The volume contains the first comprehensive explorations of misère games. It includes a tutorial for the very successful approach to analyzing misère impartial games and the first attempt at using it for misère partisan games. It also includes an updated version of Unsolved Problems in Combinatorial Game Theory and the Combinatorial Games Bibliography. The well-known normal-play games of Hex and Go are featured as well as new games: Toppling Dominoes has already spawned several papers and graduate theses; Maze extends the analysis of option-closed games; the question of Nim-dimension is introduced and new regularities are seen in take-and-break games.

Mathematical Sciences Research Institute
Publications

63

Games of No Chance 4

Mathematical Sciences Research Institute Publications

For a complete list, see http://library.msri.org/books

5 Blackadar: *K-Theory for Operator Algebras*, second edition
9 Moore/Schochet: *Global Analysis on Foliated Spaces*, second edition
27 Carlsson/Cohen/Hsiang/Jones (eds.): *Algebraic Topology and Its Applications*
28 Clemens/Kollár (eds.): *Current Topics in Complex Algebraic Geometry*
29 Nowakowski (ed.): *Games of No Chance*
30 Grove/Petersen (eds.): *Comparison Geometry*
31 Levy (ed.): *Flavors of Geometry*
32 Cecil/Chern (eds.): *Tight and Taut Submanifolds*
33 Axler/McCarthy/Sarason (eds.): *Holomorphic Spaces*
34 Ball/Milman (eds.): *Convex Geometric Analysis*
35 Levy (ed.): *The Eightfold Way*
36 Gavosto/Krantz/McCallum (eds.): *Contemporary Issues in Mathematics Education*
37 Schneider/Siu (eds.): *Several Complex Variables*
38 Billera/Björner/Green/Simion/Stanley (eds.): *New Perspectives in Geometric Combinatorics*
39 Haskell/Pillay/Steinhorn (eds.): *Model Theory, Algebra, and Geometry*
40 Bleher/Its (eds.): *Random Matrix Models and Their Applications*
41 Schneps (ed.): *Galois Groups and Fundamental Groups*
42 Nowakowski (ed.): *More Games of No Chance*
43 Montgomery/Schneider (eds.): *New Directions in Hopf Algebras*
44 Buhler/Stevenhagen (eds.): *Algorithmic Number Theory: Lattices, Number Fields, Curves and Cryptography*
45 Jensen/Ledet/Yui: *Generic Polynomials: Constructive Aspects of the Inverse Galois Problem*
46 Rockmore/Healy (eds.): *Modern Signal Processing*
47 Uhlmann (ed.): *Inside Out: Inverse Problems and Applications*
48 Gross/Kotiuga: *Electromagnetic Theory and Computation: A Topological Approach*
49 Darmon/Zhang (eds.): *Heegner Points and Rankin L-Series*
50 Bao/Bryant/Chern/Shen (eds.): *A Sampler of Riemann–Finsler Geometry*
51 Avramov/Green/Huneke/Smith/Sturmfels (eds.): *Trends in Commutative Algebra*
52 Goodman/Pach/Welzl (eds.): *Combinatorial and Computational Geometry*
53 Schoenfeld (ed.): *Assessing Mathematical Proficiency*
54 Hasselblatt (ed.): *Dynamics, Ergodic Theory, and Geometry*
55 Pinsky/Birnir (eds.): *Probability, Geometry and Integrable Systems*
56 Albert/Nowakowski (eds.): *Games of No Chance 3*
57 Kirsten/Williams (eds.): *A Window into Zeta and Modular Physics*
58 Friedman/Hunsicker/Libgober/Maxim (eds.): *Topology of Stratified Spaces*
59 Caporaso/McKernan/Mustaţă/Popa (eds.): *Current Developments in Algebraic Geometry*
60 Uhlmann (ed.): *Inverse Problems and Applications: Inside Out II*
61 Breuillard/Oh (eds.): *Thin Groups and Superstrong Approximation*
62 Eguchi/Eliashberg/Maeda (eds.): *Symplectic, Poisson, and Noncommutative Geometry*
63 Nowakowski (ed.): *Games of No Chance 4*
64 Bellamy/Rogalski/Schedler/Stafford/Wemyss: *Noncommutative Algebraic Geometry*
65 Deift/Forrester (eds.): *Random Matrix Theory, Interacting Particle Systems and Integrable Systems*

Games of No Chance 4

Edited by

Richard J. Nowakowski
Dalhousie University

CAMBRIDGE
UNIVERSITY PRESS

Richard J. Nowakowski
Department of Mathematics and Statistics
Dalhousie University
Halifax, NS, Canada
rjn@mathstat.dal.ca

Silvio Levy (*Series Editor*)
Mathematical Sciences Research Institute
levy@msri.org

The Mathematical Sciences Research Institute wishes to acknowledge support by the National Science Foundation and the *Pacific Journal of Mathematics* for the publication of this series.

CAMBRIDGE
UNIVERSITY PRESS

32 Avenue of the Americas, New York, NY 10013-2473, USA

Cambridge University Press is part of the University of Cambridge.

It furthers the University's mission by disseminating knowledge in the pursuit of education, learning and research at the highest international levels of excellence.

www.cambridge.org
Information on this title: www.cambridge.org/9781107011038

First published 2015

A catalog record for this publication is available from the British Library.

ISBN 978-1-107-01103-8 Hardback

Games of No Chance 4
MSRI Publications
Volume **63**, 2015

Contents

Games of No Chance 4
MSRI Publications
Volume **63**, 2015

Preface

This volume has its roots in the Banff International Research Station (BIRS) Workshop on Combinatorial Games (2008) organized by Elwyn Berlekamp, Tristan Cazenave, Aviezri Fraenkel, Martin Mueller and Richard J. Nowakowski. The research presented and collaboration started at BIRS has had great impact on the field. Some important seminal papers are in this volume.

Misère games are well represented. Aaron Siegel continues his work on impartial misère quotients in "The structure and classification of misère quotients" and gives a walk-through of the techniques required to analyze these games in "Misère canonical forms of partizan games". Mike Weimerskirch's article offers another approach. Of even greater impact is Siegel's "Misère canonical forms of partizan games", which has already spawned several papers, while Meghan R. Allen's "Peeking at partizan misère quotients" gives a hint of the new directions being taken.

In "Nimbers in partizan games", Carlos Pereira dos Santos and Jorge Nuno Silva expand on an early question of Berlekamp and introduce the nim-dimension measure.

The early analysis (prior to 1960) of Wythoff's game opened up a window into relationships between some games, enumerations schemes, partitions and sturmian words. This work continues in Aviezri Fraenkel and Udi Peled's "Harnessing the unwieldy MEX function", Fraenkel's "The Rat Game and the Mouse Game" and Urban Larsson's "Restrictions of m-Wythoff Nim and p-complementary Beatty sequences".

Advances on well-known games are represented by articles on Dots-and-Boxes by Sébastien Collette, Erik D. Demaine, Martin L. Demaine and Stefan Langerman; on Hex by Philip Henderson and Ryan Hayward; and on a variant thereof, Bidding Hex, by Sam Payne and Elina Robeva. Teigo Nakamura extends his work on Go in "Evaluating territories of Go positions with capturing races". The analysis of Sprouts is extended by Julien LeMoine and Simon Viennot.

Tristan Cazenave's "Monte-Carlo approximation of temperature" is about the method that has revolutionized the approach to games taken by computer scientists. J.P. Grossman and Richard J. Nowakowski show the existence of a new type of regularity in hexadecimal games.

New games with interesting structures are also introduced: MAZE, by Neil McKay, Nowakowski and Angela Siegel; Toppling Dominoes by Alex Fink,

Nowakowski, Aaron Siegel and David Wolfe; and Woodpush by Cazenave and Nowakowski. Variants of Clobber are surveyed by Laurent Beaudou, Eric Duchêne and Sylvain Gravier.

The book concludes with an updated list (with discussion) of unsolved problems in combinatorial game theory, and an update to Fraenkel's bibliography of articles published on combinatorial games.

Thanks to all who made the workshop a success. A special thanks to the BIRS organization and staff. Their help leaves the participants and organizers free to concentrate on the job (fun) of research.

Lastly, a big thanks to Silvio Levy for the final preparation of this volume.

Richard J. Nowakowski (Dalhousie University)

Games of No Chance 4
MSRI Publications
Volume **63**, 2015

Peeking at partizan misère quotients

MEGHAN R. ALLEN

1. Introduction

In two-player combinatorial games, the last player to move either wins (normal play) or loses (misère play). Traditionally, normal play games have garnered more attention due to the group structure which arises on such games. Less work has been done with games played under the misère play convention, Just as in normal play, misère games can be placed in equivalence classes, where two games G and H are equivalent if the outcome class of $G+K$ is the same as the outcome class of $H+K$ for all games K. However, Conway showed that, unlike in normal play, these misère equivalence classes are sparsely populated, making the analysis of misère games under such equivalence classes far less useful than their normal play counterparts [ONAG]. Even though these equivalence classes are sparse, Conway developed a method, called genus theory, for analyzing impartial games played under the misère play convention [Allen 2006; WW; ONAG]. For years, this was the only universal tool available for those studying misère games.

In [Plambeck 2009; 2005; Plambeck and Siegel 2008; Siegel 2006; 2015b], many results regarding impartial misère games have been achieved. These results were obtained by taking a game, restricting the universe in which that game was played, and obtaining its misère quotient. However, while, as Siegel [2015a] says "a partizan generalization exists", few results have been obtained regarding the structure of the misère quotients which arise from partizan games.

For a game $G = \{G^L \mid G^R\}$, we define $\overline{G} = \{\overline{G^R} \mid \overline{G^L}\}$. Those familiar with normal play will notice that under the normal play convention rather than $\overline{G}$, we would generally write $-G$. In normal play, this nomenclature is quite sensible as $G+(-G)=0$ [Albert et al. 2007], giving us the Tweedledum–Tweedledee principle; the second player can always win the game $G+(-G)$ by mimicking the move of the first player, but in the other component. However, in misère play, not only does the Tweedledum–Tweedledee strategy often fail, $G+\overline{G}$ is not necessarily equivalent to 0. For example, $*_2+\overline{*_2} = *_2+*_2$ is not equivalent to 0 [Allen 2006; WW]. However, having the property that $G+\overline{G}$ is equivalent to

0 is much desired, as it gives a link to which partizan misère games may behave like their normal counterparts.

To this end, this paper shows that

(1) $*+*$ is indistinguishable from 0 in the universe of all-small games, and

(2) there exists a set of games with the property that $G+\overline{G}$ is always equivalent to 0 relative to all-small games.

Using these results, the misère quotients of two nontrivial partizan examples are calculated. One such example has cardinality nine, a cardinality not found within impartial misère quotients [Plambeck and Siegel 2008]. As well, the partially ordered outcome set of one this example is given. This paper concludes with a list of six open problems of varying depth and scope in the area of partizan misère quotients.

While some elementary definitions are reviewed, this paper assumes the reader has a basic familiarity with the impartial misère quotient construction developed by Plambeck and Siegel.

2. Indistinguishability

This section contains a brief review of the indistinguishability definitions developed by Plambeck and Siegel.

Let G be a game (impartial or partizan). Then we use $o^-(G)$ to denote the misère play outcome of G, keeping the minus sign so as to not forget that we are dealing with misère play games rather than normal play ones.

We say that a set of games Υ is *closed* if it is

(1) closed under addition, i.e., if $G, H \in \Upsilon$, then $G+H \in \Upsilon$, and

(2) option closed, i.e., if $G \in \Upsilon$, then every option of G is also in Υ.

Frequently, the set of games over which we want to work is not closed. As such, we are required to take the *closure of the set* where for Υ a set of games, $c\ell\,(\Upsilon)$ is the smallest closed set such that $\Upsilon \subseteq c\ell\,(\Upsilon)$.

Suppose Υ to be a closed set of games with $G, H \in \Upsilon$. Then *G and H are indistinguishable over* Υ if

$$o^-(G+K) = o^-(H+K) \text{ for all } K \in \Upsilon,$$

and we write

$$G \equiv H \pmod{\Upsilon}.$$

Indistinguishability (mod Υ) is both an equivalence relation compatible with addition, and so, $\Upsilon/_{\equiv_\Upsilon}$ is well-defined and forms a monoid [Plambeck and Siegel 2008], which is the *misère quotient of* Υ. We denote this monoid by

$\mathscr{Q}(\Upsilon)$. Moreover, $\mathscr{Q}(\Upsilon)$ is partitioned into four disjoint outcome sets, $\mathscr{N}$, $\mathscr{P}$, $\mathscr{L}$, and $\mathscr{R}$, meaning Next, Previous, Left, and Right respectively, where, for example, $[G]_{\equiv_\Upsilon} \in \mathscr{N}$ if and only if $o^-(G) = \mathscr{N}$.

For a more detailed discussion of misère quotients, their development, and results on the monoid structures obtained, this paper refers the reader to the work of Plambeck and Siegel, most notably [Plambeck 2009; 2008; Plambeck 2005; Siegel 2015b; 2006].

3. All-small games and $* + *$

Suppose that Υ is a closed set of *impartial games* with $* \in \Upsilon$. Then we have the following result:

Proposition 3.1. $* + * \equiv 0 \pmod{\Upsilon}$ [WW].

However, if Υ contains certain partizan games, Proposition 3.1 fails.

Proposition 3.2. *Let* $1 = \{0|\cdot\}$ *and suppose* $1, * \in \Upsilon$, *a closed set of games. Then* $* + * \not\equiv 0 \pmod{\Upsilon}$.

Proof. It is easy to show that while $o^-(1) = \mathscr{R}$, Left can force a win if Right moves first in $1 + * + *$. □

Thus, while we cannot extend Proposition 3.1 to all partizan games, we can extend the result to all-small games, as shown in the following theorem.

Theorem 3.3. *Let* Υ *be a closed set of all-small games with* $* \in \Upsilon$. *Then* $* + * \equiv 0 \pmod{\Upsilon}$.

The proof of this theorem extends that of a similar result for impartial games given in [Siegel 2006].

Proof. Take $G \in \Upsilon$. We want $o^-(G + * + *) = o^-(G)$. Proceed by induction on the options of G.

We know that $o^-(0) = \mathscr{N}$, and $o^-(* + *) = \mathscr{N}$, so the base case holds.

Now suppose true for all options of G and consider G.

Since G is nonzero and all-small, Left must have a move from G. Suppose Left wins moving first in G. Then Left wins moving first in $G + * + *$ by moving to $G^L + * + *$, where G^L is a winning position for Left moving second. Since G^L is an option of G, by induction, $o^-(G^L + * + *) = o^-(G^L)$. Therefore Left wins moving second in $G^L + * + *$, and so Left wins moving first in $G + * + *$.

Suppose Left wins moving second in G. Right has two possible starting moves in $G + * + *$. Ge may move to either $G^R + * + *$ or to $G + *$. Suppose Right moves to $G^R + * + *$. By induction, $o^-(G^R + * + *) = o^-(G^R)$, where Left has a winning move moving first in G^R. Thus Left has a winning move moving first in $G^R + * + *$, and so, this is not a good opening move for Right. If Right moves to $G+*$, then Left responds with G, leaving Right to make the next move in G, and so Left wins.

Therefore, if Left moving first (or second) wins G, then Left moving first (or second) wins $G+*+*$. A symmetric argument works for Right.

Therefore $o^-(G)=o^-(G+*+*)$, and so $*+*\equiv 0 \pmod{\Upsilon}$ when Υ contains only all-small games. □

Corollary 3.4. *Let Υ be the set of all all-small games. Then* $*+*\equiv 0 \pmod{\Upsilon}$.

Proof. The set of all all-small games is closed. Thus the result follows from Theorem 3.3. □

The importance of this result is two-fold. Not only does it extend a result for impartial games, it also allows us to reduce misère monoid calculations when examining closed sets of all-small games, as we need only consider positions which contain at most one $*$.

4. Conjugation and equivalence with 0

As reviewed in the introduction, $G+\overline{G}$ is not necessarily equivalent to 0 for G played under the misère convention. However, this does raise an interesting area of investigation. For what G is it true that $G+\overline{G}\equiv 0 \pmod{c\ell(G,\overline{G})}$? This section gives an infinite set of games for which this is true.

Definition 4.1. Let G be a game. Then G is a *binary game* if at any point, a player has either no moves available or exactly one move available.

Definition 4.2. A position G is called ab$n$ if

(1) G is all-small,

(2) G is binary,

(3) each alternating path in the game tree of G is of length n or less.

Consider the games given in Figure 1. Then $*$ is ab1, G_1, G_2, and G_3 are ab2, and H_1 and H_2 are ab4.

Note that if G is ab$n$, then G is ab$m$ for all $m>n$. Also note that if G is ab$n$, then all of G's options are also ab$n$.

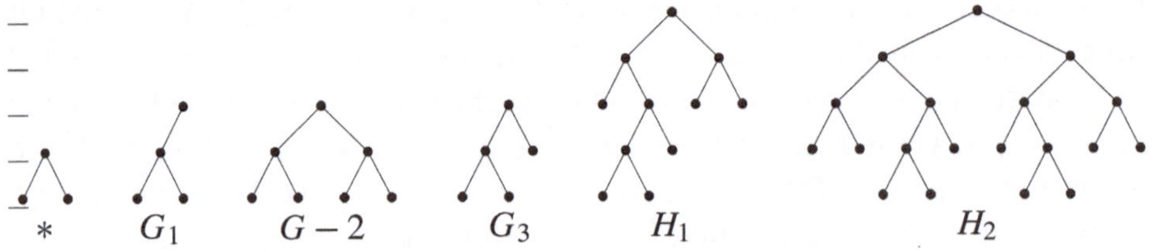

Figure 1. The games $*$, G_1, G_2, G_3, H_1, and H_2.

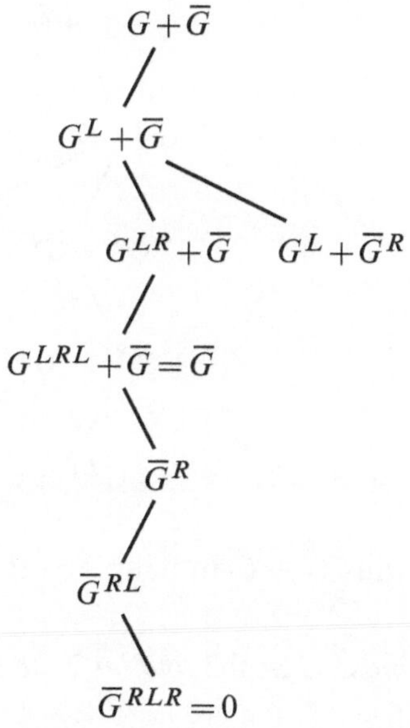

Figure 2. Left wins $G+\overline{G}$ moving first if G is ab3 and $G^{LRL}=0$.

We restrict ourselves first to examining games which are ab3. We will show that if G is ab3, then $G+\overline{G}\equiv 0 \pmod{c\ell(G, \overline{G})}$. We first require the following proposition.

Proposition 4.3. *Let G be* ab3. *Then* $o^-(G+\overline{G})=\mathscr{N}$.

Proof. Proceed by induction on the birthday of G.

Suppose $G=0$. Then $o^-(0+0)=o^-(0)=\mathscr{N}$, as required.

Suppose true for all K which are ab3 and which have smaller birthday than G. Consider G.

Suppose $G^L=0$. Then $\overline{G}^R=0$. Left moves first in $G+\overline{G}$ to $G^L+\overline{G}=\overline{G}$. Right's only response is to $\overline{G}^R=0$, and so Left wins.

Now suppose $G^{LRL}=0$. Then $\overline{G}^{RLR}=0$. Figure 2 shows how Left moving first can win $G+\overline{G}$, noting that $\overline{G^L}=\overline{G}^R$, and that the birthday of G^L is strictly less than the birthday of G, so $o^-(G^L+\overline{G}^R)=\mathscr{N}$ by induction.

Suppose that $G^{LR}=0$. Then $\overline{G}^{RL}=0$. If $G^R=0$ or $G^{RLR}=0$, then repeat one of the above arguments to get that Left wins moving first in $G+\overline{G}$. Otherwise, suppose that $G^{RL}=0$. Figure 3 shows how Left moving first can win $G+\overline{G}$.

A symmetric argument shows how Right wins moving first in $G+\overline{G}$, and so the result holds. □

We can now prove our main result.

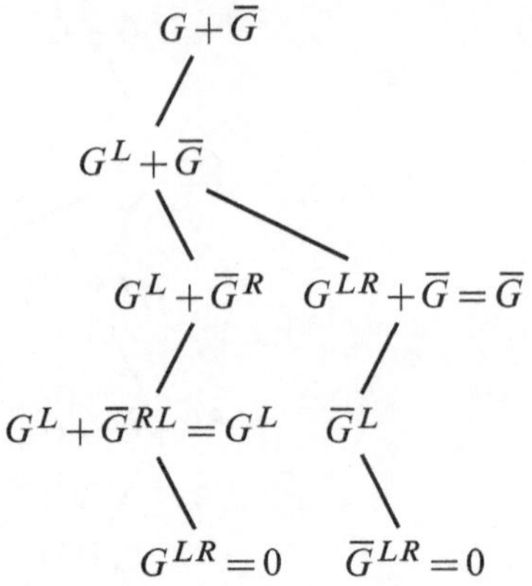

Figure 3. Left wins $G+\overline{G}$ moving first if G is ab3 and $G^{LR}=G^{RL}=0$.

Theorem 4.4. *Suppose G is* ab3 *and let Υ be the set of all all-small games. Then* $G+\overline{G}\equiv 0 \pmod{\Upsilon}$.

Proof. This proof is very similar to that of Theorem 3.3.

Proceed by induction on the birthday of G.

Suppose $G=0$. Then clearly $0+0\equiv 0 \pmod{\Upsilon}$.

Suppose true for all L which are ab3 and which have smaller birthdays than G. Consider $G+\overline{G}$. We want for any all-small H, $o^-(H)=o^-(H+G+\overline{G})$. Proceed by induction on H.

Suppose $H=0$. Then $o^-(0)=\mathscr{N}$, and, by Proposition 4.3, $o^-(G+\overline{G})=\mathscr{N}$. This shows the base case for the induction on H.

Now suppose $o^-(K)=o^-(K+G+\overline{G})$ for all all-small K with lesser birthday than that of H.

Since H is all-small and nonzero, we know that some H^L must exist. Suppose Left moving to H^L is a winning move for Left moving first in H. Claim that Left can win moving first in $H+G+\overline{G}$ with the move $H^L+G+\overline{G}$. Since the birthday of H^L is strictly less than the birthday of H, have $o^-(H^L+G+\overline{G})=o^-(H^L)$. Since Left wins moving first in H, this means $o^-(H)=\mathscr{P}$ or $\mathscr{L}$, so $o^-(H^L+G+\overline{G})=\mathscr{N}$ or $\mathscr{L}$, so Left wins moving first in $H+G+\overline{G}$.

Suppose Left wins moving second in H. Consider Right's three possible first moves in $H+G+\overline{G}$, given in Figure 4. Suppose Right makes the first move to $H^R+G+\overline{G}$. Since Left wins moving second in H, this gives $o^-(H^R)=\mathscr{N}$ or $\mathscr{L}$. Since the birthday of H^R is strictly less than the birthday of H, by induction, $o^-(H^R+G+\overline{G})=\mathscr{N}$ or $\mathscr{L}$, so Left wins $H+G+\overline{G}$ moving second if Right's first move is to $H^R+G+\overline{G}$.

Suppose Right makes the first move to $H+G^R+\overline{G}$. Left responds by moving to $H+G^R+\overline{G}^L$. Since $\overline{G^R}=\overline{G}^L$ and G^R is ab3, by induction,

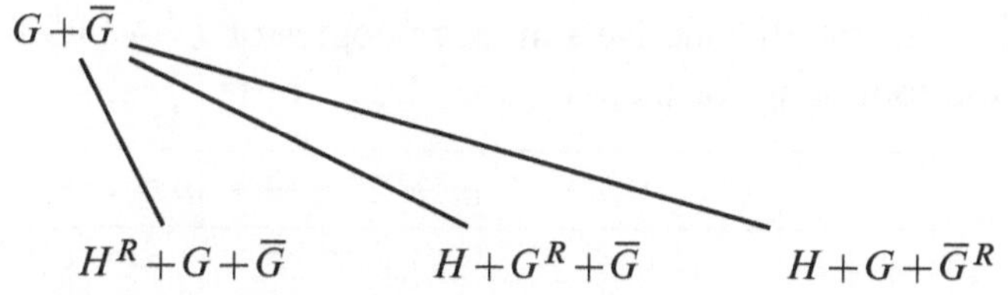

Figure 4. Right's possible opening moves in $H+G+\overline{G}$.

$o^-(H+G^R+\overline{G}^L)=o^-(H)$. Since Left wins moving second in H, by induction, Left wins moving second in $H+G^R+G$. Similarly, if Right's first move is to $H+G+\overline{G}^R$, then Left will also win.

Therefore, if Left moving first (or second) wins H, then Left moving first (or second) wins $H+G+\overline{G}$. A symmetric argument works for Right.

Therefore $o^-(H)=o^-(H+G+\overline{G})$, and so $G+\overline{G}\equiv 0 \pmod{\Upsilon}$. □

Corollary 4.5. *Let Υ be a closed set of all-small games, not necessarily all all-small games. Suppose G is* ab3. *Then $G+Gb\equiv 0 \pmod{c\ell(\Upsilon)}$.*

Proposition 4.3, Theorem 4.4, and Corollary 4.5 are surprising results. The proposition gives pairs of games $(G, \overline{G})$ in which we always know the outcome class of their sum under the misère play convention. Other than for tame games, impartial games whose misère quotients are the same as that of sums of Nim heaps, very little is known about how to deal with disjunctive sum of misère games. See [Allen 2006] and [ONAG] for a discussion on the sums of tame games, and [Mesdal and Ottaway 2007] for a discussion on some difficulties arising with the disjunctive sum on arbitrary misère games. Theorem 4.4 parallels the result discussed at the beginning of this section, namely that under normal play, $G+(-G)$ always equals 0.

A natural question is how far can Theorem 4.4 be extended? Some simple leg work shows that, for H_2 given in Figure 1, $o^-(H_2+\overline{H_2})=\mathscr{P}$, and so $H_2+\overline{H_2}\not\equiv 0 \pmod{c\ell(H_2, \overline{H}_2)}$. Hence, Theorem 4.4 does not extend to all ab$n$ games for $n\geq 4$. However, it would still be worth investigating for which ab$n$ games with $n\geq 4$ have the result given in Theorem 4.4.

5. Two examples of partizan misère quotients

Consider the game $\downarrow=\{*\mid 0\}$. In this section, we will calculate $\mathscr{Q}(c\ell(\downarrow))$ and $\mathscr{Q}(c\ell(\downarrow, \overline{\downarrow}))$.

5.1. ***The partizan misère quotient of $c\ell(\downarrow)$.*** The positions in $c\ell(\downarrow)$ are 0, $*$, and $\downarrow$. Since $\downarrow$ is all-small, Theorem 3.3 gives that $*+*\equiv 0 \pmod{c\ell(\downarrow)}$. Thus, every position in $c\ell(\downarrow)$ is indistinguishable from one of the form $m\downarrow$ or $*+m\downarrow$, where $m\downarrow$ denotes the disjunctive sum of m copies of $\downarrow$. A bit of work [Allen 2009] gives the following for the outcome classes of the positions $n*+m\downarrow$:

	$m=0$	$m=1$	$m=2$	$m=3$	$m\geq 4$
$n\equiv 0$	$\mathscr{N}$	$\mathscr{L}$	$\mathscr{P}$	$\mathscr{R}$	$\mathscr{R}$
$n\equiv 1$	$\mathscr{P}$	$\mathscr{N}$	$\mathscr{N}$	$\mathscr{N}$	$\mathscr{R}$

Moreover, we can see that

(1) $4\downarrow\equiv u\downarrow \pmod{c\ell(\downarrow)}$ for any $u\geq 4$,

(2) $4\downarrow\equiv *+u\downarrow \pmod{c\ell(\downarrow)}$ for any $u\geq 4$.

Enumerating the elements then gives us:

$$0,\quad *,\quad \downarrow,\quad 2\downarrow,\quad 3\downarrow,\quad 4\downarrow,\quad *+\downarrow,\quad *+2\downarrow,\quad *+3\downarrow,$$

all of which are pairwise distinguishable. Note that if $o^-(G)\neq o^-(H)$, then the two elements are distinguished by 0. Table 1 shows the distinguishing elements in $c\ell(\downarrow)$ when $o^-(G)=o^-(H)$.

With the mappings

$$0\mapsto 1,\quad *\mapsto a,\quad \downarrow\mapsto d,$$

the following monoid is achieved:

$$\begin{aligned}
\mathscr{Q}(c\ell(\downarrow)) &= \langle 1, a, d \mid a^2=1, d^4=d^5=ad^4\rangle,\\
\mathscr{N} &= \{1, ad, ad^2, ad^3\},\\
\mathscr{P} &= \{a, d^2\},\\
\mathscr{L} &= \{d\},\\
\mathscr{R} &= \{d^3, d^4\},
\end{aligned}$$

Position 1	Position 2	distinguishing element
0	$*+\downarrow$	$*$
0	$*+2\downarrow$	$\downarrow$
0	$*+3\downarrow$	$\downarrow$
$*+\downarrow$	$*+2\downarrow$	$*$
$*+\downarrow$	$*+3\downarrow$	$*$
$*+2\downarrow$	$*+3\downarrow$	$*$
$*$	$2\downarrow$	$2\downarrow$
$3\downarrow$	$4\downarrow$	$*$

Table 1. Positions of $c\ell(\downarrow)$ and the elements which distinguish them.

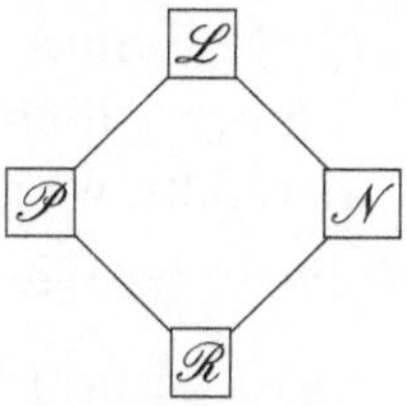

Figure 5. Outcome class partial order.

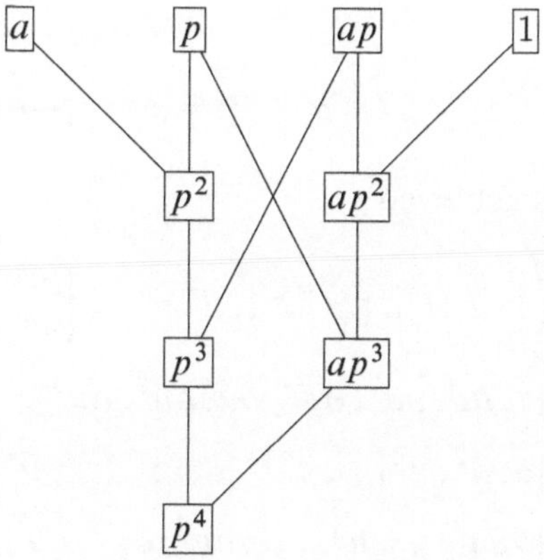

Figure 6. The partially ordered set of $\mathscr{Q}(\mathrm{c}\ell\,(\downarrow))$.

with the additive notation in $\mathrm{c}\ell\,(\downarrow)$ becoming multiplicative notation in $\mathscr{Q}(\mathrm{c}\ell\,(\downarrow))$.

This example demonstrates an important difference between partizan and impartial misère quotients. In impartial games, every finite indistinguishability quotient has either cardinality one or is of even cardinality [Plambeck and Siegel 2008]. Contrast this with the cardinality of $\mathscr{Q}(\mathrm{c}\ell\,(\downarrow))$, which is nine.

Another possible area for investigation regarding partizan misère quotients is on the partial order of the elements. Recall that, under monoid multiplicative notation,

$$x \geq y \text{ if } o^-(xz) \geq o^-(yz) \qquad \text{for all monoid elements } z,$$

and that, in terms of outcomes, the outcome lattice is given in Figure 5.

The partially ordered set of $\mathscr{Q}(\mathrm{c}\ell\,(\downarrow))$ is given in Figure 6. However, while these sets can be calculated, no general results on such partially ordered sets have been obtained.

5.2. ***The partizan misère quotient of $\mathrm{c}\ell\,(\downarrow, \uparrow)$.*** Note that $\overline{\downarrow} = \{0 \mid *\} = \uparrow$. Thus $\mathscr{Q}(\mathrm{c}\ell(\downarrow, \overline{\downarrow})) = \mathscr{Q}(\mathrm{c}\ell\,(\downarrow, \uparrow))$.

We will now calculate $\mathcal{Q}(\mathrm{c}\ell\,(\downarrow, \uparrow))$. Since $\mathrm{c}\ell\,(\downarrow, \uparrow)$ is a set of all-small games, we can apply Theorem 3.3. Since $\downarrow$ is ab2, and hence ab3, we can apply Corollary 4.5. Combining these two results, we get that all positions in $\mathrm{c}\ell\,(\downarrow, \uparrow)$ are indistinguishable from one of the following:

$$0, \quad *, \quad m\downarrow, \quad \ell\uparrow, \quad *+m\downarrow, \quad \text{or} \quad *+\ell\uparrow.$$

Moreover, it can be shown that any two positions of the above form are distinguishable ([Allen 2009]).

With the mappings

$$0 \mapsto 1, \quad a \mapsto a, \quad \downarrow \mapsto d, \quad \uparrow \mapsto u,$$

the following monoid is achieved:

$$\begin{aligned}
\mathcal{Q}(\mathrm{c}\ell\,(\downarrow, \uparrow)) &= \left\langle 1, a, d, u \;\middle|\; a^2 = 1,\, d^m u^n = \begin{cases} d^{m-n} & \text{if } m > n, \\ u^{n-m} & \text{if } m \le n, \end{cases} \right\rangle, \\
\mathcal{N} &= \{1, ad, ad^2, ad^3, au, au^2, au^3\}, \\
\mathcal{P} &= \{a, d^2, u^2\}, \\
\mathcal{L} &= \{d, u^3, u^4, u^5, \dots, au^4, au^5, au^6 \cdots\}, \\
\mathcal{R} &= \{u, d^3, d^4, d^5, \dots, ad^4, ad^5, ad^6, \dots\},
\end{aligned}$$

with the additive notation in $\mathrm{c}\ell\,(\downarrow, \uparrow)$ having become multiplicative notation in $\mathcal{Q}(\mathrm{c}\ell\,(\downarrow, \uparrow))$ (the outcome class calculations can be seen in [Allen 2009]). Notice that $\mathcal{Q}(\mathrm{c}\ell\,(\downarrow, \uparrow))$ is an infinite monoid. Thus, as in the impartial case, infinite misère monoids exist.

The partially ordered set of this monoid is particularly unpleasant, but is calculated in full in [Allen 2009].

It should also be noted that if we consider the misère monoid simply as a monoid without the outcome tetrapartitions, than it is isomorphic to the group $\mathbb{Z}_2 \oplus \mathbb{Z}$, just as in normal play.

6. Conclusion

Theorems 3.3 and 4.4 give a good starting base for further investigation of partizan misère quotients. While, as Theorem 3.3 shows, all-small games share some results with impartial games, the misère quotient of $\mathrm{c}\ell\,(\downarrow)$, which has nine elements, shows that even restricting ourselves to all-small games can yield results which do not appear for impartial misère quotients. This paper concludes with some possibilities for further research in the area of partizan misère quotients:

(1) For what $n \in \mathbb{N}$ does there exist games (or sets of games) G such that the cardinality of $\mathscr{Q}(G)$ is n?

(2) For what games G is it true that $\mathscr{Q}(G)$ is infinite?

(3) How can we extend Theorem 4.4 to other sets of partizan games?

(4) What properties of impartial misère quotients always hold for partizan misère quotients? Are there properties of impartial misère quotients which never hold for partizan misère quotients or vice versa?

(5) What can we say about the partially ordered sets that arise under the partizan misère quotient construction?

(6) Joyal [1977] constructed a category out of normal play games under disjunctive sum. The objects of his category were normal play games G, where there exists an arrow $G \to H$ if Left moving second can win the game $H + (-G)$. That this entity satisfied the further conditions required to form a category relied heavily on the use of the fact that $G + (-G)$ is equivalent to 0 and the Tweedledum–Tweedledee principle to show both the existence of an identity and compositions. In 2010, Cockett, Cruttwell, and Saff used this idea to define a combinatorial game category, that is a category with extra axioms which make it behave like Joyal's category of normal play games [Cockett et al. 2010]. However, these combinatorial game categories are very normal play focused and investigation must be undertaken to determine how, or even if, such work can be applied to the misère case.

Acknowledgements

The author would like to acknowledge NSERC, the Killam Trust, and Dalhousie University for their financial support while writing this paper. She would also like to acknowledge Dr. Richard Nowakowski of Dalhousie University and Dr. Geoff Cruttwell of the University of Calgary for their suggestions and input regarding this paper.

References

[Albert et al. 2007] M. Albert, R. J. Nowakowski, and D. Wolfe, *Lessons in play: an introduction to the combinatorial theory of games*, A K Peters, Natick, MA, 2007.

[Allen 2006] M. R. Allen, *Impartial combinatorial misère games*, Master's thesis, Dalhousie University, 2006, http://www.reluctantm.com/thesis/meghan.pdf.

[Allen 2009] M. R. Allen, *An investigation of misère partizan games*, Ph.D. thesis, Dalhousie University, 2009.

[Cockett et al. 2010] J. Cockett, G. Cruttwell, and K. Saff, "Combinatorial game categories", preprint, 2010, http://pages.cpsc.ucalgary.ca/~gscruttw/CGC.pdf.

[Joyal 1977] A. Joyal, “Remarques sur la théorie des jeux à deux personnes”, *Gazette des Sciences Mathématiques du Québec* **4** (1977), 46–52.

[Mesdal and Ottaway 2007] G. A. Mesdal and P. Ottaway, “Simplification of partizan games in misère play”, *Integers* **7** (2007), G06, 12.

[ONAG] J. H. Conway, *On numbers and games*, 2nd ed., A K Peters, Natick, MA, 2001.

[Plambeck 2005] T. E. Plambeck, “Taming the wild in impartial combinatorial games”, *Integers* **5**:1 (2005), G5, 36.

[Plambeck 2009] T. E. Plambeck, “Advances in losing”, pp. 57–89 in *Games of No Chance 3*, Cambridge University Press, 2009.

[Plambeck and Siegel 2008] T. E. Plambeck and A. N. Siegel, “Misère quotients for impartial games”, *J. Combin. Theory Ser. A* **115**:4 (2008), 593–622.

[Siegel 2006] A. N. Siegel, “Misère games and misère quotients”, lecture notes, 2006. arXiv math.CO/0612616

[Siegel 2015a] A. N. Siegel, “Misère canonical forms of partizan games”, pp. (pages to be supplied by publisher) in *Games of no chance* 4, edited by R. Nowakowski, Cambridge University Press, New York, 2015.

[Siegel 2015b] A. N. Siegel, “The structure and classification of misère quotients”, pp. 241–266 in *Games of No Chance 4*, edited by R. Nowakowski, Mathematical Sciences Research Institute Publications **63**, Cambridge University Press, New York, 2015.

[WW] E. R. Berlekamp, J. H. Conway, and R. K. Guy, *Winning ways for your mathematical plays, II*, 2nd ed., A K Peters, Natick, MA, 2003.

meghanrose@gmail.com *Sackville, NB, Canada*

Games of No Chance 4
MSRI Publications
Volume **63**, 2015

A survey of Solitaire Clobber

LAURENT BEAUDOU, ERIC DUCHÊNE AND SYLVAIN GRAVIER

Solitaire Clobber is a one-player variant of the 2-player board game Clobber introduced by Albert et al. in 2002. According to simple rules, the objective of Solitaire Clobber is to capture the maximum number of stones from a given graph. Two versions of Solitaire Clobber were recently investigated: a partisan and an impartial one. In this survey, we give an overview of the major results about Solitaire Clobber, more especially about the impartial version. In particular, the game is considered on grids, trees, and hypercubes. Two new results are provided: when playing on a tree, we show that the minimum number of remaining stones can be computed in polynomial time. We also assert that any game position on a "large" grid can be reduced to 1 or 2 stones. Note that in each part of this survey, we propose several open problems related to Solitaire Clobber.

1. Introduction

1.1. ***Two-player versions.*** In 2001, Albert, Grossman, Nowakowski and Wolfe investigated a new 2-player partisan game called Clobber; they developed the first results in [Albert et al. 2005]. In terms of game values, it turns out that Clobber is difficult even when played on basic positions. This complexity explains the author's motivation for studying Clobber. The description of the game follows below.

2-player Clobber. Black and white stones are placed on the vertices of an undirected graph, at most one per vertex. The first player moves only black stones and the second player the white ones. A player moves by picking up one of his stones and "clobbering" an adjacent stone of the opposite color (vertically or horizontally). The clobbered stone is deleted and replaced by the one that was moved. The last player to move wins.

Clobber is usually played on a grid where the initial position is the one of a checkerboard, as depicted by Figure 1.

In the last few years, several events were organized around Clobber like the first international Clobber tournament. It was held at the 2002 Dagstuhl seminar on algorithmic and combinatorial game theory (see the report in [Grossman 2004]). Since 2005 the game has been one of the events of the Computer Olympiad.

Figure 1. Initial Clobber position on a 5×6 grid.

A short article for the general public was also written in *Science News* (see [Peterson 2002]).

Albert et al. [2005] proved that determining the winner of a Clobber game is an NP-hard question. Moreover, even when the game is played on one or two rows, there are no proved results. Nevertheless the following two conjectures made in [Albert et al. 2005] seem to hold:

Conjecture 1. *The game position* $(\bullet\circ)^n$ *is winning for the first player if and only if* $n \neq 3$.

Conjecture 2. *For n odd, the position* $\binom{\bullet}{\circ}^n$ *is a first player win.*

When n is even, it is easy to see that the position $\binom{\bullet}{\circ}^n$ is a second player win by invoking symmetry arguments.

Problem P1 (2-player Impartial Clobber). Since Clobber appears to be a hard game even on regular positions, what about the impartial version? What can we say about 2-player Impartial Clobber, where the rules are the same, except that each player can indifferently move a black or a white stone. For example start by studying this game on $1 \times n$ and $2 \times n$ checkerboard positions. Can we provide the same kind of conjecture as for the partisan version?

1.2. *Solitaire versions.* E. Demaine, L. Demaine and R. Fleisher [Demaine et al. 2004] investigated the *Solitaire Clobber* game. It can be seen as the solitaire version of the 2-player partisan Clobber. The way to move the stones is the same as for Clobber, but the objective is adapted for a unique player. Indeed, the goal is to minimize the number of remaining stones on the graph. Here is a full description of this solitaire variant.

Partisan Solitaire Clobber. Black and white stones are placed on the vertices of an undirected graph, at most one per vertex. The player moves by picking up alternately a black and a white stone and "clobbering" an adjacent stone of the opposite color. The clobbered stone is replaced by the one that was moved. The objective consists in minimising the number of remaining stones on the graph.

Demaine et al. [2004] investigated this game on rows and grids with initial "checkerboard position".

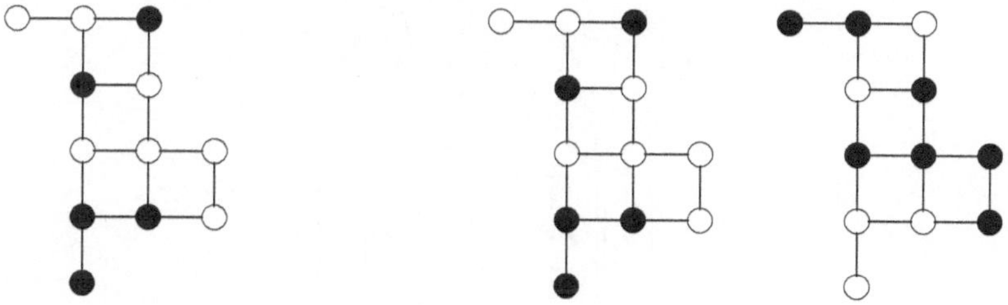

Figure 2. Left: play on a grid graph G. Right: play on complementary copies of G.

Proposition 3 [Demaine et al. 2004]. *In Partisan Solitaire Clobber, the reducibility value of a checkerboard row of size n is* $\lceil n/4 \rceil$ *if* $n \not\equiv 3 \mod [4]$, *and* $\lceil n/4 \rceil + 1$ *otherwise.*

The results about checkerboard grids will be developed in Section 2 of this survey. In particular, Demaine et al. showed that such grids are 2-reducible.

Dorbec, Duchêne and Gravier [Dorbec et al. 2008] studied a second version of Solitaire Clobber, which corresponds to an impartial 1-player variant. We call it *Impartial Solitaire Clobber*. The only difference with the above solitaire game is that the player is not forced to alternate black and white captures. Without this constraint, it turns out that Solitaire Clobber becomes easier to work on. Clearly, 1-reducible positions do not need to have the same number of black and white stones (e.g. ●○○○○○, which is 1-reducible for Impartial Solitaire Clobber and not for the partisan version).

There exists a natural correlation between the two versions of Solitaire Clobber. Indeed, playing the impartial game on a graph G is equivalent to playing the partisan version on two complementary copies of G. This property is illustrated by Figure 2, where G is a grid graph (i.e., an induced subgraph of the grid). Optimal play on the Impartial Solitaire Clobber on G (left) is equivalent to optimal main on the partisan game on two copies of G (right).

1.3. ***Definitions and notations for Solitaire Clobber.*** We here introduce several definitions and notations that are available for both versions of Solitaire Clobber.

A *game position* on a graph $G = (V, E)$ is a mapping $\Phi : V \to \{\bullet, \circ\}$.

Given a game position Φ of Solitaire Clobber on a graph G, we say that the *reducibility value* of (G, Φ) is the minimum number of stones that can be left on the graph from initial position Φ. This value is denoted by $\mathrm{rv}(G, \Phi)$. We also say that a position on a graph is *k-reducible* (for a positive integer k) if there exists a succession of moves that leaves at most k stones on the graph. In particular, as for the "classical" Solitaire game, we are interested in Solitaire Clobber positions that are 1-reducible.

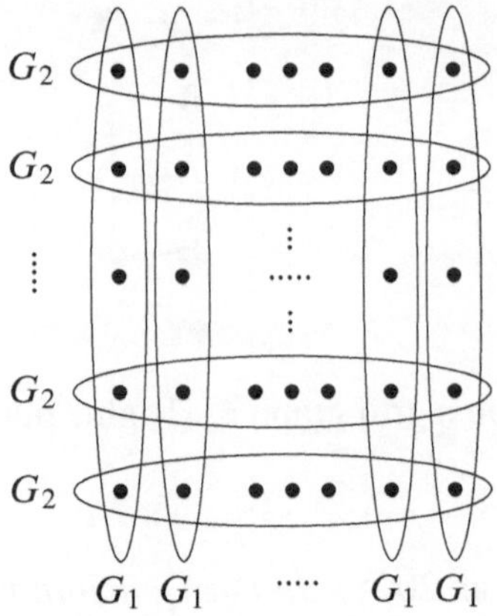

Figure 3. The Cartesian product of two graphs G_1 and G_2.

For short if there is no confusion, we may say that G or Φ is k-reducible instead of (G, Φ) is k-reducible. Similarly, we may confound a vertex with the stone it supports.

If all the vertices of the graph G have the same color, we say that G is *monochromatic*. If there exists a vertex v such that $G \setminus v$ is monochromatic and G is not, then G is said to be *quasimonochromatic*.

Given two graphs $G_1 = (V_1, E_1)$ and $G_2 = (V_2, E_2)$, the *Cartesian product* $G_1 \Box G_2$ is the graph $G = (V, E)$ where $V = V_1 \times V_2$ and $(u_1u_2, v_1v_2) \in E$ if and only if $u_1 = v_1$ and $(u_2, v_2) \in E_2$, or $u_2 = v_2$ and $(u_1, v_1) \in E_1$. One generally depicts such a graph with $|V_2|$ vertical copies of G_1, and $|V_1|$ horizontal copies of G_2, as shown on Figure 3.

A *grid graph* is an induced subgraph of the $k \times n$ grid $P_k \Box P_n$, where P_k and P_n are two paths of respective lengths k and n.

1.4. *Complexity results.* Whatever the type of the game (2-player or Solitaire) and the way of moving (Impartial or Partisan), it turns out that Clobber is a hard game. The proofs of the NP-hardness often rely on the Hamiltonicity problem on graphs. The correlation between these two problems appears when playing Clobber on a quasimonochrome position.

The first complexity result was proved by Albert et al. [2005]. Their proof uses a reduction to the *Hamiltonian path* problem.

Theorem 4 [Albert et al. 2005]. *Determining who wins from a 2-player Clobber position is NP-hard.*

The same kind of result was provided by Demaine et al. [2004] for Solitaire Clobber.

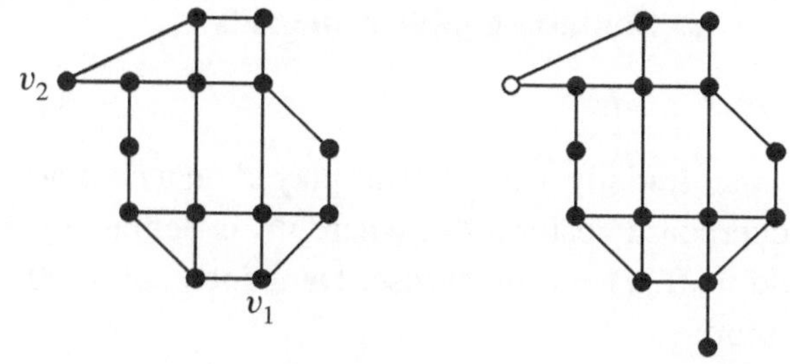

Figure 4. Reduction to a Hamiltonian path.

Theorem 5 [Demaine et al. 2004]. *Deciding whether or not a position of Solitaire Clobber is* 1*-reducible is NP-complete.*

Their proof can be adapted for both impartial and partisan versions. As for the 2-player game, it requires a reduction to the Hamiltonian path problem. On Figure 4 below it is easy to see that the graph on the left admits an Hamiltonian path between vertices v_1 and v_2 if and only if the position of the right is 1-reducible for Impartial Solitaire Clobber. A small gadget (see [Demaine et al. 2004]) leads to the same kind of reduction for the partisan version.

Demaine et al. [2004], using a result of A. Itai, C. H. Papadimitriou and J. L. Szwarcfiter [Itai et al. 1982], showed that the NP-completeness of Solitaire Clobber remains true on grid graphs. In Section 2, we give some elements to solve the complexity on grids. Indeed, we show that for Impartial Solitaire Clobber, nonmonochromatic positions on grids are 2-reducible. Nevertheless the complexity remains open.

1.5. ***Organization of the paper.*** A major part of this survey will be devoted to the investigation of Impartial Solitaire Clobber. Section 2 deals with the game on grids and compares the different results about both versions of Solitaire Clobber. In particular, it is showed that grids are always 2-reducible for Impartial Solitaire Clobber, whatever the (nonmonochromatic) initial position is. In this section, some of the stuff introduced by Demaine et al. will be reinvestigated for the study of the impartial game. In Conjecture 16 we present algorithms to compute the reducibility value of game positions on paths, cycles and trees. On such families of graph, this computation can be done in polynomial time. In Section 4 we give constant upper bounds on the reducibility value when playing on Hamming graphs and multipartite complete graphs. Section 5 is devoted to more general results, such as positions that minimize or maximize the reducibility value on an arbitrary graph. Moreover in each section we propose several other combinatorial optimization problems related to Solitaire Clobber.

2. Solitaire Clobber on grids

2.1. *Special positions on grids.*

Checkerboard positions. Initially, Clobber was played on a complete grid ($k \times n$ rectangle) with checkerboard position Φ_c, where Φ_c is defined by $\Phi_c(i, j) = \bullet$ if $(i + j)$ is even, and $\Phi_c(i, j) = \circ$ otherwise. Demaine et al. [2004] solved the game on such a position:

Theorem 6 [Demaine et al. 2004]. *The $k \times n$ checkerboard is 2-reducible. Moreover, the $k \times n$ checkerboard is 1-reducible if and only if $k.n \not\equiv 0$ mod [3].*

The proof of 2-reducibility is done by induction through local rules. In order to prove that a position is not 1-reducible, Demaine et al. define an invariant δ available for both versions of Solitaire Clobber on bipartite graphs. The description of δ follows.

Definition 7. Let Φ be a game position on a bipartite graph $G = (V, E)$. Denote by S_0 and S_1 two disjoined independent sets of G such that $S_0 \cup S_1 = V$. We allocate the color white the to set S_0, and the color black to S_1. A stone of G is said to be *clashing* if its color differs from the color of the independent set to which it belongs. Define the quantity

$$\delta(G, \Phi) = \text{number of stones} + \text{number of clashing stones}.$$

The following result found by Demaine et al. motivates the introduction of this invariant.

Proposition 8. *Let Φ be a game position of Solitaire Clobber on a bipartite graph G. Then $\delta(G, \Phi)$ mod [3] keeps the same value during the game.*

Observe that if $k.n \equiv 0$ mod [3] then the value of δ on a checkerboard position on a $k \times n$ grid is a multiple of 3. Nevertheless, a position consisting of a single stone has value 1 or 2; therefore when $k.n \equiv 0$ mod [3], the $k \times n$ checkerboard is not 1-reducible.

Theorem 6 was proved for Partisan Solitaire Clobber and Proposition 8 asserts that one can not do better for the Impartial version.

Quasimonochromatic positions for Impartial Solitaire Clobber. One of the motivation to investigate the Impartial version of Solitaire Clobber is to consider game positions that are not balanced in the numbers of black and white stones. In this way, a natural game position is the quasimonochromatic position. One may hope to solve this special position on a more general graph than a grid. Unfortunately, given a graph and a quasimonochromatic position (with a unique white stone v), determining if this position is 1-reducible is equivalent to determine if G admits

Figure 5. Forbidden configurations.

an Hamiltonian path starting from v. Therefore, the problem of 1-reducibility of a quasimonochromatic position on a graph G is NP-complete.

Itai et al. [1982] proved an even stronger result claiming that finding an Hamiltonian path in a grid graph is still NP-complete. They complete their result by a nice structural property of Hamiltonian paths in the $k \times n$ grid. Before stating their result we need to introduce some notations. Given a vertex A of the grid, we denote by $x(A)$ (respectively $y(A)$) the row (resp. column) of A.

Theorem 9 [Itai et al. 1982]. *Let G be a $k \times n$ grid and A, B be two vertices of G. Let c be the 2-coloring of the grid. There exists an Hamiltonian path from A to B if and only if A and B satisfy the following conditions*:

- ***Coloring condition***: *If $k.n$ is even then $c(A) \neq c(B)$. Otherwise $c(A) = c(B) = c(1, 1)$.*
- ***Connectivity condition***: *If $k = 1$ then A and B are the extremities of the grid. If $k = 2$ then either $(A = (1, 1)$ and $B = (2, 1))$ or $(A = (1, n)$ and $B = (2, n))$ or (A and B belong to distinct columns).*
- ***Forbidden configurations***: *If $k = 3$, n is even and $c(A) \neq c(1, 1)$ then if $x(A) = 2$ then $y(A) \geq x(B)$ otherwise $y(A) \geq y(B) - 1$* (see Figure 5).

Observe that this last result solves the 1-reducibility of quasimonochromatic positions on the $k \times n$ grid. Of course, one may provide a simpler proof since we need to fix only one extremity of the Hamiltonian path.

Corollary 10. *Let G be a $k \times n$ grid with $k \geq n \geq 2$, c be the 2-coloring of G and Φ be the quasimonochromatic position with the unique white stone on v. Then (G, Φ) is 1-reducible if and only if either $c(v) = c(1, 1)$ or $k.n$ is even.*

Proof. First remark that if $c(v) \neq c(1, 1)$ and $k.n$ is odd then any pair $(A = v, B)$ does not satisfies the coloring condition of Theorem 9. Therefore there is no Hamiltonian path from A to B which implies that Φ is not 1-reducible.

Now, suppose that $c(v) = c(1, 1)$ or $k.n$ is even. One may assume that $v \neq (1, 1)$ and set $A = v$. If $k.n$ is even and $c(v) = c(1, 1)$ then set $B = (2, 1)$, otherwise $B = (1, 1)$. Now, by Theorem 9, G admits an Hamiltonian path from A to B which gives a way to 1-reduce Φ. □

By Corollary 10, the only non-1-reducible positions are the ones for which $c(v) \neq c(1, 1)$ and $k.n$ is odd. It is worth to notice that in that case, $\delta \equiv 0 \bmod [3]$.

Moreover, as proved in [Gravier and Parreau 2006], in highest dimension $d \geq 3$, the coloring condition is sufficient to insure the existence of an Hamiltonian path from A to B in the d-dimensional grid. Therefore it is easier to play Impartial Solitaire Clobber on quasimonochromatic positions in grids of large dimension.

Theorem 9 suggests to investigate the structural properties of spanning trees in the $k \times n$ grid. In [Gravier and Parreau 2006], the authors generalized the coloring and connectivity conditions and add some forbidden configurations, in order to determine a necessary and sufficient condition on 3 vertices insuring that there exists a spanning tree of the grid having precisely these vertices as leaves. Nevertheless they were not able to solve the following general problem:

Problem P2 (spanning tree of the grid with p fixed leaves (p-STG)). Given a $k \times n$ grid G and p vertices $v_1, \ldots, v_p$, does there exist a spanning tree of G having $v_1, \ldots, v_p$ as leaves?

Using an inductive proof based on the result of [Itai et al. 1982] (for $p = 2$), Gravier and Parreau [2006] proved that for any fixed integer $p \geq 2$, an analogue of the p-STG problem in grid graphs remains NP-Complete.

2.2. *General case.* In this section, we will present a recent result due to Derouet-Jourdan and Gravier which solves Impartial Solitaire Clobber on "large enough" grids.

Theorem 11 [Derouet-Jourdan and Gravier 2008]. *Let Φ be a nonmonochromatic position on a $k \times n$ grid with $k \geq n \geq 4$. Then Φ is 2-reducible.*

Sketch of proof. Let Φ be a nonmonochromatic position on a $k \times n$ grid G. Let $k = 4p_1 + r_1$ and $n = 4p_2 + r_2$ with $0 \leq r_i < 4$. Denote by R_i (respectively C_j) the restriction of (G, Φ) to the i-th row (resp. j-th column) of G. For $1 \leq i < p_1$ and $1 \leq j < p_2$, let

$$B(i, j) = \left(\bigcup_{s=i}^{i+3} R_s\right) \cap \left(\bigcup_{s=j}^{j+3} C_s\right).$$

For all $1 \leq i \leq p_1$, let

$$B(i, p_2) = \left(\bigcup_{s=i}^{i+3} R_s\right) \cap \left(\bigcup_{s=p_2}^{p_2+3+r_2} C_s\right).$$

Similarly, for all $1 \leq j \leq p_2$, let

$$B(p_1, j) = \left(\bigcup_{s=p_1}^{p_1+3+r_1} R_s\right) \cap \left(\bigcup_{s=j}^{j+3} C_s\right).$$

Finally, let

$$B(p_1, p_2) = \left(\bigcup_{s=p_1}^{p_1+3+r_1} R_s\right) \cap \left(\bigcup_{s=p_2}^{p_2+3+r_2} C_s\right).$$

Observe that the family $B(i, j)$ is a partition of G with $(p_1 - 1).(p_2 - 1)$ grids of size 4×4, $(p_1 - 1)$ grids of size $(4 + r_1) \times 4$, $(p_2 - 1)$ grids of size $4 \times (4 + r_2)$ and one grid of size $(4 + r_1) \times (4 + r_2)$.

In the remainder of the proof, $\genfrac{}{}{0pt}{}{\bullet}{\circ}/\genfrac{}{}{0pt}{}{\circ}{\bullet}$ will denote indifferently one of the patterns $\genfrac{}{}{0pt}{}{\bullet}{\circ}$ or $\genfrac{}{}{0pt}{}{\circ}{\bullet}$.

We will say that a position on a $a \times b$ grid Q (with $a \geq b$) is *nice* if it can be reduced to one of the following patterns on the first column of Q:

- either $B_1 = 1$ or 2 copies of $\genfrac{}{}{0pt}{}{\bullet}{\circ}/\genfrac{}{}{0pt}{}{\circ}{\bullet}$,
- or both ($B_2 =$ one copy of $\genfrac{}{}{0pt}{}{\bullet}{\circ}/\genfrac{}{}{0pt}{}{\circ}{\bullet}$ plus $\circ$ on some vertex v) and ($B_2' =$ one copy of $\genfrac{}{}{0pt}{}{\bullet}{\circ}/\genfrac{}{}{0pt}{}{\circ}{\bullet}$ plus $\bullet$ on v).

This property is interesting. Indeed, consider two consecutive blocks $B(i, j)$ and $B(i, j+1)$. We claim that if $B(i, j)$ is nice, then we can move the stones of B_t to $B(i, j+1)$ insuring that the resulting $B'(i, j+1)$ is nonmonochromatic. To prove this claim, first assume that $B(i, j)$ can be reduced to B_1 with one copy of $\genfrac{}{}{0pt}{}{\circ}{\bullet}$:

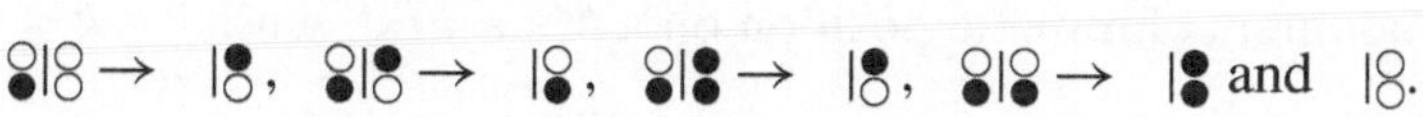

In the last case, we can choose either $\genfrac{}{}{0pt}{}{\circ}{\circ}$ or $\genfrac{}{}{0pt}{}{\bullet}{\bullet}$ in order to insure that the resulting $B'(i, j+1)$ is not monochromatic.

If $B(i, j)$ can be reduced to B_1 with two copies of $\genfrac{}{}{0pt}{}{\bullet}{\circ}/\genfrac{}{}{0pt}{}{\circ}{\bullet}$, then proceed as previously for each copy.

Now, if $B(i, j)$ can be reduced to B_2 and B_2' then choose one of them according to the color of the stone placed on the neighbour of v in $B(i, j+1)$. For the copy of $\genfrac{}{}{0pt}{}{\bullet}{\circ}/\genfrac{}{}{0pt}{}{\circ}{\bullet}$ in B_2 or B_2' proceed as previously.

With the help of a computer to check a large number of cases, we proved the following lemma:

Lemma 12 [Derouet-Jourdan and Gravier 2008]. *Each nonmonochromatic game position on a* $4 \times n$ *grid with* $n = 4, \ldots, 7$ *is nice except the* 4×4 *position namely* Bad *depicted below*:

To settle the case of Bad's and monochromatic positions, we consider the concatenation of blocks. Firstly we consider the two following possible reductions of a Bad block:

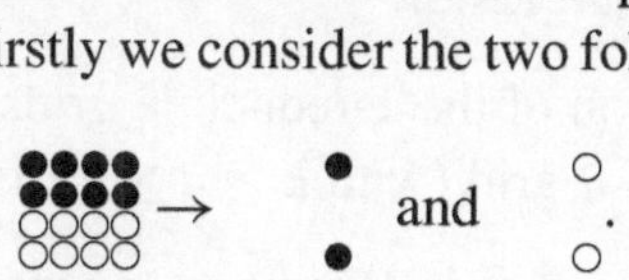

Secondly, we check by computer that:

Lemma 13 [Derouet-Jourdan and Gravier 2008]. *For every game position on a* 4×4 *grid A, one of the following properties is true*:

To settle the monochromatic case, we explore again by computer a large number of cases:

Lemma 14 [Derouet-Jourdan and Gravier 2008]. *For every nonwhite monochromatic position on a* $4 \times n$ *grid* A *with* $n = 4, \ldots, 7$, *we have that* A *plus* 2 *white columns is nice.*

From Lemma 14, one can prove by induction that $M + A$ with M being a white monochromatic position on $4 \times p$ (with $p \geq 4$) and A being a nonwhite monochromatic position on $4 \times q$ (with $4 \leq q \leq 7$), is nice.

Now to prove Theorem 11, consider an Hamiltonian path of blocks starting from $B(1, 1)$ if p_1 is odd, otherwise from $B(1, p_2)$ and ending at $B(p_1, p_2)$. Applying Lemmas 12–14 along this path, we get a nonmonochromatic position on the grid $k \times n$ with $4 \leq k \leq n \leq 7$. To finish the proof, we check again by computer that every nonmonochromatic position on a $k \times n$ grid, with $4 \leq k \leq n \leq 7$, is 2-reducible.

Since the Cartesian product of two (path-)Hamiltonian graphs contains as a subgraph a 2-dimensional grid, one may mention the following corollary of Theorem 11.

Corollary 15. *Given two (path-)Hamiltonian graphs* H *and* G *of order at least* 4, *and* Φ *a nonmonochromatic position on* $G \square H$, *we have that* Φ *is* 2*-reducible.*

This result "almost" implies the result on hypercubes (Theorem 26 below). Nevertheless the proof is quite different; it uses a computer to check a huge number of cases and leads to a weaker result.

The remaining questions left by Theorem 11 can be formulated as follows:

Problem P3 (2-reducibility of small grids). Does there exist a nonmonochromatic position on the $2 \times n$ or the $3 \times n$ grid (with $n \geq 2$) which is not 2-reducible?

Our feeling is that, maybe, there exist such positions for the $2 \times n$ grid, but none for the $3 \times n$.

The next problem is maybe easier:

Problem P4 (characterization of the 1-reducible grids). Given a nonmonochromatic position Φ of the $k \times n$ grid (with $k \geq n \geq 4$), is Φ 1-reducible?

A careful analysis of the proof in [Derouet-Jourdan and Gravier 2008] may solve Problem P4. For the special positions studied previously (checkerboard and quasimonochromatic), we remarked that if Φ is not 1-reducible then $\delta \equiv 0 \bmod [3]$. Therefore one may conjecture:

Conjecture 16. *A position* Φ *on a "large enough" grid* G *is not* 1*-reducible if and only if* Φ *is monochromatic or* $\delta(G, \Phi) \equiv 0 \bmod [3]$.

3. Algorithms on Impartial Solitaire Clobber

In this section, we only deal with Impartial Solitaire Clobber. We know that computing the reducibility value of a given game position is NP-hard in general. But for some specific cases, it can be done efficiently.

3.1. ***Paths and cycles.*** A game position on a path (with a stone on each vertex) will be considered as a word on the alphabet {●, ○}. We use the classical notations for words and languages: $●^+$ represents all nonzero repetitions of ●, the notation $●^\star$ represents the words $●^+$ or the empty word. For example, $●○^+●$ defines all the game positions that begin with ●, continue with a sequence of ○ (at least one), and end with ●.

We now give a characterization of the 1-reducible game positions on a path:

Lemma 17. *A* 1*-reducible path is either a single stone or in the form* $●○^\star●^\star○$*, or its symmetric* $○●^\star○^\star●$*.*

Proof. One easily sees that such paths are 1-reducible. We now prove that 1-reducible paths are necessarily in this form. First note that on a 1-reducible path, one must play from the extremities at each step. A different move would split the game into two subpaths, which would yield at least two stones in the end.

The proof is done by induction on the size of the path. Let P be a 1-reducible path of length strictly greater than 1. Without loss of generality, one may assume that $P = ●○P'$ and the first move gives $●P'$. Since the length of $●P'$ is less than the length of P, by induction, $●P' = ●○^\star●^\star○$ or P' is empty. Therefore $P = ●○○^\star●^\star○$ or $P = ●○$. Both are in the form $●○^\star●^\star○$. □

The set of words $\{●○^\star●^\star○, ○●^\star○^\star●, ●, ○\}$ is a regular language, which means that it can be recognized by a finite automaton. The automaton associated to 1-reducible paths is depicted by Figure 6. The initial state is 0, and the final ones are 1, 3, 5. The set of 1-reducible paths can then be recognized in linear time.

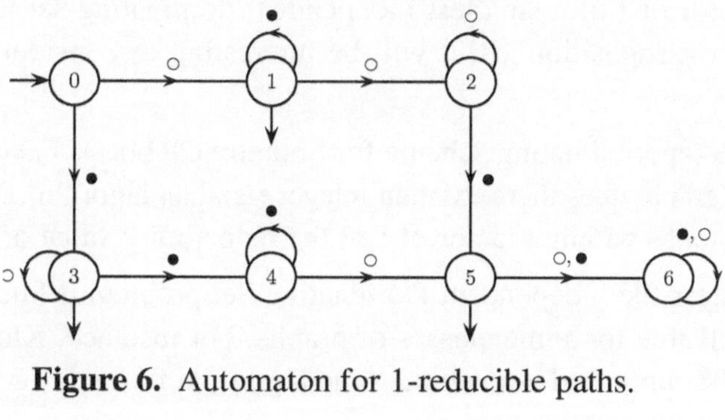

Figure 6. Automaton for 1-reducible paths.

Deciding whether a game position is k-reducible on a graph G consists in finding a partition of G into k 1-reducible subgraphs. We here give a way to see the reducibility value of a game position as the solution of a well-known problem in graph theory. When playing on paths and cycles, we show that the reducibility value can be computed in polynomial time by using this auxiliary problem.

Given a game position Φ on a graph G, consider the graph $H(G, \Phi)$ as the intersection graph of all 1-reducible subgraphs of (G, Φ) (i.e., the vertices of $H(G, \Phi)$ are the 1-reducible subgraphs of (G, Φ), and a pair of vertices (u, v) of $H(G, \Phi)$ is an edge if and only if the subgraphs u and v of G have a common vertex).

Proposition 18. *A game position Φ on a graph G is k-reducible if and only if $H(G, \Phi)$ admits a maximal independent set of cardinality k.*

Proof. Since Φ is k-reducible, there exists a partition $\mathcal{U} = \{U_1, \dots, U_t\}$ of G into $t \le k$ 1-reducible subgraphs. Each stone u of Φ must belong to exactly one 1-reducible subgraph U_i of the partition. This means that in $H(G, \Phi)$, $\mathcal{U}$ must be independent (at most one U_i per stone) and maximal (each stone must be covered by at least one U_i).

Let $\mathcal{U} = \{U_1, \dots, U_k\}$ be a maximal independent set of $H(G, \Phi)$. Since $\mathcal{U}$ is independent and each U_i is a 1-reducible subgraph, then the restriction of Φ to $\cup U_i$ is k-reducible. Now since each vertex of G is a vertex of $H(G, \Phi)$, by maximality of $\mathcal{U}$, the position Φ is k-reducible. □

In terms of graph theory, the maximal independent set problem is known as:

Independent dominating set. Given an integer k and a graph G, does there exist an independent set D of G of cardinality $\le k$ such that each vertex v of G either belongs to D or is adjacent to a vertex of D?

This problem is NP-complete for arbitrary graphs (see [Garey and Johnson 1979; Chang and Nemhauser 1984; Hochbaum and Shmoys 1985]). R. Irving [1991] showed that even the problem of finding an independent dominating set within a factor of t of a smallest independent dominating set is still NP-hard. According to Proposition 18, it will be interesting to consider the following problem:

Problem P5 (approximating scheme for Solitaire Clobber). Given a game position Φ on a graph, does there exist an integer t and an algorithm reducing Φ to a number of stones within a factor of t of the reducibility value of Φ?

Nevertheless the Independent Dominating Set problem is known to be polynomially solvable for some classes of graphs. For instance, Klostermeyer and Eschen [2000] proposed a quadratic time algorithm for arc-circular graphs.

The class of arc-circular graphs contains the intersection graphs of paths (known as *interval graphs*) and cycles. Moreover, since the number of 1-reducible subgraphs of a path (resp. cycle) on n vertices is at most n^2, one may use the algorithm of Klostermeyer et al. to determine the reducibility value of paths (resp. cycles) in $O(n^4)$ time. However, Blondel, Hendrickx and Jungers [Blondel et al. 2008] suggest another approach to compute this value in linear time. They encode a position by words, but letters refer to separations between stones: s means that it is a separation between two stones with same color and a means it is a separation between opposite colors. For example, the position ●●●○○●○○● is represented by the word $ssasasa$. This breaks the symmetry between the colors since each word encodes two opposite positions.

Theorem 19 [Blondel et al. 2008]. *Given a game position Φ on a path on n vertices, the reducibility value of Φ can be computed in $O(n)$ time.*

By removing an edge of a cycle, we get a path and we can compute its reducibility value in linear time. This provides an algorithm running in quadratic time. This complexity can be improved to linear time:

Corollary 20 [Blondel et al. 2008]. *The reducibility value of a game position on a cycle can be computed in linear time.*

Proof. Let $v_1, \ldots, v_n$ be the n vertices of a cycle C of size n with edges (v_i, v_{i+1}) where the subscript is taken modulo n. Let Φ be a nonmonochromatic position on C. Suppose that $\Phi(v_1) = \bullet$. We consider all the different ways of playing v_1. There are five cases.

- No move involves v_1 (rightmost diagram in Figure 7). Then

$$\mathrm{rv}(C, \Phi) \leq \mathrm{rv}(C - v_1, \Phi) + 1.$$

- The vertex v_1 is clobbered by v_n (middle diagram in Figure 7). This is possible only if the first white vertex v_j at the left of v_1 moves to v_n. Let

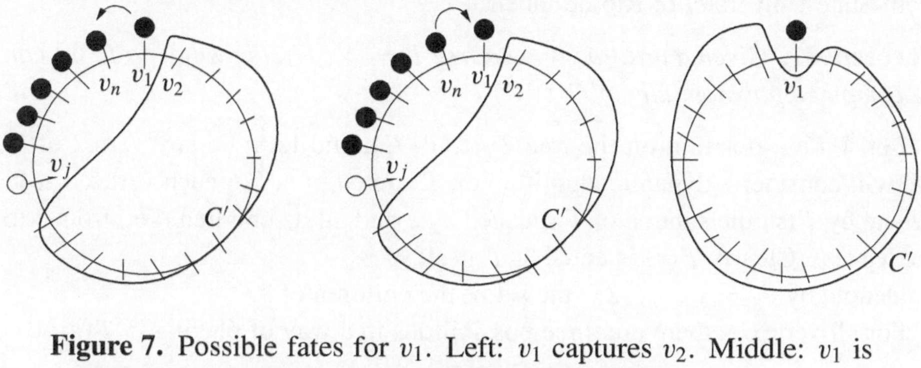

Figure 7. Possible fates for v_1. Left: v_1 captures v_2. Middle: v_1 is captured by v_2. Right: v_1 is not involved in the move.

C' be the path induced by $v_1, \ldots, v_{j-1}$ and Φ' be a position on C' defined by $\Phi'(v_1) = \circ$ and $\Phi'(v_i) = \Phi(v_i)$ for all $i = 2, \ldots, j-1$. Therefore, $\mathrm{rv}(C, \Phi) \leq \mathrm{rv}(C', \Phi')$.

- The vertex v_1 is clobbered by v_2. This case is similar to the previous one.
- The vertex v_1 clobbers v_n (leftmost diagram in Figure 7). This is possible only if the first white vertex v_j at the left of v_1 moves to v_n. If $j = n$, let $C' = C - v_1$ and Φ' be a position on C' such that $\Phi'(v_i) = \Phi(v_i)$ for all $i \neq 1, n$ and $\Phi'(v_n) = \bullet$. Then $\mathrm{rv}(C, \Phi) \leq \mathrm{rv}(C', \Phi')$. Otherwise let C' be the path induced by $v_2, \ldots, v_{j-1}$ and Φ' be the restriction of Φ on C'. We thus have $\mathrm{rv}(C, \Phi) \leq \mathrm{rv}(C', \Phi') + 1$.
- The vertex v_1 clobbers v_2. This case is similar to the previous one.

Now, since we explored all the possibilities to play v_1, we get $\mathrm{rv}(C, \Phi)$ by taking the minimum value among all the five possibilities. Thanks to Theorem 19, we obtain a linear time algorithm to compute $\mathrm{rv}(C, \Phi)$. □

3.2. ***Trees.*** Farber [1981/82] proved that the Independent Dominating Set problem is polynomial on chordal graphs. A graph G is chordal if and only if there exist a tree T and a family $\mathcal{F}$ of subtrees of T such that G is the intersection graph of $\mathcal{F}$. Therefore, one may use Proposition 18 in order to find a polynomial time algorithm for Impartial Solitaire Clobber on trees. Unfortunately, given a tree T and a game position on T we don't know whether the number of 1-reducible induced subtrees of T is polynomial.

However, deciding whether a game position on a tree is 1-reducible can be computed in linear time: it suffices to play recursively from the hanging vertices whenever it is possible (a hanging vertex is a vertex of degree 1). If there remains a unique stone in the end, then the tree is 1-reducible. Otherwise, it is not.

In this section, we propose a dynamic algorithm to solve Impartial Solitaire Clobber on trees. This result was obtained in collaboration with Luerbio Faria from State University of Rio de Janeiro.

Theorem 21. *Given a position Φ on a tree $T = (V, E)$, the value* $\mathrm{rv}(T, \Phi)$ *can be computed polynomially in $O(|V|^3)$.*

Let Φ be a position on the tree $T = (V, E)$, and let r be any vertex of T. We will consider a dynamic algorithm on T rooted in r. For each vertex $s \in V$, denote by $T(s)$ the subtree of T induced by s and all its children (i.e., from s to the leaves). Clearly $T(r)$ is equal to T itself.

Denote by $J = \{s_1, \ldots, s_p\}$ the set of the children of s.

For all vertex s, there are three possibilities in a way of playing in $T(s)$:

- in the end s is $\circ$.

- in the end s is $\bullet$.
- in the end s supports no stone.

Remark that in the third case, the stone on s moves to some vertex s_i at some step during the game. Therefore it corresponds to playing a *joker* move in the subtree $T(s_i)$.

Hence for each vertex s of T scanned from the leaves to the root, we compute five *k-values*:

(1) $k_{-1}(s)$, which is the reducibility value of $T(s)$ when a stone $\circ$ is left on s in the end.

(2) $k_1(s)$, which is the reducibility value of $T(s)$ when a stone $\bullet$ is left on s in the end.

(3) $k_0(s)$, which is the reducibility value of $T(s)$.

(4) $k_{bj}(s)$, which is the reducibility value of $T(s)$ when the *black joker* is played. The *black joker* consists in turning a white stone $\circ$ on s into a black one $\bullet$ whenever during the play. It must be used exactly once.

(5) $k_{wj}(s)$, which is the reducibility value of $T(s)$ when the *white joker* is played. The *white joker* consists in turning a black stone $\bullet$ on s into a white one $\circ$ whenever during the play. It must be used exactly once.

When computing a k-value, if $T(s)$ is monochromatic $\circ$ (resp. $\bullet$), then we assign the value ∞ to $k_1(s)$ and $k_{wj}(s)$ (resp. $k_{-1}(s)$ and $k_{bj}(s)$).

Note that $k_0(r)$ is exactly the value $\mathrm{rv}(T, \Phi)$. Then we also have for all vertex s, $k_0(s) \leq \min(k_{-1}(s), k_1(s))$.

Now consider a way of playing on $T(s)$ such that there is no move from s to some s_i. Let

$$J_{-1} = \{s_i \mid \text{there is a move of a stone } \circ \text{ from } s_i \text{ to } s\},$$

$$J_1 = \{s_i \mid \text{there is a move of a stone } \bullet \text{ from } s_i \text{ to } s\},$$

$$J_0 = \{s_i \mid \text{there is no move between } s_i \text{ to } s\}.$$

Remark that by definition of J_1 and J_{-1}, we have

$$-1 \leq |J_1| - |J_{-1}| \leq 1.$$

The value of this way of playing (i.e., the number of remaining stones) is thus greater or equal to

$$W(J_{-1}, J_1, J_0) = \sum_{i \in J_{-1}} (k_{-1}(s_i) - 1) + \sum_{i \in J_1} (k_1(s_i) - 1) + \sum_{i \in J_0} k_0(s_i) + 1.$$

A part of the resolution of Impartial Solitaire Clobber on trees relies on the following problem:

***J*-tripartition.** Given an integer a, a set $J = \{s_1, \dots, s_p\}$ and three mappings k_i (with $i \in \{-1, 0, 1\}$) from J to $\mathbb{N} \cup \{\infty\}$, find a partition $J_{-1} \cup J_0 \cup J_1$ of J which minimizes

$$W = \sum_{i \in J_{-1}} (k_{-1}(s_i) - 1) + \sum_{i \in J_1} (k_1(s_i) - 1) + \sum_{i \in J_0} k_0(s_i) + 1$$

under the constraint

$$|J_1| - |J_{-1}| = a. \tag{1}$$

Lemma 22. *J-tripartition can be solved in $O(|J|^2)$ time.*

Proof. For all $s_i \in J$, set $x_i = k_1(s_i) - k_0(s_i) - 1$ and $y_i = k_{-1}(s_i) - k_0(s_i) - 1$. Let $X = \{i \mid s_i \in J_1\}$ and $Y = \{i \mid s_i \in J_{-1}\}$. Thus, x_i represents the relative gain of putting s_i in J_1 and y_i the relative gain of putting s_i in J_{-1}. The overall gain is then

$$W(X, Y) = \sum_{i \in X} x_i + \sum_{i \in Y} y_i.$$

We want to minimize this quantity with respect to the constraint $|X| - |Y| = a$. Without loss of generality, assume that $a \geq 0$.

Additionally one may assume that $a \geq p$, otherwise there is no solution.

Let (X_n^*, Y_n^*) be an optimal solution, i.e., a solution such that $W(X_n^*, Y_n^*)$ is minimum with respect to the fact that $|X_n^*| = n$ and $|X_n^*| - |Y_n^*| = a$.

Let (X_n, Y_n) be the solution produced by Algorithm 1 with $|X_n| = n$.

If $n = a$, then clearly $W(X_n, Y_n) = W(X_n^*, Y_n^*)$. Now suppose $n > a$, and assume $W(X_k, Y_k) = W(X_k^*, Y_k^*)$ for all $k < n$, where the pair (X_k, Y_k) is obtained by Algorithm 1 and (X_k^*, Y_k^*) are two optimal sets with respect to the fact that $|X_k^*| = k$ and $|X_k^*| - |Y_k^*| = a$.

Let Δ denote the set $(X_n^* \cup Y_n^*) \setminus (X_{n-1} \cup Y_{n-1})$. Since $|X_n^*| > |X_{n-1}|$ and $|Y_n^*| > |Y_{n-1}|$, we have $|\Delta| \geq 2$.

Case I: There exist $i, j \in \Delta$ such that $i \in X_n^*$ and $j \in Y_n^*$.

By the induction hypothesis, we have $W(X_n^* \setminus \{i\}, Y_n^* \setminus \{j\}) \geq W(X_{n-1}, Y_{n-1})$. Moreover, by the first item of Algorithm 1 we get $W(X_{n-1} \cup \{i\}, Y_{n-1} \cup \{j\}) \geq W(X_n, Y_n)$. Thus $W(X_n^*, Y_n^*) \geq W(X_n, Y_n)$.

Case II: $\Delta \subset X_n^*$. Let i, j be two distinct elements of Δ. Then, since $|Y_n^*| > |Y_{n-1}|$, there exists $l \in Y_n^* \setminus Y_{n-1}$. As $\Delta \subset X_n^*$, it turns out that $l \notin \Delta$ and thus $l \in X_{n-1}$. By the induction hypothesis we have $W(X_n^* \cup \{l\} \setminus \{i, j\}, Y_n^* \setminus \{l\}) \geq W(X_{n-1}, Y_{n-1})$. Morevoer, by the second item of Algorithm 1, we obtain $W(X_{n-1} \cup \{i, j\} \setminus \{l\}, Y_{n-1} \cup \{l\}) \geq W(X_n, Y_n)$. Thus $W(X_n^*, Y_n^*) \geq W(X_n, Y_n)$.

Case III: $\Delta \subset Y_n^*$. Symmetric to case II, by considering the third item of Algorithm 1.

Algorithm 1. J-tripartition

Require: (x_i, y_i) for $i \in \{1, 2, \dots, p\}$ and a

Ensure: $\sum_{i \in X^*} x_i + \sum_{i \in Y^*} y_i$ is minimum and $|X^*| - |Y^*| = a$

put the a elements i with the smallest values of x_i in X_a and set $Y_a = \varnothing$
$n \leftarrow a$

repeat
$X^* \leftarrow X_n$
$Y^* \leftarrow Y_n$
$n \leftarrow n + 1$
For all triples (i', j', j), take the minimum value of $W(X_n, Y_n)$ among the three following cases:

(1) For $i', j' \notin \{X_{n-1}, Y_{n-1}\}$, set $X_n = X_{n-1} \cup \{i'\}$, $Y_n = Y_{n-1} \cup \{j'\}$.

(2) For $i', j' \notin \{X_{n-1}, Y_{n-1}\}$ and $j \in Y_{n-1}$, set $X_n = X_{n-1} \cup \{j\}$ and $Y_n = (Y_{n-1} \setminus \{j\}) \cup \{i', j'\}$.

(3) For $i', j' \notin \{X_{n-1}, Y_{n-1}\}$ and $j \in X_{n-1}$, set $Y_n = Y_{n-1} \cup \{j\}$ and $X_n = (X_{n-1} \setminus \{j\}) \cup \{i', j'\}$.

until $(W(X_n, Y_n) \geq W(X^*, Y^*))$ or $(n > a + \lfloor \frac{p-a}{2} \rfloor)$
if $W(X_n, Y_n) < W(X^*, Y^*)$ **then**
$X^* \leftarrow X_n$
$Y^* \leftarrow Y_n$
end if

Complexity analysis. The initialization step can be done in $O(p \log p)$. In the loop, the considered sets need at most $O(p^3)$ operations to be enumerated. The update of best sets costs $O(1)$. In the end, we get a $O(p^4)$-time algorithm.

However, in Algorithm 1 we do not need to consider all the possible triples (i', j', j), but only one that minimizes W. To do this, we start by sorting the sets $A = \{x_1, \dots, x_p\}$, $B = \{y_1, \dots, y_p\}$ and $C = \{x_1 - y_1, \dots, x_p - y_p\}$ by increasing order. This can be done in $O(p \log p)$ time. Additionally consider the function $\xi : \{1, \dots, p\} \to \{x, y, z\}$ such that

$$\xi(i) = \begin{cases} x & \text{if } i \in X_{n-1}, \\ y & \text{if } i \in Y_{n-1}, \\ z & \text{otherwise.} \end{cases}$$

Let

$$\begin{aligned}
i'_1 &= \text{index of } \min\{x_i \mid \xi(i) = z\},\\
j'_1 &= \text{index of } \min\{y_j \mid \xi(j) = z \text{ and } j \neq i'_1\},\\
i'_2 &= \text{index of } \min\{x_i \mid \xi(i) = z \text{ and } i \neq i'_1\},\\
j_1 &= \text{index of } \min\{x_i - y_i \mid \xi(i) = y\},\\
j_2 &= \text{index of } \max\{x_i - y_i \mid \xi(i) = x\}.
\end{aligned}$$

Each one of the indices above can be computed in $O(p)$ time thanks to the sorts of A, B and C. Now for each step in the loop of Algorithm 1, it is sufficient to consider the triple (i', j', j) such that $(i', j') = (i'_1, j'_1)$ or (i'_2, i'_1). For step 2, set $j = j_1$, and for step 3, set $j = j_2$. Finally, since updating ξ can be done in constant time, we get a $O(p^2)$ time algorithm. □

We now have enough ingredients to prove Theorem 21.

Proof of Theorem 21. By induction from the leaves to the root r of T.

Let l be a leaf of T. We have straightforwardly $k_0(l) = 1$. If l is white, then $k_{-1}(l) = k_{bj} = 1$ and $k_1(l) = k_{wj} = \infty$. Conversely if l is black.

We now suppose that for some vertex s of T, the five k-values of each child s_i for $1 \leq i \leq p$ are known. Then the k-values of s can be computed as follows:

(1) $k_{-1}(s)$. Since s must be white in the end, there is no move from s to some s_i. Thus an optimal way of playing on $T(s)$ is a solution of J-tripartition with $a = 0$ if s is initially ○, otherwise $a = -1$.

(2) $k_1(s)$. Identical to the previous case with $a = 0$ or 1 according to the initial color of s.

(3) $k_0(s)$. Let us consider first the computation of $k_\varnothing(s)$, which is the reducibility value of $T(s)$ when there remains no stone on s in the end. If there remains no stone on s in the end, one must play from s to some child s_j. Assume that s_j is fixed. There are two possible ways of playing to compute $k_\varnothing(s)$ (depending on whether $s \to s_j$ is a black or white move): as for $k_1(s)$, compute the reducibility value of $T(s) \setminus T(s_j)$ when a stone ● is left on s in the end. Denote by W_b this value. Play then on $T(s_j)$ by considering the $s \to s_j$ move as a black joker. These operations leave $W_b + k_{bj}(s_j) - 1$ stones on $T(s)$. Do the same by yielding a stone ○ on s and then considering the white joker.

Apply this method by fixing each child $s_j \in J$, and take the minimum value among these $2|J|$ computations.

Now set $k_0(s) = \min(k_{-1}(s), k_1(s), k_\varnothing(s))$.

(4) $k_{bj}(s)$. Proceed as for the computation of $k_O(s)$. But since we have one additional move (the joker one $\bullet \to \circ$), we need an additional child in J_{-1}. Hence replace a by $a-1$ in the previous cases.

(5) $k_{wj}(s)$. Identical to the computation of $k_{bj}(s)$, except that we replace a by $a+1$ in the previous cases.

When considering the complexity of this algorithm, each vertex is scanned exactly once to compute its five k-values. The computation of $k_O(s)$ is the highest one, since it requires to apply $|J|$ times the J-tripartition algorithm. Therefore $k_O(s)$ needs $O(p^3)$ operations. Since p is at most the degree of a vertex, we get a total complexity in $O(n^3)$. □

Remark that if we apply the algorithm of Theorem 21 for a path where the root is a leaf, then $p = 1$ at each time we apply J-tripartition. Hence this algorithm also solves Impartial Solitaire Clobber on paths in linear time.

It will be interesting to develop algorithms for other classes of graphs. Since the structure of chordal graphs is closed to the one of trees, one may ask the following problem:

Problem P6 (Impartial Solitaire Clobber on chordal graphs). Given a position Φ on a chordal graph G, can we compute $\mathrm{rv}(G, \Phi)$ in polynomial time?

4. Other classes of graphs

In this section we determine the reducibility value for some classes of graphs.

4.1. ***Impartial Solitaire Clobber on Hamming graphs.*** Playing Solitaire Clobber on cliques is not a tricky activity. It is straightforward that all nonmonochromatic cliques are 1-reducible. A more detailed analysis even lead to the fact that one can choose the color and the location of the final stone.

Hamming graphs are multiple Cartesian product of cliques. The clique K_n, $K_2 \square K_3$ or $K_4 \square K_5 \square K_2$ are examples of Hamming graphs. Hypercubes, defined by $Q_n = \square^n K_2$, constitute the most famous class of Hamming graphs. It is well-known that any Hamming graph admit an Hamiltonian cycle.

For these reasons, it is not surprising that the reducibility value of a position on a Hamming graph is small. From Corollary 15, it turns out that Hamming graphs are 2-reducible. Note that this result was already contained in [Dorbec et al. 2008], where the case of Hamming graphs is fully investigated.

As for the case of cliques, one can reduce "strongly" the family of Hamming graphs, in the sense that one can choose the location and the color of the final stones. This property is developed through the two following definitions:

Definition 23. A graph G is strongly 1-reducible if: for any vertex v of G, for any position Φ on G (provided $(G \setminus v, \Phi)$ is not monochromatic), for any color c ($\bullet$ or $\circ$), there exists a way of playing that yields a unique stone of color c on v.

Definition 24. A graph G is strongly 2-reducible if: for any vertex v of G, for any position Φ on G (provided $(G \setminus v, \Phi)$ is not monochromatic), for any two colors c and c' (provided there exist two different vertices u and u' such that $\Phi(u) = c$ and $\Phi(u') = c'$), there exists a way of playing that yields a stone of color c on v, and (possibly) a second stone of color c' somewhere else.

The following result asserts that almost all the Hamming graphs are strongly 1-reducible.

Theorem 25 [Dorbec et al. 2008]. *Any Hamming graph that is neither* $K_2 \square K_3$ *nor an hypercube is strongly* 1*-reducible.*

The case of hypercubes and the graph $K_2 \square K_3$ are a bit different.

In the case of hypercubes, there are some game positions for which the reducibility value equals 2. For example, the position  on the hypercube Q_2 is not 1-reducible. Dorbec et al. [2008] showed that for all integer $n > 1$, there exists a nonmonochromatic position on Q_n which is not 1-reducible. Nevertheless, hypercubes were proved to be 2-reducible, and even more since the location and the color of the final stones can be fixed (under some conditions). It corresponds to Theorem 26.

Theorem 26 [Dorbec et al. 2008]. *Hypercubes are strongly* 2*-reducible.*

Note that this result is not sufficient to quickly decide whether or not a game position on an hypercube is 1-reducible. In [Dorbec et al. 2008] Conjecture 27 is proposed. Since hypercubes are bipartite graphs, the invariant δ of Definition 7 and Proposition 8 is available.

Conjecture 27. *The nonmonochromatic game positions* Φ *on the hypercube* Q_n *that are* 1*-reducible are exactly those for which the invariant*

$$\delta(Q_n, \Phi) \not\equiv 0 \bmod [3].$$

We conclude this part with the Hamming graph $K_2 \square K_3$. It was proved in [Dorbec et al. 2008] that nonmonochromatic positions on this graph are always 1-reducible. However the conditions for strong 1-reducibility are not fulfilled.

One may ask whether there are strong 1-reducible graphs which are not Hamming graphs, and more generally:

Problem P7 (strongly 1-reducible graphs). Characterize all the graphs that are strongly 1-reducible.

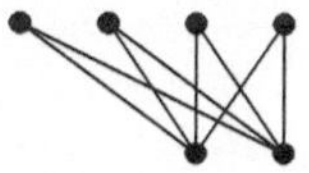

Figure 8. The complete bipartite graph $K_{4,2}$.

4.2. ***Impartial Solitaire Clobber on complete multipartite graphs.*** Given $k \geq 2$ integers $n_1, \ldots, n_k$, the complete k-partite graph $K_{n_1,\ldots,n_k}$ is the graph $G = (V, E)$ such that V can be partitioned into k independent sets S_i with $|S_i| = n_i$ for $1 \leq i \leq k$, and for all $u \in S_i$ and $v \in S_j$ with $i \neq j$ we have $(u, v) \in E$. See Figure 8 for an example.

For convenience, the edges of the complete k-partite graphs will not be drawn in the other figures.

Complete bipartite graphs. As shown in the previous sections, the invariant $\delta(G, \Phi)$ defined by Demaine et al. is a powerful tool to compute the reducibility value on several families of bipartite graphs. It is also the case of balanced complete bipartite graphs $K_{n,n}$, as proved in [Duchêne et al. 2009]. For such graphs we have a result similar to the one of checkerboard grids:

Proposition 28 [Dorbec et al. 2008]. *A position Φ on a balanced complete bipartite graph $K_{n,n}$ is* 1*-reducible if $\delta(K_{n,n}, \Phi) \not\equiv 0 \bmod [3]$. Otherwise, the reducibility value is equal to* 2.

When the complete bipartite graph is unbalanced, the reducibility value may be higher. The invariant δ is thus not sufficient to compute the exact reducibility value.

Let us denote by $K_{n,m}$ with $n > m > 0$ an unbalanced complete bipartite graph, and Φ a game position on $K_{n,m}$.

The values n_b and n_w denote respectively the numbers of black and white stones in the independent set of size n. Obviously we have $n = n_b + n_w$. Similarly, m_b and m_w denote respectively the numbers of black and white stones in the independent set of size m.

Without loss of generality, we consider positions satisfying $n_b \leq n_w$. Under this condition, we define a nonnegative integer $q = n - 2n_b$. We also define the function $f(\Phi) = q - m - m_b$.

Theorem 29 [Duchêne et al. 2009]. *Let Φ be a game position on $K_{n,m}$ with $n > m > 0$ and $n_b \leq n_w$.*

If $f(\Phi) < 0$ then $\mathrm{rv}(K_{n,m}, \Phi) \leq 2$ *and* $\mathrm{rv}(K_{n,m}, \Phi) = 1$ *if and only if* $\delta(K_{n,m}, \Phi) \not\equiv 0 \bmod [3]$.

If $f(\Phi) \geq 0$ then $\mathrm{rv}(K_{n,m}, \Phi) = f(\Phi) + 2$.

The ground of Theorem 29 can be explained as follows. When $f(\Phi) < 0$, it is always possible to reduce Φ to a nonmonochromatic position on a balanced

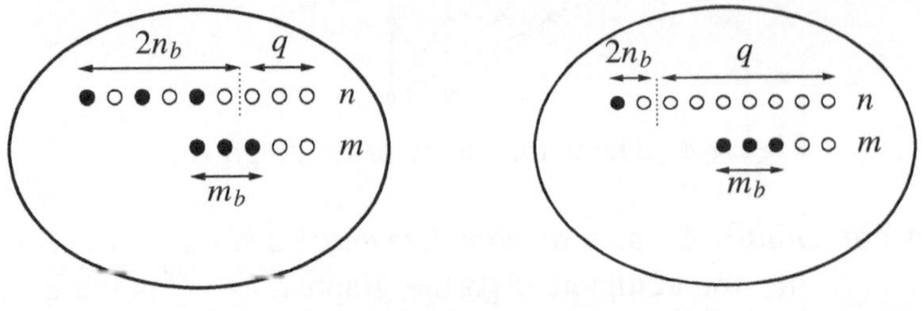

Figure 9. Two examples of positions satisfying $f(\Phi) < 0$.

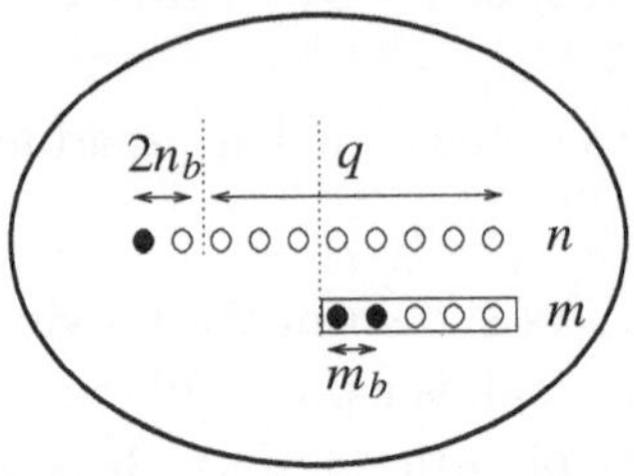

Figure 10. A position satisfying $f(\Phi) \geq 0$.

complete bipartite graph $K_{m,m}$. This operation can be done by alternating black and white moves to reduce the size of the larger independent set, or by clobbering at most $m_b - 1$ black stones of the smaller independent set. Two examples of such positions are given by Figure 9. Once the graph is reduced to a balanced $K_{m,m}$, it suffices to use Proposition 28.

When $f(\Phi) \geq 0$, it means that the number of white stones in the larger independent set is too big comparing to the number of black stones in the smaller independent set. A reduction to a balanced position is not possible. Figure 10 illustrates this situation. In such cases, the reducibility value equals $f(\Phi) + 2$. One key of the proof is to show that when playing any move from Φ, the function f never decreases.

Complete k-partite graphs with $k \geq 3$. In the case of complete k-partite graphs with $k \geq 3$, the invariant δ is no more available. Nevertheless, the introduction of the function f for complete bipartite graphs is the key for the other k-partite graphs. We distinguish two cases, depending on the number of maximum independent sets of the graph.

Theorem 30 [Duchêne et al. 2009]. *Let Φ be a nonmonochromatic position on a k-partite graph G with $k \geq 3$. If G has at least two independent sets of maximum size, then Φ is 1-reducible.*

An example of such a 1-reducible position is given by Figure 11 (left).

When the k-partite graph G has a unique maximum independent set M_0 (as on Figure 11 (right), then the reducibility is once again fixed by the function f.

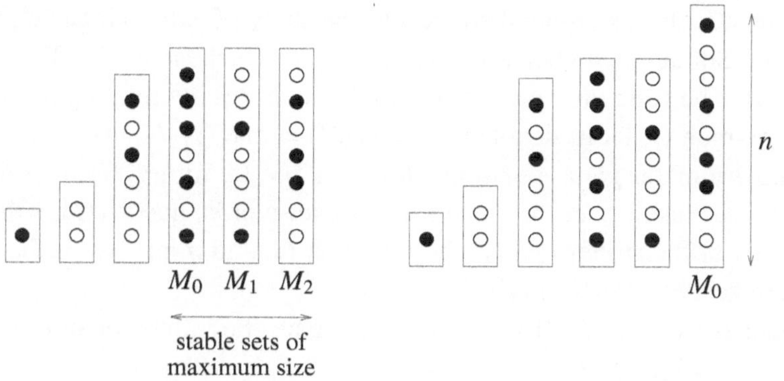

Figure 11. Left: 6-partite complete graph with three maximum independent sets. Right: 6-partite complete graph with a unique maximum independent set.

The parameters introduced for complete bipartite graphs are extended for k-partite graphs : the value n is the size of the maximum independent set M_0, the quantities n_b and n_w are the numbers of black and white stones in M_0, the parameter m is the size of $G \setminus M_0$, and q is still equal to $n - 2n_b$. The function f is unchanged (i.e., equal to $q - m - m_b$). Then the reducibility value is similar as in Theorem 29:

Theorem 31 [Duchêne et al. 2009]. *Let Φ be a nonmonochromatic position on a k-partite graph G with $k \geq 3$. If $f(\Phi) < 0$ then Φ is* 1*-reducible. If $f(\Phi) \geq 0$ then* $\mathrm{rv}(K_{n,m}, \Phi) = f(\Phi) + 2$.

Problem P8 (general invariant). In view of the previous result and the use of the function f, could we provide a general invariant that extends δ to nonbipartite graphs?

5. General results about Impartial Solitaire Clobber

If you play Impartial Solitaire Clobber on a random graph with a random position, you have great chance to leave a unique stone in the end. This result is due to Ruszinkó (personal communication) and can be formulated in the following way in terms of random graphs:

Proposition 32 (Ruszinkó). *Almost all graphs are* 1*-reducible.*

In other words, it means that for any fixed $p \in]0, 1[$, the random graph $G_{n,p}$ is 1-reducible with probability tending to 1 as n tends to infinity — for details about the Erdős–Rényi random graph model $G_{n,p}$, see [Bollobás 2001].

In contrast with this probabilistic result, the study of extremal game positions of Impartial Solitaire Clobber was investigated in [Duchêne et al. 2009; Blondel et al. 2008]. The following part deal with the main results relating to this topic.

Given a graph G, let us denote by $\text{maxrv}(G) = \max(\text{rv}(G, \Phi))$ taken over all non-monochromatic game positions Φ on G, and by $\text{minrv}(G)$ the minimum of these values taken over the same set. In other words, $\text{maxrv}(G)$ is the worst arrangement of the stones that can be found on G, and $\text{minrv}(G)$ is the the one that minimizes the reducibility value.

In [Duchêne et al. 2009] we tried to estimate the values of $\text{minrv}(G)$ and $\text{maxrv}(G)$ for all graph G.

The first result asserts that it is always possible to find a 1-reducible position on any connected graph.

Proposition 33. *Let G be a connected graph. Then we have* $\text{minrv}(G) = 1$.

Proof. Let $G = (V, E)$ be a connected graph. We will prove a stronger result: for any vertex u of G and any color $c \in \{\circ, \bullet\}$, there exists a position Φ on G such that (G, Φ) is 1-reducible with the last stone located on u and with color c.

The proof works by induction. It is obvious if $|V| = 1$. Now suppose that $|V| \geq 2$. Let u be a vertex of G and $c \in \{\bullet, \circ\}$. Let $G_1, \dots, G_t$ be the t connected components of $G - u$. For each $i = 1, \dots, t$, let u_i be a neighbour of u in G_i. Without loss of generality, assume that $c = \bullet$. Set $c_i = \bullet$ for $1 \leq i \leq \lceil t/2 \rceil$, and $c_j = \circ$ for $\lceil t/2 \rceil < j \leq t$. By the induction hypothesis, there exists a position Φ_i on G_i such that there is a way of playing that yields a unique stone of color c_i on u_i. Now consider the position Φ on G defined as follows: $\Phi(v) = \Phi(u)$ for all $v \neq u$, $\Phi(u) = \bullet$ if t is even, and $\circ$ otherwise. If t is even, then play alternately a pair $(u_j \to u, u_i \to u)$ with $i \leq \lceil t/2 \rceil$ and $j > \lceil t/2 \rceil$. If t is odd, play first $u_1 \to u$ and then as the case where t is even. □

The discussion is more tricky when estimating the value of $\text{maxrv}(G)$. Upper bounds are given in [Duchêne et al. 2009]. They involve the minimum degree d of the graph.

Theorem 34 [Duchêne et al. 2009]. *Given a graph G of minimum degree d, we have* $\text{maxrv}(G) \leq n - d$.

Theorem 34 asserts that it is always possible to find a strategy with at least d moves. Note that a greedy strategy is provided to do so.

When d is fixed, Duchêne et al. [2009] give in a characterization of the graphs G satisfying exactly $\text{maxrv}(G) = n - d$. Roughly speaking, they correspond to stars of cliques of size $d + 1$. This family of graphs is fully defined below and is denoted by $\mathcal{G}_d$.

For any $d \geq 1$, let us define $\mathcal{G}_d$, a set of connected graphs of degree minimum d as follows:

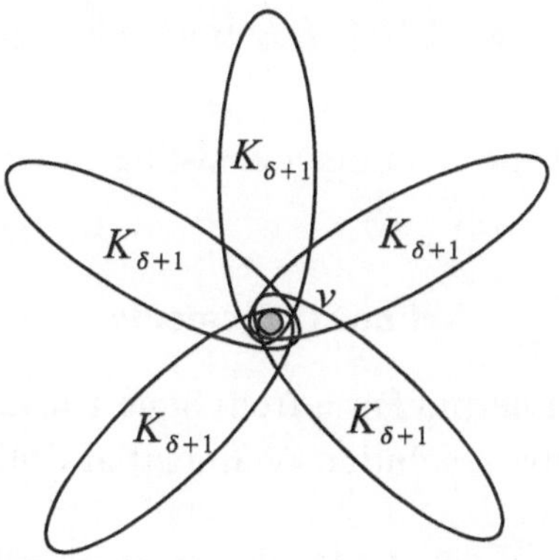

Figure 12. The graph $S_k(K_{d+1})$.

- The complete graph on $d+1$ vertices belongs to $\mathcal{G}_d$ for all $d \geq 1$.
- For any integer $k \geq 2$, let us define $S_k(K_{d+1})$ as the graph obtained by k disjoined copies of K_d plus one additional vertex v adjacent to all the vertices of each K_d.
- For $d = 2$, add the cycle on 4 vertices C_4.
- No other graph belongs to $\mathcal{G}_d$.

Theorem 35 [Duchêne et al. 2009]. *For all $d \geq 1$, the set of connected graphs G having minimum degree d and such that* $\operatorname{maxrv}(G) = n - d$ *is exactly $\mathcal{G}_d$.*

For example, a position Φ on a graph $G \in \mathcal{G}_d$ that satisfies $\operatorname{rv}(G, \Phi) = n - d$ can be built as follows:

- If $G = K_{d+1}$, then choose any nonmonochromatic position Φ on G.
- If $G = S_k(K_{d+1})$, then choose $\Phi(v) \neq \Phi(u)$ for all vertex $u \neq v$.
- If $G = C_4 = (v_1, \ldots, v_4)$, then choose a position Φ such that $\Phi(v_1) = \Phi(v_2) = \bullet$ and $\Phi(v_3) = \Phi(v_4) = \circ$.

Given a graph G, it seems that the computation of $\operatorname{maxrv}(G)$ is a hard problem.

Problem P9 ($\operatorname{maxrv}(G)$). Given a graph G and an integer k, does there exist a polynomial time algorithm deciding whether $\operatorname{maxrv}(G) \leq k$?

Blondel et al. [2008] give in a closed formula for $\operatorname{maxrv}(G)$ when G is a path or a cycle. Their results are summed up below:

Proposition 36 [Blondel et al. 2008]. *If P is a path of size n, then* $\operatorname{maxrv}(P) = \lceil n/2 \rceil$.

Note that on a path P of size n, examples of game positions Φ satisfying $\operatorname{rv}(P, \Phi) = \lceil n/2 \rceil$ are those in the form $\bullet^k \circ \bullet^k$.

Proposition 37 [Blondel et al. 2008]. *If C is a cycle of size n, then* $\operatorname{maxrv}(C) = \lceil n/3 \rceil$.

This value is reached for positions consisting in repetitions of the pattern (●●○).

Acknowledgements

We would like to thank Luerbio Faria from State University of Rio de Janeiro for his involvement in the resolution of Impartial Solitaire Clobber on trees (Theorem 21 in particular).

This work is supported by the Fédération Recherche Maths à Modeler.

References

[Albert et al. 2005] M. H. Albert, J.P. Grossman, R. J. Nowakowski, and D. Wolfe, "An introduction to clobber", *Integers* **5**:2 (2005), A1, 12.

[Blondel et al. 2008] V. D. Blondel, J. M. Hendrickx, and R. M. Jungers, "Solitaire Clobber as an optimization problem on words", *Integers* **8** (2008), G04, 12.

[Bollobás 2001] B. Bollobás, *Random graphs*, 2nd ed., Cambridge Studies in Advanced Mathematics **73**, Cambridge University Press, Cambridge, 2001.

[Chang and Nemhauser 1984] G. J. Chang and G. L. Nemhauser, "The k-domination and k-stability problems on sun-free chordal graphs", *SIAM J. Algebraic Discrete Methods* **5**:3 (1984), 332–345.

[Demaine et al. 2004] E. D. Demaine, M. L. Demaine, and R. Fleischer, "Solitaire clobber", *Theoret. Comput. Sci.* **313**:3 (2004), 325–338.

[Derouet-Jourdan and Gravier 2008] A. Derouet-Jourdan and S. Gravier, "Solitaire Clobber sur la Grille", 2008. Unpublished manuscript.

[Dorbec et al. 2008] P. Dorbec, E. Duchêne, and S. Gravier, "Solitaire Clobber played on Hamming graphs", *Integers* **8** (2008), G3, 21.

[Duchêne et al. 2009] E. Duchêne, S. Gravier, and J. Moncel, "New results about Solitaire Clobber", *RAIRO–Operations Research* **43** (2009), 463–482.

[Farber 1981/82] M. Farber, "Independent domination in chordal graphs", *Oper. Res. Lett.* **1**:4 (1981/82), 134–138.

[Garey and Johnson 1979] M. R. Garey and D. S. Johnson, *Computers and intractability: a guide to the theory of NP-completeness*, W. H. Freeman and Co., San Francisco, CA, 1979.

[Gravier and Parreau 2006] S. Gravier and A. Parreau, "Hamiltonicité dans une grille, un jeu d'enfant ?", internal report, École Normale Supérieure de Lyon, 2006, http://enslyon.free.fr/rapports/info/Aline_Parreau_1.pdf.

[Grossman 2004] J.P. Grossman, "Appendix A: report on the first international clobber tournament", *Theoretical Computer Science* **313**:3 (2004), 533–537.

[Hochbaum and Shmoys 1985] D. S. Hochbaum and D. B. Shmoys, "A best possible heuristic for the k-center problem", *Math. Oper. Res.* **10**:2 (1985), 180–184.

[Irving 1991] R. W. Irving, "On approximating the minimum independent dominating set", *Inform. Process. Lett.* **37**:4 (1991), 197–200.

[Itai et al. 1982] A. Itai, C. H. Papadimitriou, and J. L. Szwarcfiter, "Hamilton paths in grid graphs", *SIAM J. Comput.* **11**:4 (1982), 676–686.

[Klostermeyer and Eschen 2000] W. F. Klostermeyer and E. M. Eschen, "Perfect codes and independent dominating sets", pp. 7–28 in *Proceedings of the Thirty-first Southeastern International Conference on Combinatorics, Graph Theory and Computing* (Boca Raton, FL, 2000), vol. 142, 2000.

[Peterson 2002] I. Peterson, "Getting clobbered", *Science News* **161** (2002).

laurent.beaudou@univ-bpclermont.fr
LIMOS, Bureau D005, Complexe scientifique des Cézeaux, 63173 Aubière, France

eric.duchene@univ-lyon1.fr *Laboratoire LIRIS, Université Claude Bernard Lyon 1, Bâtiment Nautibus (ex 710), 843, Bd. du 11 novembre 1918, 69622 Villeurbanne, France*

sylvain.gravier@ujf-grenoble.fr *Institut Fourier, 100 rue des Maths - BP 74, 38402 Saint Martin d'Hères, France*

Games of No Chance 4
MSRI Publications
Volume **63**, 2015

Monte-Carlo approximation of temperature

TRISTAN CAZENAVE

Monte-Carlo tree search is a powerful paradigm for the game of Go. We propose to use Monte-Carlo tree search to approximate the temperature of a game, using the mean result of the playouts. Experimental results on the sum of five 7×7 Go games show that it improves much on a global search algorithm.

1. Introduction

Monte-Carlo Go has recently improved to compete with the best Go programs [Coulom 2007; Gelly et al. 2006; Gelly and Silver 2007]. We are interested in the use of Monte-Carlo methods when there are independent games. In such cases it might be interesting to analyze the games independently instead of considering them as a unified game.

Section 2 describes related works. Section 3 presents the Monte-Carlo algorithms we have tested. Section 4 details experimental results. Section 5 concludes.

2. Related works

In this section we expose related works on Monte-Carlo Go. We first explain basic Monte-Carlo Go as implemented in Gobble in 1993. Then we address the combination of search and Monte-Carlo Go, followed by the UCT algorithm, and previous works on the approximation of temperature.

2.1. ***Monte-Carlo Go.*** The first Monte-Carlo Go program is Gobble [Bruegmann 1993]. It uses simulated annealing on a list of moves. The list is sorted by the mean score of the games where the move has been played. Moves in the list are switched with their neighbor with a probability dependent on the temperature. The moves are tried in the games in the order of the list. At the end, the temperature is set to zero for a small number of games. After all games have been played, the value of a move is the average score of the games it has been played in first. Gobble-like programs have a good global sense but lack of tactical knowledge. For example, they often play useless Ataris, or try to save captured strings.

2.2. ***Search and Monte-Carlo Go.*** A very effective way to combine search with Monte-Carlo Go has been found by Rémi Coulom with his program Crazy Stone [Coulom 2007]. It consists in adding a leaf to the tree for each simulation. The choice of the move to develop in the tree depends on the comparison of the results of the previous simulations that went through this node, and of the results of the simulations that went through its sibling nodes.

2.3. ***UCT.*** The UCT algorithm has been devised recently [Kocsis and Szepesvári 2006], and it has been applied with success to Monte-Carlo Go in the program Mogo [Gelly et al. 2006; Gelly and Silver 2007] among others.

When choosing a move to explore, there is a balance between exploitation (exploring the best move so far), and exploration (exploring other moves to see if they can prove better). The UCT algorithm addresses the exploration/exploitation problem. UCT consists in exploring the move that maximizes $\mu_i + C\sqrt{\log(t)/s}$. The mean result of the games that start with the c_i move is μ_i, the number of games played in the current node is t, and the number of games that start with move c_i is s.

The C constant can be used to adjust the level of exploration of the algorithm. High values favor exploration and low values favor exploitation.

2.4. ***Thermography.*** Thermography [Berlekamp et al. 1982] can be used to play in a sum of combinatorial games. In Go endgames, it has already been used to find better than professional play [Spight 2002], relying on a computer assisted human analysis. A simple and efficient strategy based on thermography is Hotstrat, it consists in playing in the hottest game. Hotstrat competes well with other strategies on random games [Cazenave 2002], but it can be improved taking into account the subgame type [Andraos et al. 2007]. Another approach used to play in a sum of hot games is to use locally informed global search [Müller and Li 2006; Müller et al. 2004]. Our previous work on evaluating the temperature evaluated goals temperature on a single board using a Monte-Carlo method [Cazenave 2009]. In this paper, we either use Hotstrat or Minimax to evaluate the subgames built for each separate board.

3. Search algorithms

In this section, we present the different algorithms we have tested. They all use the score of a playout for UCT instead of the usual probability of winning because the difference in points is meaningful to approximate the temperature.

3.1. ***Global search.*** The direct application of UCT to playing on several boards is to do a global search. A move can be made on any board, the normal UCT tree is developed, and the playout is played separately on each separate board. In our implementation of global search, the color to play after the UCT tree descent starts the playout on the first board, then when the game on the first board is over, the other player starts the playout of the second board, and so on up to the completion of all the playouts on all boards.

3.2. *Dual search.* Dual search consists in performing two UCT searches on each separate board. The first one always starts with Black, and the second one always starts with White. In each search, after the descent of the UCT tree, a normal playout is played on the separate board. Each search is allocated the same number of playouts. At the end of the searches, the program knows the mean value of the playouts starting with a Black move (μ_{Black}), and the mean value of the playouts starting with a White move (μ_{White}). The temperature of the board is approximated with $(\mu_{\text{Black}} + \mu_{\text{White}} - \text{size} \times \text{size})/2$.

The player to move chooses to play the best UCT move of the board with the greatest approximated temperature.

3.3. *Threat search.* Threat search consists in performing four UCT searches on each separate board. The first one always starts with Black, and the second one always starts with White. After the first search is completed, Black knows the best UCT move. The third search always starts with the best Black move and it is followed by the descent of an UCT tree that also starts with a Black move (so all the playouts starts with two Black moves). The fourth search is the equivalent for White of the third search. Each search is allocated the same number of playouts. At the end of the searches, the program knows the mean value of the playouts starting with a Black move, the mean value of the playouts starting with a White move, the mean value of the playouts starting with two Black moves, and the mean value of the playouts starting with two White moves. It either use these values to compute the temperature with HotStrat, or to perform a Minimax search on all the values of all the boards.

The player to move chooses to play the best UCT move of the board with the greatest approximated temperature, or the best UCT move returned by Minimax.

4. Experimental results

The random games are played using the same policy as in Mogo [Gelly et al. 2006]. We tested the algorithm on a game composed of five 7×7 boards. The komi is set to 7.5 points. Therefore the maximum number of points is 252.5 for White, and 245.0 for Black.

In Table 1 the results of games between the global search algorithm and the dual search algorithm are given. The algorithms use the 0.3 UCT constant. For each algorithm, the table gives the mean number of points against the approximation algorithm.

Size	Playouts	Black	Mean number of Black points	White	Mean number of White points
5 × 7×7	1,000	Dual	199.21	Global	53.28
5 × 7×7	1,000	Global	47.13	Dual	205.37

Table 1. Results of the global search program against the dual search program.

Size	Playouts	Black	Mean number of Black points	White	Mean number of White points
5 × 7×7	1,000	Threat	127.03	Dual	125.45
5 × 7×7	1,000	Dual	120.50	Threat	131.96

Table 2. Results of the threat search program against the dual search program.

In Table 2 the results of games between the threat search algorithm and the dual search algorithm are given.

The threat search algorithm is twice slower as the dual search algorithm. We played the dual search algorithm against another dual search algorithm that is twice slower in order to compare with the threat search algorithm. Table 3 gives the results of games between the two dual search algorithms. Results of the dual threat algorithm with 2,000 playouts are similar to the results of the threat search algorithm with 1,000 playouts and they take the same time.We also give in Table 3 the result of the dual algorithm with 2,000 playouts against the threat algorithm with 1,000 playouts. The dual algorithm has a clear win. Using the threat search algorithm is more complicated and gives worse results than the more simple dual search algorithm.

In the previous experiment, the threat search algorithm uses HotStrat to choose the board to play, in order to test if HotStrat was a potential problem, we replaced it with a Minimax on the tree composed of the first two moves for Black and for White for all the boards. Minimax gave results very similar to HotStrat. The behavior of the threat search algorithm is not due to HotStrat.

Size	Black	Mean number of Black points	White	Mean number of White points
5 × 7×7	Dual (1,000)	118.39	Dual (2,000)	134.10
5 × 7×7	Dual (2,000)	128.13	Dual (1,000)	124.34
5 × 7×7	Threat (1,000)	118.58	Dual (2,000)	133.86
5 × 7×7	Dual (2,000)	131.79	Threat (1,000)	120.68

Table 3. Results with different numbers of playouts.

5. Conclusion

When a game is composed of independent games, it is better to approximate the temperature using separate Monte-Carlo tree searches on each game than using a global Monte-Carlo search. When we tested the algorithms that evaluate threats, we obtained results comparable to the more simple dual search algorithm.

References

[Andraos et al. 2007] C. R. S. Andraos, M. M. Zaky, and S. A. Ghoneim, "Comparative study of approximate strategies for playing sum games based on subgame types", pp. 212–219 in *Computers and games* (Turin, Italy, 2006), edited by H. J. van den Herik et al., Lecture Notes in Computer Science **4630**, Springer, Berlin, 2007.

[Berlekamp et al. 1982] E. Berlekamp, J. H. Conway, and R. K. Guy, *Winning ways for your mathematical plays*, Academic, New York, 1982.

[Bruegmann 1993] B. Bruegmann, "Monte–Carlo Go", 1993, http://www.ideanest.com/vegos/MonteCarloGo.pdf.

[Cazenave 2002] T. Cazenave, "Comparative evaluation of strategies based on the value of direct threats", 2002, http://www.lamsade.dauphine.fr/~cazenave/papers/ts.pdf.

[Cazenave 2009] T. Cazenave, "Goal threats, temperature and Monte–Carlo go", pp. 135–150 in *Games of no chance 3*, Math. Sci. Res. Inst. Publ. **56**, Cambridge, 2009.

[Coulom 2007] R. Coulom, "Efficient selectivity and back-up operators in Monte–Carlo tree search", pp. 72–83 in *Computers and games* (Turin, Italy, 2006), edited by H. J. van den Herik et al., Lecture Notes in Computer Science **4630**, Springer, Berlin, 2007.

[Gelly and Silver 2007] S. Gelly and D. Silver, "Combining online and offline knowledge in UCT", pp. 273–280 in *Proceedings of the 24th international conference on machine learning (ICML 2007)* (Corvalis, Oregon), edited by Z. Ghahramani, ACM, New York, 2007.

[Gelly et al. 2006] S. Gelly, Y. Wang, R. Munos, and O. Teytaud, "Modification of UCT with patterns in Monte–Carlo Go", INRIA, 2006, http://hal.inria.fr/docs/00/12/15/16/PDF/RR-6062.pdf.

[Kocsis and Szepesvári 2006] L. Kocsis and C. Szepesvári, "Bandit based Monte–Carlo planning", pp. 282–293 in *Machine learning: ECML 2006*, edited by H. J. van den Herik et al., Lecture Notes in Comput. Sci. **4212**, Springer, Berlin, 2006.

[Müller and Li 2006] M. Müller and Z. Li, "Locally informed global search for sums of combinatorial games", pp. 273–284 in *Computers and games*, edited by H. J. van den Herik et al., Lecture Notes in Comput. Sci. **3846**, Springer, Berlin, 2006.

[Müller et al. 2004] M. Müller, M. Enzenberger, and J. Schaeffer, "Temperature discovery search", pp. 658–663 in *Proceedings of the 19th national conference on artifical intelligence* (San Jose, CA, 2004), AAAI, Menlo Park, CA, 2004.

[Spight 2002] W. L. Spight, "Go thermography: The 4/21/98 Jiang–Rui endgame", pp. 89–105 in *More games of no chance* (Berkeley, CA, 2000), edited by R. Nowakowski, Math. Sci. Res. Inst. Publ. **42**, Cambridge Univ. Press, 2002.

LAMSADE, Université Paris-Dauphine, Place du Maréchal de Lattre de Tassigny, 75775 Paris Cedex 16, France

cazenave@lamsade.dauphine.fr

Games of No Chance 4
MSRI Publications
Volume **63**, 2015

Retrograde analysis of Woodpush

TRISTAN CAZENAVE AND RICHARD NOWAKOWSKI

Retrograde analysis enables to solve games, starting from the end and moving toward the beginning. For games with repetition of positions and the ko rule, the standard retrograde analysis algorithm has to be adapted. We present an algorithm to solve games with the ko rule by retrograde analysis. The game of Woodpush has the ko rule and its limited state space enables its complete analysis with a modified retrograde analysis algorithm. The resulting Woodpush program plays perfect moves instantly.

1. Introduction

The game of Woodpush is a recent game, developed by the second author, that involves ko. (See [Nowakowski 2015, B2] this volume.) In this game, moves that repeat the previous position are forbidden. It is a game more simple than Go but it retains the complexity of managing ko situations. As yet there is no accepted starting position but

```
|L| |L| ... |L| | | |R| |R| ... |R| |R|
```

is a popular starting board where Left is represented with L, and Right with R. An example of a starting position for a game of width 9 is

```
|L| |L| | | |R| |R|
```

A Left move consists in sliding one of his pieces to the right as in the following example:

```
|L| |L| | | |R| |R|   =>   |L| | |L| | |R| |R|
```

If some pieces are on the way of the sliding piece, they are jumped over. For example, the Right move in the following position jumps over the Left piece contiguous to it:

```
|L| | | |L|R| | |R|   =>   |L| | |R|L| | | |R|
```

When a piece has an opponent piece behind it, it can move backward and push all the pieces behind, provided it does not repeat the previous position (i.e., creates a ko):

Keywords: none.

```
|L| | |R|L| | | |R|  =>  |L| |R|L| | | | |R|
```

For example, after this last Left move that moves backward, Right cannot move backward because it is a direct ko that repeats the previous position.

The game is won when the opponent has no more pieces on the board. Here is an example of a winning move by Right:

```
| | | | | | | |R|L|  =>  | | | | | | | | |R|
```

The Left piece falls of the board and Right has won.

Retrograde analysis has been applied to many problems. It enables to generate databases of positions or databases of patterns. For each possible position or pattern it enables to find the status of the position and other information such as the minimal number of moves required to win in the position. Once generated, databases enable to control, reduce or even replace search.

Retrograde analysis was first used to solve chess endgames [van den Herik and Herschberg 1985; Thompson 1986; Stiller 1996; Thompson 1996] containing up to six pieces. Chess endgame databases enable to play endgames perfectly and even discovered new chess knowledge about endgames [Nunn 1993].

Another successful application of retrograde analysis is the computation of Checkers endgames by Chinook [Lake et al. 1994; Schaeffer 1997] which is an important part of the program that solved Checkers [Schaeffer 2007]. Retrograde analysis has also been used in single player games such as the 16 puzzle. It consisted in computing an admissible heuristics involving only some of the pieces [Culberson and Schaeffer 1998]. Pattern database can also be combined and improve on single pattern databases [Korf and Felner 2002]. Another application of pattern databases is Rubik's cube [Korf 1997] where separate databases for corner and side cubes can be computed and improve much the admissible heuristic. Pattern databases can also be used for the game of Go, computing for example databases on eyes or on life [Cazenave 1993; Cazenave 1996b; Cazenave 1996a]. Improvements include associating patterns to abstract conditions such as external liberties [Cazenave 2001] and reducing memory requirements using metarules [Cazenave 2003].

Some complex games such as Awari have been completely solved with retrograde analysis [Romein and Bal 2003].

In his thesis [Fraser 2002], Bill Fraser describes the BruteForce program that searches an endgame region in Go to calculate thermographs for every position. It enables his program to find means, temperatures, and orthodox lines of play. Our work is related since we use a brute force approach that takes ko into account, however we simply compute the values of positions and not the associated thermograph. Moreover we deal with long loops in the game graph, long loops only very rarely occur in Go positions.

Section 2 presents the retrograde analysis algorithms we have tested. Section 3 details experimental results. Section 4 concludes.

2. Retrograde analysis algorithms

In this section, the different algorithms we have tested are presented. The standard Retrograde analysis is first detailed. Then the adaptation of the algorithm to manage direct ko is explained. Retrograde analysis that scores positions with loops is then presented in order to evaluate positions that involve loops in the state space. Eventually, the combination of retrograde analysis with direct ko and of retrograde analysis with avoidance of loop enables to solve the game. In order to produce correct scores, retrograde analysis is used again at the end of this process.

2.1. ***Standard retrograde analysis.*** At the start of the algorithm the database is initialized with the scores of terminal positions (i.e., positions where only one player has pieces). Other positions are marked as unknown. The score of a terminal position is positive if Left has won and equals the number of moves Left can play in a row, it is the opposite of the number of moves in a row Right can play if Right has won the position.

The retrograde analysis algorithm uses two tables, one table for the positions where Left is to move and one table for the positions where Right is to move. Each entry of the table is an integer that contains the score of the position.

The algorithm iteratively calls two functions one after another. The first function finds new values for Left given the values for Right. The second function finds new values for Right given the values for Left. When no new values are found, the algorithm terminates. The pseudocode for the function that finds new values for Left is given in Algorithm 1.

Algorithm 1. Step for Left of the retrograde analysis algorithm.

stepForLeft ()
for *position* in all possible positions **do**
  if value for Left to play of *position* is unknown **then**
    if no children of *position* is unknown for Right **then**
      value for Left to play of *position* ← value for Right to play of the best child of *position*
    end if
  end if
end for

A problem with this standard algorithm is that it does not take the possible ko status of a position into account. A position can either have a move that is forbidden by ko or the same move allowed because there is no ko. Therefore the same position can have different values according to its associated ko moves. We show in the next subsection how we solve this problem.

2.2. ***Retrograde analysis with direct ko.*** To take ko into account, every possible position is associated to a set of possibly forbidden moves. The evaluation of a position now takes its ko status into account in algorithm 2. This algorithm searches through all possible positions and all possible ko moves in these positions. It does not consider the child with the selected ko move to compute the value of a position associated to a ko move. Note that the children of *position* for Left have an associated ko move for Right and that this ko move is used to find the relevant value for Right of each child.

Algorithm 2. Step for Left of the retrograde analysis algorithm with ko moves.

---

stepKoForLeft ()
for *position* in all possible positions **do**
  for *ko* in all possible ko moves of *position* **do**
    if value for Left to play of *position* with the *ko* move forbidden is unknown **then**
      let *children* be the children of *position* except the child for *ko*
      if no position in *children* is unknown for Right to play **then**
        value for Left to play of *position* with *ko* ← value for Right to play of the best child in *children*
      end if
    end if
  end for
end for

This algorithm correctly makes the difference between the different ko moves of a position. However it does only score a small subset of all positions. This is due to loops in the state space. If a position is inside a loop, it cannot be scored since a position is scored only if all its descendants are scored. Moreover the child in the loop cannot be scored since the position to score is also one of its descendant and is not scored. We give in the next subsection an algorithm that enables these positions to be scored.

2.3. ***Retrograde analysis with avoidance of loops.*** In order to evaluate positions that are in a loop, we adopt the following policy: if there is a winning move for the color to play, and if another move leads to a position in a loop, the policy decides to ignore the loopy move and to evaluate the position with the value of the best winning move. This is described in Algorithm 3.

Algorithm 3. Step for Left of the retrograde analysis algorithm with ko moves and avoidance of loops.

---

stepKoForLeftAvoidLoop ()
for *position* in all possible positions **do**
  for *ko* in all possible ko moves of *position* **do**
    if value for Left to play of *position* with the *ko* move forbidden is unknown **then**
      let *children* be the children of *position* except the child for *ko*
      if at least one child in *children* is won for Left **then**
        value for Left to play of *position* with *ko* ← value for Right to play of the best child in *children*
      end if
    end if
  end for
end for

This algorithm is only to be applied once the retrograde analysis with direct ko is finished and cannot find new positions. When it is the case, all unknown positions are either in a loop or have a descendant that is in a loop. The policy that avoids loops is necessary to find new values.

However applying the function that avoids loops until no new values are found is not enough. Because these new values enable new losing positions to be evaluated for the player to play. Therefore new passes of the direct ko algorithms are needed to evaluate these positions.

The function that avoids loops is not guaranteed to find the exact values for the positions, but as it only scores positions that are won for the color to play, it cannot evaluate a position as lost if it is won. The score may be underestimated but the sign of the score is correct.

2.4. ***Interleaved retrograde analysis.*** In order to evaluate the maximum number of positions, retrograde analysis with direct ko and retrograde analysis with avoidance of loops are interleaved and consecutively called until they are both unable to find new values. When it is the case, the remaining unknown positions are considered to lead either to a loop or to a loss. Instead of losing both colors will choose to loop, making these positions a draw. The algorithm that interleaves retrograde analysis is given in Algorithm 4.

Algorithm 4. Interleaving retrograde analysis with ko moves and with avoidance of loops.

```
interleavedAnalysis ()
score all terminal positions
while new positions with ko or with avoidance of loops are scored do
  while new positions with ko are scored do
    stepKoForLeft ()
    stepKoForRight ()
  end while
  while new positions with avoidance of loops are scored do
    stepKoForLeftAvoidLoop ()
    stepKoForRightAvoidLoop ()
  end while
end while
score unknown positions as draw
```

2.5. ***Retrograde analysis again.*** When interleaved retrograde analysis is finished, all positions have the correct sign but not always the correct score. In order to find the correct scores, another algorithm is required which reevaluates non terminal positions according to the scores of their children. The program alternates Left and Right passes until there is no more changes.

3. Experimental results

For size 7, retrograde analysis uses 128 passes. All positions are scored. Right, the second player, wins by four. Here is the winning line with perfect play from both players with succeeding positions written one after another:

```
|L| |L| |R| |R|
| |L|L| |R| |R|
| |L|L|R| | |R|
| | |L|R|L| |R|
| | |L|R|L|R| |
| |L|R|L| |R| |
| |L|R|L|R| | |
|L|R|L| |R| | |
|L|R|L|R| | | |
| |R|L|R|L| | |
| | |R|L|R|L| |
| |R|L| |R|L| |
```

```
| |R|L| | |R|L|
|R|L| | | |R|L|
|R|L| | | | |R|
|L| | | | | |R|
|L| | | | |R| |
| |L| | | |R| |
| |L| | |R| | |
| | |L| |R| | |
| | |L|R| | | |
| | | |R|L| | |
| | | | |R|L| |
| | | | |R| |L|
| | | |R| | |L|
| | | |R| | | |
```

For size 9, our algorithm uses 366 passes. 933 positions cannot be scored by the algorithm. They are given value 0 which means a draw. The initial position, scored with the rule of direct ko, is a draw due to a superko (repetition of a position after more than two moves):

```
|L| |L| | | |R| |R|
|L| | |L| | |R| |R|
|L| | |L| |R| | |R|
|L| | | |L|R| | |R|
|L| | | |L|R| |R| |
| |L| | |L|R| |R| |
| |L| | |L|R|R| | |
| | |L| |L|R|R| | |
| | |L|R|L|R| | | |
| |L|R|L| |R| | | |
| |L| |R|L|R| | | |
| | |L|R|L|R| | | |
| | |L| |R|L|R| | |
| | | |L|R|L|R| | |
| | | |L| |R|L|R| |
| | | |L|R|L| |R| |
| | | |L|R|L|R| | |
| | |L|R|L| |R| | |
| | |L| |R|L|R| | |
| | | |L|R|L|R| | |
```

For size 11, our algorithm uses 860 passes. 8,563 positions are scored 0 by the algorithm. The initial position is also a draw:

```
|L| |L| | | | | |R| |R|
|L| | |L| | | | |R| |R|
|L| | |L| | | |R| | |R|
|L| | | |L| | |R| | |R|
|L| | | |L| |R| | | |R|
|L| | | | |L|R| | | |R|
|L| | | | |L|R| | |R| |
| |L| | | |L|R| | |R| |
| |L| | | |L|R| |R| | |
| | |L| | |L|R| |R| | |
| | |L| | |L|R|R| | | |
| | | |L| |L|R|R| | | |
| | | |L|R|L|R| | | | |
| | |L|R|L| |R| | | | |
| | |L| |R|L|R| | | | |
| | | |L|R|L|R| | | | |
| | | |L| |R|L|R| | | |
| | | | |L|R|L|R| | | |
| | | | |L| |R|L|R| | |
| | | | |L|R|L| |R| | |
| | | | |L|R|L|R| | | |
| | | |L|R|L| |R| | | |
| | | |L| |R|L|R| | | |
| | | | |L|R|L|R| | | |
```

4. Conclusion

An algorithm that evaluates Woodpush positions using retrograde analysis has been presented. It correctly evaluates the score of positions and scores positions that lead to a superko as a draw. The resulting program plays perfect moves instantaneously.

For size 7 Woodpush, the game is a loss by 4 for the first player. For size 9 and 11 the game is a draw with the direct ko rule due to a repetition of positions.

An extension of our work is to extend our results to games with superko. Either using superko together with retrograde analysis, or using the direct ko database to solve superko with search.

Retrograde analysis algorithms that undo moves instead of browsing all possible positions are usually more rapid. It is certainly possible to adapt our algorithm in this way even though speed of the retrograde analysis algorithm is not a matter for the sizes that were solved.

References

[Cazenave 1993] T. Cazenave, "Apprentissage de le résolution de problèmes de vie et de mort au jeu de Go", Rapport de DEA, Université Paris 6, 1993.

[Cazenave 1996a] T. Cazenave, "Automatic acquisition of tactical Go rules", pp. 10–19 in *Game Programming Workshop in Japan '96* (Hakone, Kanagawa, Japan), edited by H. Matsubara, Computer Shogi Association, Kanagawa, Japan, 1996.

[Cazenave 1996b] T. Cazenave, *Système d'apprentissage par auto-observation: Application au jeu de Go*, Ph.D. thesis, Université Paris 6, 1996, http://www.lamsade.dauphine.fr/~cazenave/papers/these.pdf.

[Cazenave 2001] T. Cazenave, "Generation of patterns with external conditions for the game of Go", pp. 275–293 in *Advance in Computer Games 9*, edited by H. van den Herik and B. Monien, Universiteit Maastricht, Maastricht, 2001.

[Cazenave 2003] T. Cazenave, "Metarules to improve tactical Go knowledge", *Inform. Sci.* **154**:3-4 (2003), 173–188.

[Culberson and Schaeffer 1998] J. C. Culberson and J. Schaeffer, "Pattern databases", *Comput. Intelligence* **14**:3 (1998), 318–334.

[Fraser 2002] W. E. Fraser, *Computer-assisted thermographic analysis of go endgames*, Ph.D. thesis, University of California, Berkeley, 2002, http://search.proquest.com/docview/304802211.

[van den Herik and Herschberg 1985] H. van den Herik and I. S. Herschberg, "The construction of an omniscient endgame database", *ICCA Journal* **8**:2 (1985), 6–87.

[Korf 1997] R. E. Korf, "Finding optimal solutions to Rubik's Cube using pattern databases", pp. 700–705 in *AAAI-97* (Providence, RI), AAAI Press, Menlo Park, CA, 1997.

[Korf and Felner 2002] R. E. Korf and A. Felner, "Disjoint pattern database heuristics", *Artificial Intelligence* **134**:1-2 (2002), 9–22.

[Lake et al. 1994] R. Lake, J. Schaeffer, and P. Lu, "Solving large retrograde-analysis problems using a network of workstations", pp. 135–162 in *Advances in computer chess*, vol. 7, edited by H. J. van den Herik et al., University of Limburg, Maastricht, 1994.

[Nowakowski 2015] R. Nowakowski, "Unsolved problems in combinatorial game theory", in *Games of no chance 4*, edited by R. Nowakowski, Mathematical Sciences Research Institute Publications **63**, Cambridge University Press, New York, 2015.

[Nunn 1993] J. Nunn, "Extracting information from endgame databases", *ICCA Journal* **16**:4 (1993), 191–200.

[Romein and Bal 2003] J. W. Romein and H. E. Bal, "Solving awari with parallel retrograde analysis", *IEEE Computer* **36**:10 (2003), 26–33.

[Schaeffer 1997] J. Schaeffer, *One jump ahead: Challenging human supremacy at checkers*, Springer, New York, 1997.

[Schaeffer 2007] J. Schaeffer, "Game over: Black to play and draw in checkers", *ICGA Journal* **30**:4 (2007), 187–197.

[Stiller 1996] L. Stiller, "Multilinear algebra and chess endgames", pp. 151–192 in *Games of no chance* (Berkeley, CA, 1994), edited by R. J. Nowakowski, Math. Sci. Res. Inst. Publ. **29**, Cambridge Univ. Press, 1996.

[Thompson 1986] K. Thompson, "Retrograde analysis of certain endgames", *ICCA Journal* **9**:3 (1986), 131–139.

[Thompson 1996] K. Thompson, "6-piece endgames", *ICCA Journal* **19**:4 (1996), 215–226.

cazenave@lamsade.dauphine.fr *LAMSADE, Université Paris-Dauphine, 75775 Paris, France*

rjn@mathstat.dal.ca *Mathematics and Statistics, Dalhousie University, Halifax B3H 3J5, Canada*

Games of No Chance 4
MSRI Publications
Volume **63**, 2015

Narrow misère Dots-and-Boxes

SÉBASTIEN COLLETTE, ERIK D. DEMAINE,
MARTIN L. DEMAINE AND STEFAN LANGERMAN

We study misère Dots-and-Boxes, where the goal is to minimize score, for narrow boards. In particular, we characterize the winner for $1 \times n$ boards with an explicit winning strategy for the first player with a score of $\lfloor (n-1)/3 \rfloor$. We also give preliminary results for $2 \times n$ and for Swedish $1 \times n$ (where the boundary is initially drawn).

1. Introduction

Recall the classic children's game Dots-and-Boxes [Berlekamp et al. 2003]. We start with an $m \times n$ square grid of dots. Players alternate drawing individual edges of the grid. If a player completes a box of the grid, s/he gets a point and must draw another edge; this process can repeat several times within a single turn. The game ends when all edges have been drawn, i.e., when all mn boxes have been completed. In normal Dots-and-Boxes, the player to receive the most points wins. In misère Dots-and-Boxes, the player to receive the fewest points wins. A draw (tie) occurs when mn is even and the players complete the same number of boxes.

Normal Dots-and-Boxes endgames are known to be NP-hard; see [Demaine and Hearn 2009]. In addition, no winning strategies are known when m or n is sufficiently large. To our knowledge, even the $1 \times n$ case is open for arbitrary n. On the other hand, misère Dots-and-Boxes may be easier to analyze.

In Section 2, we give a winning strategy for the first player in $1 \times n$ misère Dots-and-Boxes that guarantees a score of at most $\lfloor (n-1)/3 \rfloor$ boxes, which is a

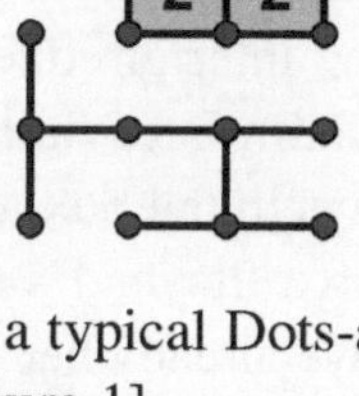

Figure 1. In the middle of a typical Dots-and-Boxes game. Based on [Berlekamp et al. 2003, Figure 1].

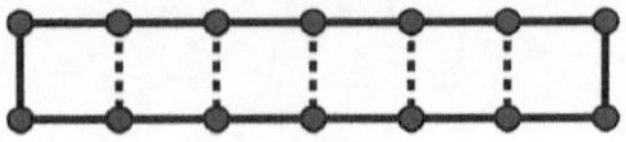

Figure 2. Boundary edges are solid; interior edges are dashed.

win by roughly $n/6$. The essence of the strategy is to avoid any parity switching of who leads the game, which we show is possible, unlike general boards. In Sections 3 and 4, we give preliminary results for the $2 \times n$ game and for the Swedish $1 \times n$ game (where the boundary is initially drawn).

Terminology. See Figure 2. A *boundary* edge is an edge of the bounding rectangle. An *interior* edge is any nonboundary edge.

2. Misère 1 × *n*

In a $1 \times n$ board, there are $n-1$ interior edges; the remaining $2n+2$ edges are boundary. We distinguish the leftmost and rightmost boxes as *end* boxes.

Theorem 1. *For all $n \geq 1$, misère $1 \times n$ Dots-and-Boxes is a first-player win.*

Proof. First we describe Player 1's strategy, which divides the game into two phases. In Phase I, some interior edges remain untaken, and Player 1 always takes such an edge. The initial choice of interior edge is any not incident to an end box, if there is one, and otherwise an arbitrary interior edge. We ignore any boxes that Player 2 takes, and instead focus on the last edge played. If Player 2 takes an interior edge, Player 1 takes another arbitrary interior edge. If Player 2 takes a boundary edge, Player 1 takes one of the two incident interior edges, if one of them is untaken, and otherwise an arbitrary interior edge. This rule may cause Player 1 to take a box, in which case Player 1 takes another, arbitrary interior edge (if any exist). In Phase II, when all interior edges are depleted, Player 1 takes any boundary edge that does not complete a box; we will show that such an edge always exists.

We show that no edge can ever complete two boxes simultaneously. During Phase I, Player 1 goes first and takes only interior edges, so the number of taken interior edges is always at least the number of taken boundary edges. Further we claim that, within each nonboundary box except possibly the one in which Player 2 just played, the number of taken interior edges is always at least the number of taken boundary edges. The claim is trivially true before either player has played in the box. If Player 1 plays in the box, the claim certainly remains true. Whenever Player 2 plays in the box, Player 1's next move will be to play in the box, unless both interior edges have already been taken; in either case, the claim remains true. For boundary boxes, the number of boundary edges can exceed the number of interior edges, but only when all (zero or one) interior edges have

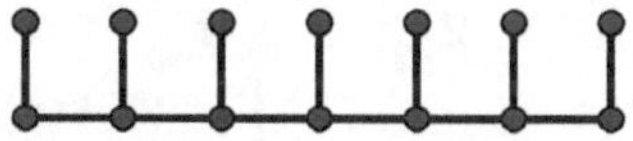

Figure 3. A spanning tree of a $1 \times n$ grid has $2n + 1$ edges.

been taken. Thus, at any time, no box except possibly the one in which Player 2 just played could be completed by an interior edge; all other boxes can be completed only by boundary edges, each of which is incident to only one box. Therefore at no time can any edge complete two boxes simultaneously.

Next we prove that Player 1 completes no boxes during Phase II. By definition, Player 1 will take a box during Phase II only if every box is either completed or one edge from being completed. At such a time, the taken edges must include a spanning tree of the grid, which consists of $2n + 1$ edges (see Figure 3), plus exactly one edge for each completed box (because the cycle formed by each completed box can be broken by a single edge removal). Because we proved that each edge completed at most one box, the number of complete turns must be the number of taken edges minus the number of completed boxes. Thus the number of complete turns must be $2n + 1$, meaning that it is Player 2's turn. Therefore Player 1 completes no boxes during Phase II.

We claim that Player 1 never completes an end box. An end box has at most one interior edge, so there is only one possible move by Player 1 that could complete the box in Phase I. But when Player 2 plays the top or bottom edge of the box, Player 1 will take the interior edge, before Player 2 could have played the opposite (bottom or top) edge of the box. Therefore this move by Player 1 did not complete the box.

If Player 1 plays first in a nonend box of the board, then we claim that Player 1 will not complete this box; refer to Figure 4. If Player 1 also plays second in this box, then the claim is obvious: Player 1 will play in the box only if it does not complete the box. If Player 2 plays second in the box, then by definition Player 1 will immediately take the remaining interior edge of the box. As this is only the third move in the box, this move does not complete the box. Player 1 will not play the final boundary edge of the box because that would complete the box.

Finally we show that Player 1 completes at most $\lfloor (n-1)/3 \rfloor$ boxes. As argued above, for Player 1 to complete a box, it must not be an end box and Player 2 must play in it first. Indeed, Player 2 must play in that box again, taking the other boundary edge, or else we would have already entered Phase II. Thus, every box taken by Player 1 can be charged to two moves by Player 2, as well as the two following interior edges taken by Player 1. Furthermore, the completed box means that Player 1 also takes another interior edge (if there is one). Thus every box completed by Player 1 corresponds to an increase in the number of taken interior by at least 3. Therefore Player 1 completes at most $(n - 1)/3$ boxes. □

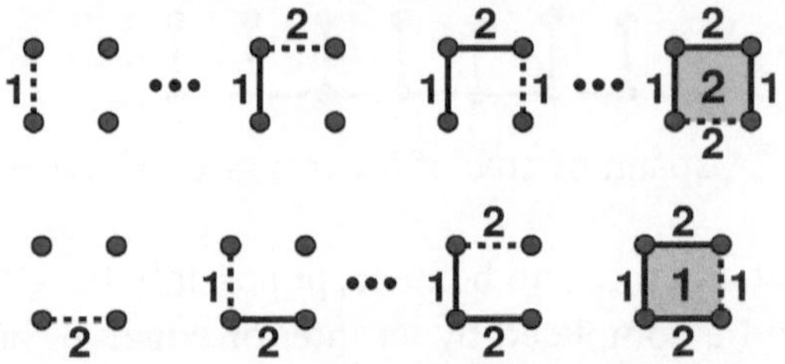

Figure 4. Behavior of a nonend box under Player 1's strategy, depending on who takes the first box edge. Top: Player 1 starts. Bottom: Player 2 starts. Nonimmediate responses are denoted by $\cdots$.

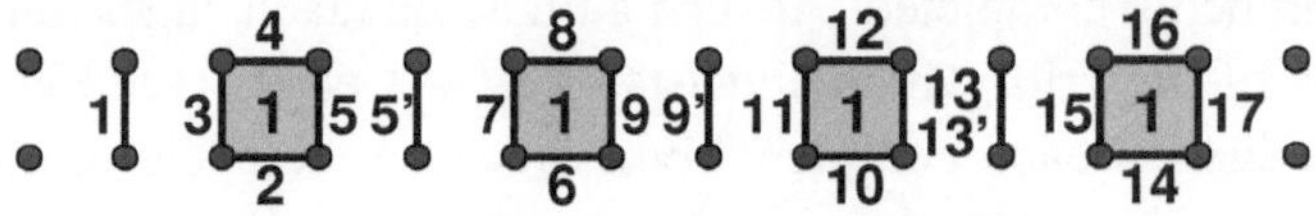

Figure 5. A strategy for Player 2 that causes Player 1's strategy to complete $\lfloor (n-1)/3 \rfloor$ boxes. Edge labels denote turn number.

Figure 5 shows an example where the strategy of Theorem 1 causes Player 1 to take $\lfloor (n-1)/3 \rfloor$ boxes. We conjecture that this strategy is optimal, at least up to additive constants.

Open Problem. *Can Player 2 force Player 1 to complete* $\lfloor (n-1)/3 \rfloor$ *boxes*?

3. Misère 2 × *n*

For misère $2 \times n$ Dots-and-Boxes, which player is the initial leader changes with n. In the absence of parity-switching moves, the first player should win for odd n and the second player should win for even n. By Theorem 1, we already know this to be the case for $n = 1$. We have also verified this claim by exhaustive computational search for $n = 2$ and $n = 3$.

A natural strategy for the leading player, generalizing the $1 \times n$ strategy, is the following. If you can complete a box, then take it. (This rule prevents the formation of larger parity-changing cycles.) Otherwise, if there is an untaken internal edge incident to the edge just taken by the other player, then take it. Otherwise, if there is an untaken internal edge, take it. Otherwise, if there is an edge that does not complete a box, take it. Otherwise, take any edge. (The last rule should not arise if the parity remains unchanged.) We have verified that this strategy works for 2×2 but not for 2×3 or larger boards.

In fact, for sufficiently large $2 \times n$ boards, it seems that the nonleading player can force the leading player to take around $3/4$ of the boxes. If this is the case, then either the nonleading player wins, or the leading player must change the parity. We wonder whether such a change in parity (perhaps just once or twice?) can let the leading player guarantee a win.

4. Misère Swedish 1 × *n*

In *Swedish Dots-and-Boxes* (see wilson.engr.wisc.edu/boxes), all boundary edges are initially drawn. In this case, $1 \times n$ misère Dots-and-Boxes has a much more complicated behavior; see Table 1. These results are based on exhaustive computational search.

This game seems particularly interesting because it is very simple, yet is all about the parity switching of who leads. Conceivably, $1 \times n$ Swedish games could also arise in the middle of a $2 \times n$ game, though it is unclear whether this happens under optimal play.

Conjecture 1. *The outcome of misère* $1 \times n$ *Swedish Dots-and-Boxes is given by Table 1, with periodic behavior starting from* $n = 22$ *and a period of* 10.

1×2	second player wins by 2 points	1×24	draw
1×3	second player wins by 3 points	1×25	second player wins by 1 point
1×4	first player wins by 4 points	1×26	draw
1×5	first player wins by 5 points	1×27	first player wins by 1 point
1×6	first player wins by 6 points	1×28	draw
1×7	first player wins by 5 points	1×29	first player wins by 1 point
1×8	first player wins by 4 points	1×30	draw
1×9	first player wins by 3 points	1×31	first player wins by 1 point
1×10	first player wins by 2 points	1×32	draw
1×11	first player wins by 1 point	1×33	second player wins by 1 point
1×12	draw	1×34	draw
1×13	second player wins by 1 point	1×35	second player wins by 1 point
1×14	second player wins by 2 points	1×36	draw
1×15	second player wins by 1 point	1×37	first player wins by 1 point
1×16	draw	1×38	draw
1×17	first player wins by 1 point	1×39	first player wins by 1 point
1×18	first player wins by 2 points	1×40	draw
1×19	first player wins by 1 point	1×41	first player wins by 1 point
1×20	first player wins by 2 points	1×42	draw
1×21	first player wins by 1 point	1×43	second player wins by 1 point
1×22	draw	1×44	draw
1×23	second player wins by 1 point	1×45	???

Table 1. Who wins in Swedish $1 \times n$ misère Dots-and-Boxes under optimal play, as computed by an exhaustive search.

1×2	any
1×3	any
1×4	cut 2
1×5	cut 2
1×6	cut 3
1×7	take all but 6 then cut 3
1×8	take all but 6 then cut 3
1×9	take all but 6 then cut 3
1×10	take all but 6 then cut 3
1×11	take all but 6 then cut 3
1×12	take all but 6 then cut 3
1×13	take all but 6 then cut 3
1×14	cut 2; or take all but 6 then cut 3
1×15	cut 2
1×16	cut 2
1×17	cut 2
1×18	cut 2
1×19	cut 2, 3, ..., or 9; or take 1 then cut 2
1×20	cut 3, 5, 6, ..., or 10
1×21	cut 3, 5, 6, ..., or 10; or take 1 then cut 3, 5, 6, ..., or 10
1×22	cut 3, 5, 6, ..., or 11; take 1 then cut 3, 5, 6, ..., or 10; or take 2 then cut 3, 5, 6, ..., or 10
1×23	cut 2, 3, ..., or 11; take 1 then cut 3, 5, 6, ..., or 11; take 2 then cut 3, 5, 6, ..., or 10; or take 3 then cut 3, 5, 6, ..., or 10
1×24	cut 4
1×25	cut 2, 3, ..., or 12; or take 1 then cut 4
1×26	cut 2 or 4
1×27	cut 2
1×28	cut 2, 3, ..., or 14; or take 1 then cut 2
1×29	cut 2, 3, ..., or 14
1×30	cut 2, 3, ..., or 15; or take 1 then cut 2, 3, ..., or 14
1×31	cut 3, 5, 6, ..., or 15
1×32	cut 2, 3, ..., or 16; or take 1 then cut 3, 5, 6, ..., or 15
1×33	cut 2, 3, ..., or 16; take 1 then cut 2, 3, ..., or 16; or take 2 then cut 3, 5, 6, ..., or 15
1×34	cut 2, 3, ..., or 17
1×35	cut 2, 3, ..., or 17; or take 1 then cut 2, 3, ..., or 17
1×36	cut 2, 3, ..., or 18
1×37	cut 2 or 4
1×38	cut 2, 3, ..., or 19; or take 1 then cut 2 or 4
1×39	cut 2, 3, ..., or 19
1×40	cut 2, 3, ..., or 20; or take 1 then cut 2, 3, ..., or 19
1×41	cut 3, 5, 7, 8, ..., or 20
1×42	cut 2, 3, ..., or 21; or take 1 then cut 3, 5, 7, 8, ..., or 20
1×43	cut 2, 3, ..., or 21; take 1 then cut 2, 3, ..., or 21; or take 2 then cut 3, 5, 7, 8, ..., or 20
1×44	cut 2, 3, ..., or 22
1×45	cut 2, 3, ..., or 22; or take 1 then cut 2, 3, ..., or 22
1×46	cut 2, 3, ..., or 23
1×47	cut 2 or 4
1×48	cut 2, 3, ..., or 24; or take 1 then cut 2 or 4
1×49	cut 2, 3, ..., or 24
1×50	cut 2, 3, ..., or 25; or take 1 then cut 2, 3, ..., or 24
1×51	cut 3, 5, 7, 9, 10, ..., or 25
1×52	cut 2, 3, ..., or 26; or take 1 then cut 3, 5, 7, 9, 10, ..., or 25

Table 2. All optimal moves for the first player from the initial configuration of the $1 \times n$ Swedish board, as computed by an exhaustive search. Here "take k" means to complete n boxes, "cut k" means to draw an edge to form a $1 \times k$ rectangle, and an ellipsis (...) denotes an interval of consecutive integers. To remove symmetric moves, we omit "cut k" when k is larger than half the current rectangle.

This conjecture is supported by different observations. First, it matches the behavior exposed by our exhaustive search, as shown in Table 1. Second, we have a detailed proof that the outcome is correct for a restricted form of the game, in which every edge drawn must form a rectangle of size 1×1, 1×2, 1×3, or 1×4, and the last case only when there is not already another 1×4 rectangle on the board. We believe that the outcome of this restricted game is equivalent to the original one:

Conjecture 2. *For misère $1 \times n$ Swedish Dots-and-Boxes, there exists an optimal play in which both players move so as to form rectangles of size 1×1, 1×2, 1×3, or 1×4 with every edge drawn, with the last case arising only when there is not already a 1×4 rectangle on the board.*

This conjecture is also supported by our exhaustive search: in the games we so analyze, there is always an optimal move of the restricted form, as shown in Table 2. If Conjecture 2 is true, our proof implies Conjecture 1. The outcome under optimal play of the restricted game (and thus also of the unrestricted game if Conjecture 2 holds) is the following eventually periodic sequence:

$$-2,-3,4,5,6,5,4,3,2,1,0,-1,-2,-1,0,1,2,1,2,1,[0,-1,0,-1,0,1,0,1,0,1]^*.$$

These numbers indicate by how many points the first player wins under optimal play, with a positive number meaning a first-player win, zero meaning a draw, and a negative number meaning a second-player win.

Acknowledgments

We thank Greg Aloupis, David Bremner, Karim Douïeb, and Vi Hart for helpful discussions about Swedish $1 \times n$ misère Dots-and-Boxes.

References

[Berlekamp et al. 2003] E. R. Berlekamp, J. H. Conway, and R. K. Guy, *Winning ways for your mathematical plays, III*, 2nd ed., A K Peters, Natick, MA, 2003.

[Demaine and Hearn 2009] E. D. Demaine and R. A. Hearn, "Playing games with algorithms: Algorithmic combinatorial game theory", pp. 3–56 in *Games of no chance 3*, edited by M. H. Albert and R. J. Nowakowski, Math. Sci. Res. Inst. Publ. **56**, Cambridge, 2009.

sebastien.collette@ulb.ac.be *Chargé de Recherches du F.R.S.-FNRS*

Computer Science Department, Université Libre de Bruxelles, CP 212, Boulevard du Triomphe, 1050 Brussels, Belgium

edemaine@mit.edu *Computer Science and Artificial Intelligence Laboratory, Massachusetts Institute of Technology, 32 Vassar Street, Cambridge, MA 02139, United States*

mdemaine@mit.edu *Computer Science and Artificial Intelligence Laboratory, Massachusetts Institute of Technology, 32 Vassar Street, Cambridge, MA 02139, United States*

stefan.langerman@ulb.ac.be *Directeur de Recherches du F.R.S.-FNRS*

Computer Science Department, Université Libre de Bruxelles, CP 212, Boulevard du Triomphe, 1050 Brussels, Belgium

Games of No Chance 4
MSRI Publications
Volume **63**, 2015

Toppling Conjectures

ALEX FINK, RICHARD NOWAKOWSKI,
AARON SIEGEL AND DAVID WOLFE

Positions of the game of TOPPLING DOMINOES exhibit many familiar combinatorial game theory values, often arranged in unusual and striking patterns. We show that for any given dyadic rational x, there is a *unique* TOPPLING DOMINOES position G equal to x, and that G is necessarily a palindrome. We also exhibit positions of value $+_x$ for each $x > 0$. We show that for each integer $m \geq 0$, there are exactly m distinct LR-TOPPLING DOMINOES positions of value $*m$ (modulo a trivial symmetry). Lastly, every infinitesimal TOPPLING DOMINOES position has atomic weight 0, 1 or -1.

1. Introduction

TOPPLING DOMINOES, introduced by Albert, Nowakowski and Wolfe [1], is a combinatorial game played with a row of dominoes, such as the one pictured in Figure 1. Here each domino is colored blue or red (black or white, respectively, when color printing is unavailable). On his turn, Left selects any bLue (black) domino and topples it either east or west (his choice). This removes the toppled domino from the game, together with all other dominoes in the chosen direction. Likewise, Right's options are to topple Red (white) dominoes east or west. For example, the Left options of  are

$A =$ 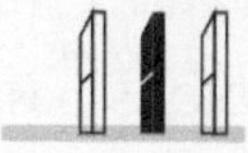, $B =$, $C =$ , $D =$ 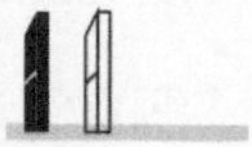.

Here A and B result from toppling the westmost domino respectively west or

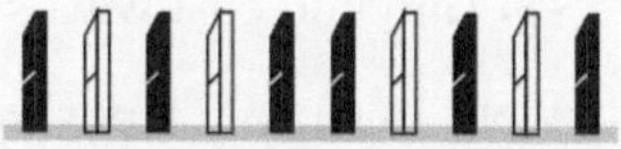

Figure 1. A typical TOPPLING DOMINOES position.

MSC2010: 05C69.
Keywords: TOPPLING DOMINOES, reduced canonical form, nimbers, dyadic rationals, palindromes, colon principle, ordinal sum, atomic weight.

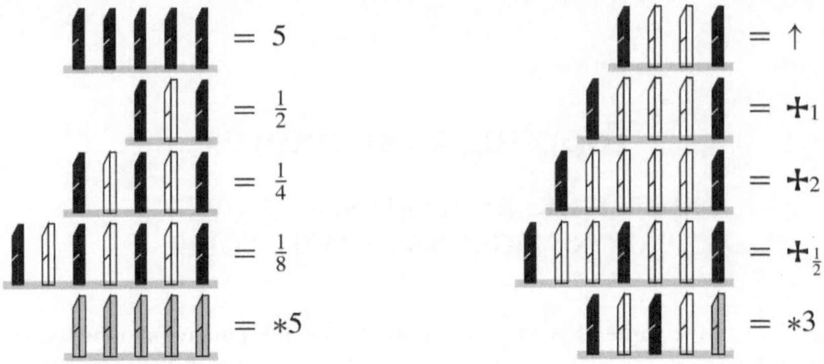

Figure 2. Some familiar combinatorial game values that arise in TOPPLING DOMINOES.

east, while C and D result from toppling the eastern black domino respectively west or east.

Positions in TOPPLING DOMINOES might also include grEen (gray) dominoes, which Either player may topple. Positions consisting entirely of black and white dominoes, such as the example above, are sometimes called LR-TOPPLING DOMINOES positions.

The reader might wish to verify some of the sample values in Figure 2. (Exercise: what is the value of Figure 1?)

The principal results of this paper are the following four theorems.

- For each number x, there is exactly one TOPPLING DOMINOES position G equal to x. Furthermore, G is necessarily a palindrome. The proof involves constructing a new LR-TOPPLING DOMINOES position $f(G)$ with value $1{:}G$ and applying this construction recursively.
- For each number $x > 0$, there exists a TOPPLING DOMINOES position G equal to $⧾_x$. It is not known whether G is unique.
- For each integer $m \geq 0$, there are exactly m distinct LR-TOPPLING DOMINOES positions of value $*m$ (modulo east/west symmetry).
- Every infinitesimal TOPPLING DOMINOES position has atomic weight 0, 1 or -1.

In Section 2, we introduce some basic facts and strategies for TOPPLING DOMINOES. The three main results are proved in Sections 3, 4 and 5, as Theorems 16, 21 and 29, respectively. Finally, in Section 6 we note that every infinitesimal TOPPLING DOMINOES position necessarily has atomic weight 0, 1 or -1.

We adopt the following notational conventions:

L	a single Left (black) domino
R	a single Right (white) domino
E	a single domino for Either (gray)
$\boldsymbol{x}, \boldsymbol{y}, \boldsymbol{z}$	an arbitrary string of dominoes

Familiarity with combinatorial game theory is assumed; the necessary background may be found in a standard reference such as [1] or [2].

2. Preliminaries

The following easy result is very useful.

Lemma 1. *A* TOPPLING DOMINOES *position's outcome class is completely determined by its end dominoes*:

- *The empty string has value* 0 (*and is the unique position of value* 0);
- $L > 0$, *and* $L\boldsymbol{x}L > 0$ *for any string* $\boldsymbol{x}$;
- $R > 0$, *and* $R\boldsymbol{x}R > 0$ *for any string* $\boldsymbol{x}$;
- $L\boldsymbol{x}R \not\gtrless 0$ *and* $R\boldsymbol{x}L \not\gtrless 0$ *for any string* $\boldsymbol{x}$.

Proof. From $L\boldsymbol{x}L$, any opening move for Right annihilates at most one of the two endpoints, and hence reverts to 0. Finally, from $L\boldsymbol{x}R$ (or $R\boldsymbol{x}L$) both players have moves to 0. □

The next results shed light on some simple winning strategies.

Lemma 2. *For any string* $\boldsymbol{x}$, *every Left move from* $L\boldsymbol{x}L$ *to some* $L\boldsymbol{y}R$ *or* $L\boldsymbol{y}E$ *is reversible through* 0.

Proof. In particular, $L\boldsymbol{x}L > 0$, and so Left's move to $L\boldsymbol{y}R$ or $L\boldsymbol{y}E$ reverses through Right's response toppling the remaining dominoes. □

Lemma 3 (sandwich lemma). *Let G be a* TOPPLING DOMINOES *position, and suppose G^L is obtained from G by toppling east* (*resp. west*). *Suppose that G^{LR} is obtained from G^L by toppling in the same direction. Then* $G^{LR} \mathrel{|\!\triangleright} G$.

Proof. Assume (by symmetry) both moves toppled east. Write $\boldsymbol{x} = G^{LR}$, so that $G^L = \boldsymbol{xy}$ for some $\boldsymbol{y}$ and $G = \boldsymbol{xyz}$ for some $\boldsymbol{z}$. Note that $\boldsymbol{x} = G^{LR}$ is available directly as a Right option of G, so that $G^{LR} = G^R \mathrel{|\!\triangleright} G$. □

The importance of Lemma 3 is that in any difference $G - H = 0$, if Left moves to any $G^L - H$, then Right's winning response cannot be to topple in the same direction in G.

Lemma 4. *For any* $\boldsymbol{x}$, $L\boldsymbol{x} > \boldsymbol{x}$.

Proof. In $\boldsymbol{Lx} - \boldsymbol{x}$ Left can win by moving to $\boldsymbol{x} - \boldsymbol{x}$. When Right moves first, Left plays the corresponding move in the other component leaving either $\boldsymbol{y} - \boldsymbol{y} = 0$ or $\boldsymbol{L}\boldsymbol{y} - \boldsymbol{y}$ which, by induction, is positive. □

Corollary 5. *Within any block of adjacent blue dominoes, the two moves that topple an end domino away from the block dominate all other moves within the block.*

3. Uniqueness of numbers

In this section, we show that for every number x there is a unique TOPPLING DOMINOES position $G = x$. The *ordinal sum* defined in [2; 3] plays a central role in this analysis. We recall the definition here.

Definition 6. Let G and H be combinatorial games. The *ordinal sum* of G and H, denoted $G{:}H$, is defined recusively by

$$G{:}H = \{G^L, G{:}H^L \mid G^R, G{:}H^R\}.$$

Intuitively, we may think of the ordinal sum as a modified disjunctive sum in which *any move on* G *annihilates the entire component* H.

The special case $1{:}G = \{0, 1{:}G^L \mid 1{:}G^R\}$ has particular importance in our analysis, owing to the following classical theorem.

Fact 7 (Berlekamp–Conway–Guy). *Let $x > 0$ be a number. Then $x = 1{:}y$ for some simpler number y.*

Corollary 8. *The set of numbers is generated from* $\{0\}$ *by the transformations* $x \mapsto -x$, $x \mapsto 1{:}x$.

With these facts in hand, the outline of our argument becomes clear. We will associate to each TOPPLING DOMINOES position G a new position $f(G)$ of value $1{:}G$. Since there is a trivial transformation $G \mapsto -G$ (replace blue dominoes with red ones and vice versa), we immediately obtain all numbers. We then show that *every* positive number G is necessarily of the form $f(H)$ for some H, after which a simple inductive argument establishes uniqueness.

For completeness, we include a proof of Fact 7 here. The theorem was originally proved in the context of BLUE-RED HACKENBUSH strings, where it was used to show that every such string encodes a unique number [2].

Proof of Fact 7. It is easily seen that $1{:}n = n + 1$ and $1{:}-n = 2^{-n}$, both of which satisfy the simplicity conclusion, so assume $x = \{x^L \mid x^R\}$ in simplest form, with $x^R > x^L > 0$. By induction, $x^L = 1{:}y^L$ and $x^R = 1{:}y^R$ for some y^L, y^R. Put $y = \{y^L \mid y^R\}$. Then

$$1{:}y = \{0, 1{:}y^L \mid 1{:}y^R\} = \{0, x^L \mid x^R\} = x,$$

since 0 is dominated by x^L. By induction we may assume that y^L and y^R are simpler than x^L and x^R, respectively, whence y is simpler than x. $\square$

With this machinery in hand, we now introduce our transformation f.

Definition 9. Let f be the morphism of strings such that, for a TOPPLING DOMINOES string $\boldsymbol{x}$, $f(\boldsymbol{x})$ is obtained by inserting one additional L in every maximal substring of Ls in $\boldsymbol{x}$, including empty substrings. For example, $f(RRLLRL) = LRLRLLLRLL$.

Equivalently, f can be described by the following transformation, where $\wedge$ denotes the "start of string" marker for $\boldsymbol{x}$:

$$\begin{array}{rclcrcl} \wedge & \to & \wedge L & \qquad & R & \to & RL \\ L & \to & L & & E & \to & EL \end{array}$$

Theorem 10. *For any* TOPPLING DOMINOES *position G, we have $f(G) = 1{:}G$.*

Proof. Let $H = \{0, f(G^L) \mid f(G^R)\}$. It suffices to show that $f(G) = H$, for then the theorem follows by induction.

Step 1: We first show that every Left (resp. Right) option of H is a Left (resp. Right) option of $f(G)$. Consider a typical G^R. We may write $G = \boldsymbol{x}R\boldsymbol{y}$ (or $G = \boldsymbol{x}E\boldsymbol{y}$) with $G^R = \boldsymbol{x}$ or $\boldsymbol{y}$. By definition of f, we have $f(G) = f(\boldsymbol{x})Rf(\boldsymbol{y})$ (or $f(\boldsymbol{x})Ef(\boldsymbol{y})$), so $f(\boldsymbol{x})$ and $f(\boldsymbol{y})$ are necessarily Right options of $f(G)$.

Likewise, for any G^L we may write $G = \boldsymbol{x}L\boldsymbol{y}$ (or $G = \boldsymbol{x}E\boldsymbol{y}$) with $G^L = \boldsymbol{x}$ or $\boldsymbol{y}$. If $G = \boldsymbol{x}E\boldsymbol{y}$, then the argument is identical to the G^R case. If $G = \boldsymbol{x}L\boldsymbol{y}$, then $f(G) = f(\boldsymbol{x})f(\boldsymbol{y})$, so $f(\boldsymbol{x})$ and $f(\boldsymbol{y})$ are necessarily Left options of $f(G)$ (note that both $f(\boldsymbol{x})$ and $f(\boldsymbol{y})$ have blue dominoes on both ends).

Finally, $f(G)$ has a move to 0, since it necessarily has a blue domino on the west end.

Step 2: To complete the proof, we show that for every option of T of $f(G)$, *either T is an option of H, or T reverses through* 0.

First of all, inspection of Definition 9 reveals that *every* Right option of $f(G)$ is obtained as in Step 1, since every red (resp. green) domino in $f(G)$ arises as the image of a red (resp. green) domino in G.

However, there are additional Left options $T = f(G)^L$ that are not of the form 0 or $f(G^L)$. In particular, they are those that expose a red or green domino, but these moves reverse out through Right's moves to 0, since $0 < f(G)$. $\square$

Corollary 11. *For every number x, there is a* TOPPLING DOMINOES *position $G = x$.*

Proof. Theorem 10 yields the mapping $G \mapsto 1{:}G$, so this result follows from Corollary 8. $\square$

We now prove uniqueness. A few more lemmas are needed.

Lemma 12. *Let G be an* LR-TOPPLING DOMINOES *position with $G > 0$. If G does not contain RR as a substring, then there exists a* TOPPLING DOMINOES *position H such that $G = f(H)$.*

Proof. H is obtained by removing a single L from every maximal substring of Ls in G. □

Lemma 13. $1{:}H \geq 1{:}K$ *if and only if* $H \geq K$.

Proof. When playing $1{:}H - 1{:}K$, a move on the base 1 taking a component to 0 can never be winning unless the other component is already 0. Hence, winning play on $1{:}H - 1{:}K$ is tantamount to play on $H - K$ until the last two moves. □

Lemma 14. *If a* TOPPLING DOMINOES *position G has consecutive non-blue dominoes, then G is not a number.*

Proof. Suppose (for contradiction) that $G = L\mathbf{x}AB\mathbf{y}L$ is a number where A and B are non-blue dominoes. Consider Right's move to $G^R = L\mathbf{x}A$. Since $G^R \mathrel{|\!\!\!>} G$, Left must have a winning move on $G^R - G$. Since $-G$ is a number and G^R is not, Left must have a winning move on the G^R component. By Lemma 3, this move must be to topple *west* on $L\mathbf{x}A$, to a position of the form $\mathbf{z}A - G$. But then Right has a response to $-G < 0$, a contradiction. □

Lemma 15. *Let G be a* TOPPLING DOMINOES *position. If G is a number and $G > 0$, then $G \cong f(H)$ for some H.*

Proof. It follows from the lemma that G is in the image of f. □

Theorem 16. *Let G and G' be* TOPPLING DOMINOES *positions, and suppose that $G = G' = x$, a number. Then $G \cong G'$.*

Proof. For $x = 0$ this is just Lemma 1. We proceed by induction on x.

First suppose $x > 0$. By the previous lemma, we have $G = f(H)$ and $G' = f(H')$ for some H, H', so by Theorem 10, $G = 1{:}H$ and $G' = 1{:}H'$. Now we know that $x = 1{:}y$ for some number y simpler than x. By Lemma 13, we have $H = H' = y$, so by induction $H \cong H'$. Therefore $G \cong G'$.

The case $x < 0$ is symmetric. □

Corollary 17. *All* TOPPLING DOMINOES *positions that are numbers are palindromes.*

Proof. If G' is the reversal of G, then certainly $G' = G$, so by Theorem 16 we have $G' \cong G$. □

4. Tinies

The main result of this section is the construction of $+_x$ and $-_x$. We present the main theorem here, but defer its proof until the end of the section.

Theorem 18. *Let* $\boldsymbol{x}$ *be the* TOPPLING DOMINOES *position of value* x, *where* $x \geq 0$ *is a number. The position* $RL\boldsymbol{x}LR$ *has value* $-_x$.

The proof of the theorem naturally entails showing that the games $\boldsymbol{x}LR + \{0 \mid -x\}$ and $RL\boldsymbol{x}LR + +_x$ are second player wins. In the arguments, Left sometimes moves twice in the TOPPLING DOMINOES component to some $\boldsymbol{x}^{LL}$, and we will need to show this is bad for Left:

Lemma 19. *Let* $\boldsymbol{x}$ *be a* TOPPLING DOMINOES *number (possibly negative). Let* $\boldsymbol{x}'$ *be a position reachable from* $\boldsymbol{x}$ *by one or more Left moves. Then* $\boldsymbol{x}' < \boldsymbol{x}$.

Proof. Recall that $\boldsymbol{x}$ consists only of L and R. Furthermore, if $\boldsymbol{x} > 0$, Lemma 14 tells us $\boldsymbol{x}$ never has consecutive R's and begins and ends in an L.

As usual, consider the difference game $\boldsymbol{x} - \boldsymbol{x}'$, and consider the case when $\boldsymbol{x} > 0$. If $\boldsymbol{x}'$ begins and/or ends in R, Left wins: Left's primary plan is to first move $-\boldsymbol{x}'$ to 0, and then move $\boldsymbol{x}$ to 0. Right cannot thwart this plan except possibly by making a first move which eliminates the only L end of $-\boldsymbol{x}'$, but Left has a good response to this opening by the next case.

In case $\boldsymbol{x}'$ begins and ends in L, then $\boldsymbol{x}'$ is in the image of f (Definition 9), and we can rewrite $\boldsymbol{x} = 1{:}\boldsymbol{y}$ and $\boldsymbol{x}' = 1{:}\boldsymbol{y}'$. Furthermore, since $\boldsymbol{x}'$ begins and ends in an L, observe that $\boldsymbol{y}'$ is reachable from $\boldsymbol{x}'$ by Left moves. So, by induction and Lemma 13, $\boldsymbol{y}' < \boldsymbol{y}$ and we have $\boldsymbol{x}' < \boldsymbol{x}$.

Finally, if $\boldsymbol{x} < 0$, $\boldsymbol{x}'$ must begin and end in R, and we can rewrite $\boldsymbol{x} - \boldsymbol{x}'$ as $(-1{:}\boldsymbol{y}) - (-1{:}\boldsymbol{y}')$ and use induction as above. □

Lemma 20. *If* $\boldsymbol{x}$ *is a* TOPPLING DOMINOES *number* $x > 0$, *then*

$$\boldsymbol{x}LR = \{x \mid 0\}.$$

Proof. Consider play on

$$\boldsymbol{x}LR + \{0 \mid -x\}.$$

If the first player moves in one component, the second player almost always moves in the other component; the only exception is when Right topples west in the first component wherein Left topples the rightmost LR, leaving $\boldsymbol{x}^R + \{0 \mid -x\}$. Right's move to $\boldsymbol{x}^R - x$ loses by Lemma 19 while when Right topples, Left can move to $\boldsymbol{x}^{RR} + 0 > 0$ since $\boldsymbol{x}^{RR}$ ends in Ls.

When the second player does respond to the first player's move in the other component, typically this leaves either $0 + 0$ or $\boldsymbol{x} - x$ to win. However, when Right is the first player and topples east, Left leaves games of the form $\boldsymbol{x}^R + 0$ where $\boldsymbol{x}^R > 0$ (since $\boldsymbol{x}$ never has consecutive Rs). When Left makes an initial move toppling a domino east, Right leaves $\boldsymbol{x}^L - x < 0$ by Lemma 19. When Left makes an initial move toppling west, Right again leaves $\boldsymbol{x}^R LR - x$ or $R - x$. The latter is clearly negative, while in the former Left's only hope is to topple east, but that leaves $\boldsymbol{x}^{RR} - x < 0$ by Lemma 19. □

Theorem 21. *If $\boldsymbol{x}$ is a positive* TOPPLING DOMINOES *number $x > 0$, then*

$$RL\boldsymbol{x}LR = \text{⧿}_x = \{x \mid 0 \,\|\, 0\}.$$

Proof. Consider play on

$$RL\boldsymbol{x}LR + \text{⧾}_x = RL\boldsymbol{x}LR + \{0 \,\|\, 0 \mid -x\}.$$

All moves outside those in $\boldsymbol{x}$ have natural winning responses. When Right moves in $\boldsymbol{x}$ or topples an end R only, Left immediately responds to $0 + \text{⧾}_x > 0$. The remaining case is Left's move to some $\boldsymbol{x}^L LR + \text{⧾}_x$ whence Right can play to

$$\boldsymbol{x}^L LR + \{0 \mid -x\}.$$

If Left now plays the second component to 0, Right can move the first component to 0. If, on the other hand, Left plays in the first component, Right replies in the second component leaving either $\boldsymbol{x}^{LL} - x$ which Right wins by Lemma 19, or $R - x$ which Right wins trivially. □

Conjecture 22. *$RL\boldsymbol{x}LR$ is the* unique TOPPLING DOMINOES *position of value ⧿_x, where $\boldsymbol{x} > 0$ is a number.*

This has been confirmed for TOPPLING DOMINOES strings of length up to 15.

5. Nimbers

In the sequel, we use $\boldsymbol{x}^a$ to mean the block of dominoes represented by $\boldsymbol{x}$ repeated a times.

Unlike dyadic rationals, nimbers have more than one representation; in LR-TOPPLING DOMINOES, $*m$ has m representations, or $2m$ counting reversals. The shortest representation is given by the following lemma.

Lemma 23. *For any non-negative integer m, $(LR)^m = *m$.*

Proof. The right followers of the form $(LR)^k L$ are positive and therefore dominated, while the ones of the form $(LR)^k$ are, by induction, $*k$. □

One position recurs as a Right follower in the arguments to come, and so we highlight it as a losing move for Right:

Lemma 24. *Let $\boldsymbol{x}$ be a* TOPPLING DOMINOES *position. Then $L\boldsymbol{x}L + *m \mathrel{|\!\rhd} 0$.*

Proof. The sum $L\boldsymbol{x}L + *m$ is a positive game added to a game equal to or fuzzy with 0. □

The main theorem of the section is that the following lemma describes the *only* forms of $*m$, and the proof of this fact will take the bulk of this section.

Lemma 25. *The position* $(LR)^a(RL)^m(LR)^a = *m$, *for* $0 \le a < m$.

Proof. The proof is routine. First, let $G = (LR)^a(RL)^m(LR)^a$ with $a < m$. We show that $G + *m = 0$. By the symmetry of G we need only consider Right going first.

If Right plays in G, all moves, except one, which topple east leave both ends L and lose by Lemma 24. The exception is the move to $(LR)^a + *m$ whence Left responds to $(LR)^a + *a = 0$.

When Right topples west, Left mirrors moves in the initial block $(LR)^a$ by moving symmetrically in the final block. Right's moves in the final block leave $(LR)^{a'} + *m$ for $a' < m$ and lose. Lastly, if Right topples west from the middle block $(RL)^m$, Right can remove the latter block leaving a position of the form $(LR)^{m'}L + *m = 1/2^{m'} + *m > 0$.

If Right plays in the second component of $G + *m$ to some $G + *m'$, there are two cases. If $m' \le a$, Left plays to $(LR)^{m'} + *m' = 0$. If $m' > a$, Left plays to $(LR)^a(RL)^m + *m'$. From here, each of Right's plays is easily seen as losing, for Right cannot leave a position of the form $LxL + *k$, yet moves to $(LR)^a + *m' = *a + *m'$ and to $(RL)^m + *m' = *m + *m'$ also lose. □

For the remainder of this section, let $G = *m$ be an LR-TOPPLING DOMINOES position. By Lemma 1, G cannot have Ls at both ends and so by symmetry we may assume $G = L\boldsymbol{x}R$.

Lemma 26. $G = *m$ *contains no substring of the form* $LL(RL)^nL$.

Proof. Say $G = \boldsymbol{x}LL(RL)^nL\boldsymbol{y}$. From $G+*m$ Left can move to $\boldsymbol{x}LL(RL)^n + *m$. By Lemma 3 if Right has a winning move in $\boldsymbol{x}LL(RL)^n$, it must be to topple a domino west. In these cases, then Left either has a move to or is left with a position of the form $L(RL)^j + *m$ for some j, which is $1/2^j + *m > 0$. If Right moves on the $*m$ then Left moves to $L(RL)^n + *k > 0$. □

The last lemma, and the symmetric one with L and R interchanged, is used to prove:

Lemma 27. $G = *m$ *is of the form* $(LR)^{a_1}(RL)^{a_2}(LR)^{a_3}\cdots(LR)^{a_n}$ *(or its reverse) for some* $n \ge 1$ *odd and* $a_1, \ldots, a_n \ge 0$.

Proof. By the previous lemma, G has no two instances of RR which aren't separated by an LL, and in particular no RRR (and the same holds with the roles of L and R reversed). Yet G cannot begin $(LR)^aLL\ldots$ (nor end $\ldots RR(LR)^a$) for Left's move to $(LR)^aL = 1/2^a$ is too strong. So if there are any pairs of consecutive like dominoes, these must alternate between RR and LL, beginning with RR and ending with LL. The form claimed in the lemma is the only possibility remaining. □

Lemma 28. *$G = *m$ is of the form $(LR)^a(RL)^m(LR)^a$ (or its reverse) for $a < m$.*

Proof. The previous lemma has G of one of the forms

$$(LR)^a \quad \text{or} \quad (LR)^a(RL)^b(LR)^c\boldsymbol{x}$$

for some string $\boldsymbol{x}$ either empty or starting in R. The first form is fine; in the second we must show that $\boldsymbol{x} = 0$, that $a = c$ and that $b > a$.

Since Left has moves from G to $0, *, \ldots, *(a-1)$, and Right has a move to $*a$, we have $m > a$. Moreover, Left has a move to $(LR)^a(RL)^b$, from which Right must respond to either $(LR)^a$ or $(RL)^b$ (since Right will lose if she leaves a double-Left-ended string). Since $m > a$, Right's move to $(LR)^a + *m$ is losing, so necessarily $(RL)^b + *m <= 0$. Therefore $b = m > a$.

Next consider c. We can't have $c > a$ since then Left would have a move from G to $(LR)^a(RL)^m(LR)^a$, which we know is equal to $*m$. Suppose instead $c < a$. Then Right can move from G to $(LR)^c(RL)^m(LR)^c\boldsymbol{x}$. If $\boldsymbol{x} = 0$ this wins immediately. Otherwise, consider Left's responses. Toppling to the west must lose, since all such moves are available to Left directly from $G + *m$. If Left topples east from within $\boldsymbol{x}$, then Right can revert to $(LR)^c(RL)^m(LR)^c$ which is equal to $*m$. Finally, if Left topples east annihilating $\boldsymbol{x}$, then Left's move must a priori be losing, since it's also available from $(LR)^c(RL)^m(LR)^c + *m$, which is equal to 0. Therefore Right's move from G to $(LR)^c(RL)^m(LR)^c\boldsymbol{x} + *m$ was a winning move, a contradiction.

This shows that $c = a$. Finally, this implies $\boldsymbol{x} = 0$, since otherwise Right could win by simply annihilating $\boldsymbol{x}$. □

Theorem 29. *The* LR-TOPPLING DOMINOES *position $G = *m$ if and only if it is of the form $(LR)^a(RL)^m(LR)^a$ for $0 \le a < m$.*

Proof. The last lemma combined with Lemma 25 proves the theorem. □

6. Atomic weights

In [1, p. 202], it is proved that any TOPPLING DOMINOES position with two gray ends is an infinitesimal with atomic weight 0, 1, or -1. In this section we generalize the result to include all TOPPLING DOMINOES infinitesimals, including those that end in L or R.

Lemma 30. *If G is an infinitesimal* TOPPLING DOMINOES *position beginning in L, then it must be of the form $(LR)^n$ or $(LR)^n E\boldsymbol{x}$ or $(LR)^n R\boldsymbol{x}$.*

Proof. If G is not of one of the above forms, fix G and rewrite it as $(LR)^m L\mathbf{y}$ where m is maximal. The string $\mathbf{y}$ cannot begin in R, for then G would either be one of the above forms or m would not be maximal. So $\mathbf{y}$ is either empty, in which case G equals some $(LR)^m L = 1/2^n$, or Left can move to $(LR)^m L$, contradicting that G is infinitesimal. □

Lemma 31. *If G is an infinitesimal* TOPPLING DOMINOES *position, then the first player has a move to a nimber. Furthermore, if the first player moves to any non-nimber, the second player can respond to a nimber.*

Proof. The proof is clear from the forms of G given by the last lemma. We first observe that the first player can move to a nimber. If G begins with an L, Left can move to 0, while Right can move to some $(LR)^n = *n$. If G begins with an E, either player can move to 0.

We now show that if the first player moves to a non-nimber, the second player can respond to a nimber. Without loss of generality, the first player topples east. When G begins with an E, the second player has a move to 0, and when G begins with an L, Left moving second has a move to 0. In the remaining case, G begins in L, and Left moved first toppling east. If Left's move was to some initial $(LR)^m$, then that already equals $*m$. If, on the other hand, the move is to some $(LR)^n E\mathbf{x}'$ or $(LR)^n R\mathbf{x}'$, then Right can move to $(LR)^n = *n$. □

Theorem 32. *All infinitesimal* TOPPLING DOMINOES *positions have atomic weight* -1, 0, *or* 1.

Proof. The proof is identical to that found in [1, pp. 202–203], and so we only provide a brief sketch here. Let G be an infinitesimal. The last lemma shows that G is of the form

$$\big\{\mathbf{0}, \{?\mid\mathbf{0}\}, \ldots \,\big|\, \mathbf{0}, \{\mathbf{0}\mid ?\}, \ldots\big\}$$

where the $\mathbf{0}$s are nimbers, and therefore have atomic weight 0. If each ? were all-small, then the rule for recursively computing atomic weights given by [1, p. 199] or [2] would yield at atomic weight of -1, 0, or 1. But if the ? are not all-small, we can replace each positive number stop in G with ⤊, and each negative number with ⇊, arriving at a new game G'. Observe that $G + H$ has the same outcome as $G' + H$ for $H \in \{\Uparrow☆, \uparrow☆, ☆, \downarrow☆, \Downarrow☆\}$, and so G has the same atomic weight as G' in $\{-1, 0, 1\}$. □

7. Conjectures

We end with a few conjectures. This first one would prove that green (gray) dominoes can introduce new values:

Conjecture 33. *The value*

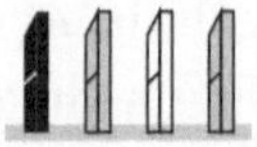

$$= \{1 \mid 0,\ \{0 \mid -1\}\}$$

does not appear in LR-TOPPLING DOMINOES.

Conjecture 34. *For all* TOPPLING DOMINOES *positions* $\boldsymbol{x}$ *and* $\boldsymbol{y}$, $\boldsymbol{x}L\boldsymbol{y} > \boldsymbol{x}\boldsymbol{y}$. *(Proving* $\boldsymbol{x}L\boldsymbol{y} \geq \boldsymbol{x}\boldsymbol{y}$ *is trivial.)*

Conjecture 35. *In* LR-TOPPLING DOMINOES, *if G is a palindrome then G's value appears uniquely.*

Next, here are a few conjectures that generalize the construction of $+_x$. In what follows, $a \geq b \geq c \geq d$ are numbers (as are x, x^L, and x^R), and are the toppling dominoes positions given by the construction of Section 3.

Theorem 36. *If* $x = \{x^L \mid x^R\}$ *is a number in canonical form, x is given uniquely by the* TOPPLING DOMINOES *position* $x^L L R x^R$.

Conjecture 37. *The game* $\{a \mid b\}$ *is given (uniquely) by the* TOPPLING DOMINOES *position* $aLRb$.

Conjecture 38. *The game* $\{a \parallel b \mid c\}$ *is given (uniquely) by the* TOPPLING DOMINOES *position* $aLRcRLb$.

Conjecture 39. *The game* $\{a \mid b \parallel c \mid d\}$ *is given (uniquely) by the* TOPPLING DOMINOES *position* $bRLaLRdRLc$.

Conjecture 40. *Some games* do not *occur in* TOPPLING DOMINOES. *(By way of comparison,* HACKENBUSH *contains no hot games.)*

References

[1] M. Albert, R. J. Nowakowski, and D. Wolfe. *Lessons in play: An introduction to combinatorial game theory.* A K Peters, Wellesley, MA, 2007.

[2] Elwyn R. Berlekamp, John H. Conway, and Richard K. Guy. *Winning ways for your mathematical plays, I.* A K Peters, Natick, MA, 2nd edition, 2001.

[3] J. H. Conway. *On numbers and games.* A K Peters, Natick, MA, 2nd edition, 2001.

finka@math.berkeley.edu — *Department of Mathematics, University of California, Berkeley, Berkeley, CA 94620, United States*

rjn@mathstat.dal.ca — *Mathematics and Statistics, Dalhousie University, Halifax B3H 3J5, Canada*

aaron.n.siegel@gmail.com

davidgameswolfe@gmail.com — *Mathematics and Computer Science, Gustavus Adolphus College, St Peter, MN 56082, United States*

Games of No Chance 4
MSRI Publications
Volume **63**, 2015

Harnessing the unwieldy MEX function

AVIEZRI S. FRAENKEL AND UDI PELED

A pair of integer sequences that split $\mathbb{Z}_{>0}$ is often — especially in the context of combinatorial game theory — defined recursively by

$$a_n = \text{mex}\{a_i, a_i : 0 \le i < n\}, \quad b_n = a_n + c_n \quad (n \ge 0),$$

where mex (**M**inimum **EX**cludant) of a subset S of nonnegative integers is the smallest nonnegative integer not in S, and $c : \mathbb{Z}_{\ge 0} \to \mathbb{Z}_{\ge 0}$. Given $x, y \in \mathbb{Z}_{>0}$, a typical problem is to decide whether $x = a_n$, $y = b_n$. For general functions c, the best algorithm for this decision problem was until now exponential in the input size $\Omega(\log x + \log y)$. We prove constructively that the problem is actually polynomial for the wide class of *approximately linear* functions c_n. This solves constructively and efficiently the complexity question of a number of previously analyzed *take-away* games of various authors.

1. Introduction

This paper is about the complexity of combinatorial games. Its main contribution is showing constructively that a large class of games whose complexity was hitherto unknown and its best winning strategy was exponential, is actually solvable in polynomial time.

For many *take-away* games on two piles, the set of *losing positions* $\{(a_n, b_n)\}$ $(n \ge 0)$, also called *P-positions*, is given by recursive formulas of the form:

$$\begin{cases} a_n = \text{mex}\{a_i, b_i : 0 \le i < n\}, \\ b_n = a_n + c_n, \quad n \ge 0, \end{cases} \tag{1}$$

where mex (**M**inimum **EX**cludant) of a subset S of nonnegative integers is the smallest nonnegative integer not in S. The mex definition implies that sequences $\{a_n\}$, $\{b_n\}$ $(n \ge 1)$ of the form (1), for $c : \mathbb{Z}_{\ge 0} \to \mathbb{Z}_{\ge 0}$, are *complementary sequences* [Fraenkel 1982]. Later on we define precisely the notions alluded to here informally.

A case in point is *Wythoff's game*, played on two piles of tokens of sizes (x, y), where we may assume $0 \le x \le y$. The two players play alternately. There are two

MSC2010: 11B39, 11B75, 68Q25, 91A46.
Keywords: combinatorial games, mex, efficient algorithm, complexity.

types of moves: (i) taking any positive number of tokens from a *single* pile (Nim rule), or (ii) removing the *same* number of tokens from both piles (Wythoff rule). In this and all games considered here, the first player unable to move (because the position is (0, 0)) loses and the opponent wins. Its P-positions are $\{(a_n, b_n)\}$, $n \geq 0$, where a_n, b_n are given by (1) with $c_n = n$ [Fraenkel 1982]. Given a game position (x, y), to decide whether $(x, y) = (a_n, b_n)$ requires the computation of $a_0, b_0, a_1, \ldots, b_{k-1}, a_k$, where k is the largest index such that $a_k \leq x$. This is an exponential computation, since the input size is $O(\log x + \log y)$.

For Wythoff's game, however, there is an explicit formula for the P-positions, namely, $a_n = \lfloor n\varphi \rfloor$, $b_n = \lfloor n\varphi^2 \rfloor$, where $\varphi = (1 + \sqrt{5})/2$ is the golden section [Fraenkel 1982]. Using this formula it is possible to efficiently compute a_n, b_n for any n, and more importantly, to decide efficiently (in polynomial time) whether a given position (x, y) of Wythoff's game is a P-position. Another efficient winning strategy is based on the Fibonacci numeration system. There is a generalization of Wythoff's game where the Wythoff move is relaxed: one can take $k > 0$ from one pile and $\ell > 0$ from the other, provided $|k - \ell| < p$, where p is a fixed positive integer parameter ($p = 1$ is Wythoff's game). The P-positions of this *p-Wythoff game* are given by (1) for $c_n = pn$ [Fraenkel 1982]. Also for this case there is an explicit formula of its P-positions: $a_n = \lfloor n\alpha \rfloor$, $b_n = \lfloor n\beta \rfloor$, where $\alpha = 1 - p/2 + \sqrt{1 + (p/2)^2}$, $\beta = \alpha + p$, leading to an efficient computation of a_n, b_n [Fraenkel 1982]. For a recent and comprehensive survey on the complexity of combinatorial games see [Demaine and Hearn 2009]. Previous surveys are in [Demaine 2001; Fraenkel 2000; 2004a]. There is an extensive literature on Wythoff's game; here we just cite two of them: [Wythoff 1906; Fraenkel 1982].

For other functions c, the present state of affairs is that ad hoc methods have to be devised for each c in order to try and decide whether an efficient winning strategy exists. For example, for $c_n = (s - 1)a_n + pn$ where $s, p \in \mathbb{Z}_{>0}$, it was shown in [Fraenkel 1998] that for $s > 1$, there are no α, β, γ, δ such that $a_n = \lfloor n\alpha + \gamma \rfloor$, $b_n = \lfloor n\beta + \delta \rfloor$. However, in this case a polynomial winning strategy can still be recovered by means of an exotic numeration system. If c is not an integer, no efficient strategy is known in general. A case in point is a game considered in [Duchêne and Gravier 2009], where $c_n = 2\lfloor n/4 \rfloor$. For the case where c is a special algebraic number, it was recently shown in [Duchêne and Rigo 2010] that an efficient winning strategy exists. But for most cases, nothing is known about the efficiency of winning.

In conclusion, the present state of affairs is that for each function c, the corresponding game has to be analyzed and ad hoc methods have to be devised which might or might not produce an efficient strategy.

The main impact of the present paper is the formulation of a constructive recursive algorithm showing that the strategy is efficient for every *approximately linear* function c_n, in particular functions of the form $c_n = k\lfloor \theta n \rfloor$, where $k \in \mathbb{Z}_{>0}$, $\theta \in \mathbb{R}_{>0}$. This solves take-away games considered previously where the complexity analysis was left open, such as in [Fraenkel 2004b; Hegarty and Larsson 2006; Duchêne and Gravier 2009]. We also prove several approximation results for certain sequences, of independent interest, which we use for formulating our recursive algorithm.

Many complementary sequences are listed in Sloane's online encyclopedia of integer sequences. They play a prominent role not only in the theory of combinatorial games, but also in combinatorics of words and spectra of numbers investigations. See [Kimberling 2007] for a survey and also [Kimberling 2008].

In the next section we formalize some of the notions and tools which will be used in the sequel. In particular, we define the notion of approximate linearity. In Section 3, which is the main technical section, we explore complementary sequences by proving Theorem 3.2, which gives a basic yet important inequality about complementary sequences. This is followed by Theorem 3.4, which establishes an important combinatorial equivalence concerning increasing sequences. Both of these results are then used to prove the fundamental Bounded Additivity Theorem 3.5. It enables one to wed together complementarity and approximate linearity in Section 4, with the aid of Fekete's lemma [1923], leading to our main result, the Approximate Linearity Theorem 4.3. This result then leads, in Section 5, to the efficient algorithm, together with an example and its complexity analysis. In Section 6 we wrap-up with a conclusion.

2. Formalizations

Definition 2.1 (P- and N-positions). (i) A position from which the player moving first has a winning strategy is called an *N-position* (**N**ext player to move wins). A position from which the player moving second has a winning strategy is called a *P-position* (**P**revious player to move wins).

(ii) A *take-away* game is played on a collection of piles of finitely many tokens. A move consists of selecting a pile and removing from it any positive number of tokens.

Note that player I, starting from a P-position, is doomed to lose the game, assuming player II plays optimally. Hence to find optimal strategies, it suffices to characterize the P-positions, and this characterization should be as simple as possible computationally. One might go about characterizing the N-positions instead, but it's more economical to characterize the P-positions, since they are rare. Intuitively, this follows from the fact that a position is a P-position if and only if *all* its direct followers are N-positions, whereas a position is an

N-position if and only if it *has* a direct P-position follower. Formally, this has been proved in [Singmaster 1981]; see also [Singmaster 1982].

Definition 2.2 (Minimal EXcludant). Let $S \subsetneq Z_{\geq 0}$. Then

$$\text{mex}(S) := \min(Z_{\geq 0} \setminus S).$$

Notice that by the mex definition, the sets $\{a_n\}_{n\geq 1}$, $\{b_n\}_{n\geq 1}$ in (1) partition the positive integers into two disjoint sets whenever $c_n \geq 1$ $(n \geq 1)$ [Fraenkel 1982]. The mex function appears frequently in the study of combinatorial games, hence it is important to understand its behavior and influence on the sequences which it defines.

The game of q-Blocking p-Wythoff's Nim ($p, q \geq 1$ integer constants) was defined by Hegarty and Larsson [2006]. It is p-Wythoff mentioned earlier, with an additional "Muller twist" (see [Smith and Stănică 2002]), namely, if the current position is (k, ℓ), then, before the next move is made, the previous player is allowed to choose up to $(q-1)$ distinct positive integers $t_1, \ldots, t_{q-1} \leq \min\{k, l\}$ and prevent the next player from moving to any position $(k - t_i, \ell - t_i)$. They proved that its P-positions are given by (1) with $c_n = p\lfloor n/q \rfloor$ $(n \geq 0)$. Alas, by a result in [Boshernitzan and Fraenkel 1981], there is no representation of the type $\lfloor n\alpha + \beta \rfloor$ for either of the sequences a_n, b_n in the general case of positive integers p, q. No numeration system seems to help, so there appears to be no efficient method for computing the sequences. This game was one of the motivations for the current work.

Duchêne and Gravier [2009] analyzed the following 2-pile extension of Wythoff's game: remove any number of tokens from a *single* pile, or remove the same positive *even* number of tokens from both piles. They proved that the P-positions (a_n, b_n) are given by $a_n = F_n$, $b_{4n} = G_{4n}$, $b_{4n+1} = G_{4n+1}$, $b_{4n+2} = G_{4n+3}$, $b_{4n+3} = G_{4n+2}$ for all $n \geq 0$, where $F_n = \text{mex}\{P_i, Q_i : 0 \leq i < n\}$, $G_n = F_n + 2\lfloor n/4 \rfloor$ for all $n \geq 0$. So F_n, G_n have the same values as a_n, b_n respectively of q-Blocking p-Wythoff for the special case $p = 2$, $q = 4$. In fact, the Duchêne and Gravier game was one of our original motivations for investigating sequences of this type. Both of the last mentioned games can be solved polynomially by our method.

Definition 2.3 (approximate linearity). We call the sequence A *approximately linear* if there exist constants $\alpha, u_1, u_2 \in \mathbb{R}$ such that

$$u_1 \leq a_n - n\alpha \leq u_2 \quad \text{for all } n \geq 0.$$

We say that u_1, u_2 are *approximation bounds*, and α is the *linearity rate*.

Remark. Notice that α is unique since $\lim_{n\to\infty} a_n/n = \alpha$, but u_1, u_2 are not.

Though there is no known explicit formula for general sequences $\{a_n\}$, $\{b_n\}$ of (1), it was conjectured that for the special case $c_n = p\lfloor n/q \rfloor$ they can be approximated by linear sequences, namely that $a_n - n\alpha$ is bounded for some constant α, and similarly for b_n. Our first result of this work was proving that indeed this approximation holds, by finding bounds depending on p, q. Hegarty [2009] independently proved the same approximation for the special case $p = 1$. We then generalized our result to the case where c is any approximately linear function.

After approximating a_n, b_n, there still remained the problem of efficiently computing the sequences, which, without an explicit formula, seemed impossible. We should mention that more often than not, for a general function c_n, there is no efficient algorithm for computing the elements of the sequences defined by (1) and deciding whether an arbitrary number belongs to it. The best known algorithms are exponential.

But quite surprisingly, after deepening our understanding of partitions created by the mex function, we indeed found an efficient algorithm. This is our most important result. The algorithm computes sequences for any approximately linear function c.

In particular, the algorithm solves efficiently the q-Blocking p-Wythoff's game, and also Duchêne's game. Moreover, since most of the results presented here are for general sequences defined by the mex function, and the idea behind the algorithm is basic, we believe that it can be generalized and implemented with some modifications for other sequences. For example, Fraenkel and Krieger [2004] and Sun and Zeilberger [2004; Sun 2005] investigated the P-positions of *Fraenkel's N-Heap Wythoff's game* [Fraenkel 2004a]. These P-positions are defined similarly to the sequences $\{a_n\}$, $\{b_n\}$, and they share the same essential properties as them. Hence the results presented here can possibly offer some new insights to understanding Fraenkel's N-Heap Wythoff's game.

3. Complementary sequences

Unless otherwise specified, any interval of the form $[a, b]$ denotes the set of integers $\{k \in \mathbb{Z} : a \le k \le b\}$. Analogously for open/half-open intervals.

Definition 3.1 (complementary sequences). We say that two integer sequences $A := \{a_n\}_{n\ge 0}$, $B := \{b_n\}_{n\ge 0}$ are *complementary sequences* if they are strictly increasing sequences which satisfy $A \cup B = \mathbb{Z}_{\ge 0}$, $A \cap B = \{a_0 = b_0 = 0\}$. We also define $c := B - A$.

We begin with a property of complementary sequences, which will be used both in the proof of Theorem 3.5, and in the efficient algorithm for computing sequences. The idea is that given an element in a sequence, we know its position between two consecutive elements of the complementary sequence.

Theorem 3.2 (complementary sequences inequality). *Suppose that* A, B *are complementary sequences. Then*

$$b_{a_k-k} < a_k < b_{a_k-k+1}, \quad a_{b_k-k} < b_k < a_{b_k-k+1}, \quad \textit{for all } k \geq 1.$$

Proof. How is the interval $I := [1, a_k]$ split between A and B? Since A is increasing, I contains precisely the first k terms of $\{a_n\}$. Hence the remaining $a_k - k$ terms of I must be the leading terms of the increasing sequence B. Therefore $b_{a_k-k} \leq a_k < b_{a_k-k+1}$. By complementarity, actually $b_{a_k-k} < a_k$. The second inequality is derived the same way by splitting $[1, b_k]$ between the two sequences. □

Remark. Denote $I = [b_r, b_s]$, by the theorem we get that the first element of $A \cap I$ is a_{b_r-r+1}, and the last element of $A \cap I$ is a_{b_s-s}, and a similar result by switching the roles of A, B.

Now for simple bounds on the growth of the smaller sequence of complementary sequences.

Corollary 3.3. *Assume that* A, B *are complementary sequences. If* $a_n < b_n$ *for all* $n \geq 1$, *then* $n \leq a_n \leq 2n - 1$ *for all* $n \geq 1$.

Proof. Since A is increasing, obviously $n \leq a_n$. From Theorem 3.2 we know that $b_{a_n-n} < a_n$, and we assumed $a_n < b_n$, so $b_{a_n-n} < b_n$. Since B is strictly increasing, the indices of the last inequality must satisfy $a_n - n < n$ as claimed. □

The following very intuitive theorem gives an equivalence between the number of elements of an increasing sequence in an arbitrary interval of length u, and an inequality about the increments of this interval. We need it for proving Theorem 3.5.

Theorem 3.4 (min-max-equivalence). *Assume that* $B = \{b_n\}_{n\geq 0}$ *is an increasing sequence, satisfying* $b_0 = 0$.

Then the following equivalence holds for any two integers x, u.

$$\min_{L\geq 0} |[L+1, L+u] \cap B| \geq x \iff \textit{for all } n \geq 0, b_{n+x} \leq b_n + u.$$

And similarly also the following equivalence holds

$$\max_{L\geq 0} |[L+1, L+u] \cap B| \leq x \iff \textit{for all } n \geq 0, b_n + u \leq b_{n+x}.$$

Proof. The implication

$$\min_{L\geq 0} |[L+1, L+u] \cap B| \geq x \implies \text{for all } n \geq 0, b_{n+x} \leq b_n + u$$

is obvious, since we can take $L = b_n$.

We'll prove the other direction: given $L \geq 0$, take the maximal n such that $b_n \leq L$. So since $b_{n+x} \leq b_n + u \leq L + u$, we get $b_{n+1}, b_{n+2}, \dots, b_{n+x} \in [L+1, L+u]$ as claimed.

The proof of the second equivalence is similar: for the first implication take $L = b_n - 1$. And for the second direction take minimal $n \geq 0$ such that $L < b_n$. □

Remark. Equivalence also holds when either x or u are negative integers.

The following result enables us to prove Theorem 4.3. The proof is by induction, and in order to better understand it, we recommend following the first steps of $m = 1, 2, 3, \dots$ with specific sequences $\{a_n\}$, $\{b_n\}$ (e.g., $b_n = a_n + \lfloor \sqrt{2}n \rfloor$).

Theorem 3.5 (bounded additivity). *Let A, B be complementary sequences. Suppose that $c = \{c_n\} = \{b_n - a_n\}_{n \geq 0}$ is a nonnegative sequence which satisfies, for all $n, m \geq 0$,*

$$c_{n+m-r} \leq c_n + c_m \leq c_{n+m+s},$$

for some integer constants $r, s \geq 0$ satisfying $r - 1 \leq s$. Then we have for all $n, m \geq 0$,

$$a_{n+m-r} - 1 \leq a_n + a_m + s - r \leq a_{n+m+s}.$$

Proof. We shall prove the inequality by induction on m. For $m = 0$, the inequality is $a_{n-r} - 1 \leq a_n + s - r \leq a_{n+s}$ for $n \geq 0$, which follows from $r - 1 \leq s$ and $a_n + s \leq a_{n+s}$ for all $n \geq 0$ (since a_n is increasing).

We assume that the following holds for all $0 \leq t \leq m$,

$$a_{n+t-r} - 1 \leq a_n + a_t + s - r \leq a_{n+t+s} \quad \text{for all } n \geq 0,$$

and we shall prove that it holds for all $0 \leq t \leq m+1$.

By the induction assumption we know

$$a_{n+t-r} - 1 \leq a_n + a_t + s - r \quad \text{for all } n \geq 0, 0 \leq t \leq m,$$

and by adding $c_{n+t-r} \leq c_n + c_t$ we get

$$b_{n+t-r} - 1 \leq b_n + b_t + s - r \quad \text{for all } n \geq 0, 0 \leq t \leq m.$$

Now, by Corollary 3.3 we have $t + 1 \leq a_{t+1} \leq 2(t+1) - 1$, hence $0 \leq a_{t+1} - (t+1) \leq t \leq m$, and we can switch t with $a_{t+1} - (t+1)$ in the above inequality to get

$$b_{n+a_{t+1}-(t+1)-r} \leq b_n + b_{a_{t+1}-(t+1)} + s - r + 1 \quad \text{for all } n \geq 0, 0 \leq t \leq m.$$

From Theorem 3.2 we know that $b_{a_{t+1}-(t+1)} < a_{t+1}$, meaning $b_{a_{t+1}-(t+1)} \leq a_{t+1} - 1$, which, concatenated to the right of the above inequality, gives

$$b_{n+a_{t+1}-(t+1)-r} \leq b_n + a_{t+1} + s - r \quad \text{for all } n \geq 0, 0 \leq t \leq m.$$

Since this holds for all $n \geq 0$, Theorem 3.4 implies that for every fixed $t \leq m$,

$$\min_{L\geq 0}|I_L \cap B| \leq a_{t+1} - (t+1) - r,$$

where the closed interval I_L is given by $I_L = [L+1, L+a_{t+1}+s-r]$. Now, since A, B are complementary, and $|I_L| = a_{t+1}+s-r$, we get

$$\max_{L\geq 0}|I_L \cap A| \leq (a_{t+1}+s-r) - (a_{t+1} - (t+1) - r) = t+1+s.$$

Again by Theorem 3.4 we get

$$a_n + a_{t+1} + s - r \leq a_{n+t+1+s} \quad \text{for all } n \geq, 0 \leq t \leq m.$$

So we established the right inequality of our goal inequality, namely

$$a_n + a_t + s - r \leq a_{n+t+s} \quad \text{for all } n \geq, 0 \leq t \leq m+1.$$

Now, for the left hand side we use similar arguments. We add $c_n + c_t \leq c_{n+t+s}$ to the last inequality, to get

$$b_n + b_t + s - r \leq b_{n+t+s} \quad \text{for all } n \geq 0, 0 \leq t \leq m+1.$$

By Corollary 3.3 we have $t \leq a_t \leq 2t-1$, hence $0 \leq a_t - t + 1 \leq t \leq m+1$, and we can switch t with $a_t - t + 1$ in the above inequality to get

$$b_n + b_{a_t - t+1} + s - r \leq b_{n+a_t-t+1+s} \quad \text{for all } n \geq 0, 0 \leq t \leq m+1.$$

From Theorem 3.2 we know that $a_t < b_{a_t-t+1}$, meaning $a_t + 1 \leq b_{a_t-t+1}$, which, concatenated to the left of the above inequality, gives

$$b_n + a_t + 1 + s - r \leq b_{n+a_t-t+1+s} \quad \text{for all } n \geq 0, 0 \leq t \leq m+1.$$

Since this holds for all $n \geq 0$, by Theorem 3.4 we get for every fixed $t \leq m+1$,

$$\max_{L\geq 0}|I_L \cap B| \leq a_t - t + 1 + s,$$

where $I_L = [L+1, L+a_t+1+s-r]$. Now, since A, B are complementary, and $|I_L| = a_t + 1 + s - r$, we get

$$\min_{L\geq 0}|I_L \cap A| \geq (a_t + 1 + s - r) - (a_t - t + 1 + s) = t - r.$$

Again by Theorem 3.4 we get

$$a_{n+t-r} \leq a_n + a_t + 1 + s - r \quad \text{for all } n \geq, 0 \leq t \leq m+1.$$

So we also established the left inequality of our goal inequality, hence the induction is complete. □

Remark. The theorem's conclusion also holds for the sequence B, that is, for all $n, m \geq 0$,

$$b_{n+m-r} - 1 \leq b_n + b_m + s - r \leq b_{n+m+s}.$$

To see this, just add the inequality assumption of the sequence c to the inequality conclusion of a_n.

We can change the indices in the theorem's conclusion to get

$$a_{n+m} \leq a_n + a_{m+r} + s - r + 1 \leq a_{n+m+\alpha+s} + 1$$

for all $n, m \geq 0$. But we do not want to be concerned with the exact indices and constants of such inequalities (though these were essential for the theorem's proof), so we state a very simple lemma.

Lemma 3.6. *Assume that $\{a_n\}_{n\geq 0}$ is a real sequence, which for some constants $r, s \in \mathbb{Z}_{\geq 0}$ and $u, v \in \mathbb{R}$, satisfies*

$$a_{n+m} \leq a_n + a_{m+r} + u \leq a_{n+m+s} + v \quad \textit{for all } n, m \geq 0.$$

Then there exist constants D, $E \in \mathbb{R}$, such that

$$a_{n+m} + D \leq a_n + a_m \leq a_{n+m} + E \quad \textit{for all } n, m \geq 0.$$

Proof. By the left inequality of the assumption we know that $a_{t+r} \leq a_t + a_{2r} + u$ for all $t \geq 0$, so by taking $t = m$ together with the assumption we get

$$a_{n+m} \leq a_n + a_{m+r} + u \leq a_n + a_m + a_{2r} + 2u.$$

We can take $D = -a_{2r} - 2u$ to fit our goal.

Now, again by the left inequality of the assumption we know that $a_{t+s} \leq a_t + a_{s+r} + u$ for all $t \geq 0$. By taking $t = n+m$ we can continue the right inequality of the assumption as $a_n + a_{m+r} + u \leq a_{n+m+s} + v \leq a_{n+m} + a_{s+r} + u + v$.

By taking $n = 0$ in the assumption we get $a_m \leq a_0 + a_{m+r} + u$, and together with the previous inequality we get $a_n + a_m - a_0 \leq a_n + a_{m+r} + u \leq a_{n+m} + a_{s+r} + u + v$. So by both ends of this inequality we can take $E = a_0 + a_{s+r} + u + v$, and we're done. □

4. Approximate linearity

We use a well known lemma from [Fekete 1923].

Lemma 4.1 (Fekete's lemma). *Let $\{s_n\}_{n\geq 0}$ be a superadditive sequence of real numbers, that is, $s_n + s_m \leq s_{n+m}$ for all $n, m \geq 0$. Then $\lim_{n\to\infty} s_n/n$ exists, and is equal to $\sup_{n\geq 1} s_n/n \in (-\infty, \infty]$.*

From this lemma we derive the following corollary, which can be applied directly to the results of Theorem 3.5.

Corollary 4.2 (Fekete's corollary). *Assume that A is a sequence of real numbers which satisfies, for some constants $u, v \in \mathbb{R}$,*

$$u \leq a_n + a_m - a_{n+m} \leq v \quad \textit{for all } n, m \geq 0.$$

Then A is approximately linear, with approximation bounds u, v, and linearity rate $\alpha = \lim a_n/n$.

Proof. We have $(a_n - v) + (a_m - v) \leq a_{n+m} - v$ for $n, m \geq 0$, so by applying Fekete's lemma to the sequence $a_n - v$ we get

$$\sup_{n\geq 1} \frac{a_n - v}{n} = \lim \frac{a_n - v}{n} = \alpha \in (-\infty, \infty].$$

Hence for $n \geq 1$ we have $a_n - v \leq n\alpha$, i.e., $a_n - n\alpha \leq v$.

Similarly, $(u - a_n) + (u - a_m) \leq u - a_{n+m}$ for $n, m \geq 0$, so by Fekete's lemma

$$\sup_{n\geq 1} \frac{u - a_n}{n} = \lim \frac{u - a_n}{n} = \alpha' \in (-\infty, \infty].$$

Hence for $n \geq 1$ $\quad u - a_n \leq n\alpha'$, i.e., $u \leq a_n + n\alpha'$.

Lastly, notice that $\alpha' = \lim (u - a_n)/n = -\lim a_n/n = -\lim (a_n - v)/n = -\alpha$, hence $\alpha \in \mathbb{R}$. Altogether we get $u \leq a_n - n\alpha \leq v$, as claimed. □

Using Theorem 3.5 together with Fekete's lemma and their corollaries, we get Theorem 4.3, which is the main theorem of this paper.

Theorem 4.3 (approximate linearity). *Let A, B be complementary sequences. Then both A and B are approximately linear if and only if $c = B - A$ is approximately linear.*

Proof. Assume that A, B are approximately linear, i.e., $u_1 \leq a_n - n\alpha \leq u_2$ and $v_1 \leq b_n - n\beta \leq v_2$ for suitable constants u_1, u_2, v_1, v_2. Then we easily get $v_1 - u_2 \leq (b_n - a_n) - n(\beta - \alpha) \leq v_2 - u_1$, which means that c is approximately linear with linearity rate $\beta - \alpha$.

In the other direction, assume c is approximately linear, i.e., $K_1 \leq c_n - n\gamma \leq K_2$ for constants K_1, K_2. We would like to use Theorem 3.5, so we first look for integer constants $r, s \geq 0$. We take an integer $r \geq 0$, which satisfies $r \geq (K_2 - 2K_1)/\gamma$. Thus,

$$c_{n+m-r} \leq (n+m-r)\gamma + K_2 \leq n\gamma + K_1 + m\gamma + K_1 \leq c_n + c_m$$

(the middle inequality is by our choice of r, and rightmost/leftmost inequalities are the approximate linearity of c).

We also take some integer $s \geq 0$ which satisfies $s \geq (2K_2 - K_1)/\gamma$. Then

$$c_n + c_m \leq n\gamma + K_2 + m\gamma + K_2 \leq (n+m+s)\gamma + K_1 \leq c_{n+m+s}.$$

But recall that the theorem also requires $s \geq r - 1$, so we choose s that also satisfies this.

In conclusion, $c_{n+m-r} \leq c_n + c_m \leq c_{n+m+s}$, so by Theorem 3.5 and Lemma 3.6 we get $u_1 \leq a_n + a_n - a_{n+m} \leq u_2$ for some constants u_1, u_2, and similarly for B. Hence Fekete's Corollary implies that A, B are approximately linear. □

Remark. If we know the approximation bounds K_1, K_2 and the linearity rate $\gamma = \lim c_n/n$, then by Theorem 3.5's conclusion and the proof of Lemma 3.6, we can compute approximation bounds for the sequences A and B after choosing — preferably minimal — r, s. Thus we regard them as known constants.

Also the linearity rates α, β are known. Since A, B are complementary sequences, and since $\lim a_n/n = \alpha$, $\lim b_n/n = \beta$, density considerations imply $\alpha^{-1} + \beta^{-1} = 1$ [Fraenkel 1982]. Also notice that $c_n = b_n - a_n$ implies $\beta - \alpha = \gamma$, so we find α, β by solving the quadratic equation $\alpha^{-1} + (\alpha + \gamma)^{-1} = 1$. For any $\theta \in \mathbb{R}_{>0}$, setting $c_n = k\lfloor \theta n \rfloor$ we have $c = k\theta$, so we obtain $a = \frac{1}{2}(2 - k\theta + \sqrt{k^2\theta^2 + 4})$, $b = \frac{1}{2}(2 + k\theta + \sqrt{k^2\theta^2 + 4})$.

If $0 < \theta < 1$ then for $0 \leq i < 1/\theta$ we get $c_i = k\lfloor \theta i \rfloor = 0$. And for $i_0 = \lceil 1/\theta \rceil$ we get $c_{i_0} = k\lfloor \theta i_0 \rfloor \geq 1$ and $b_i = a_i = i$; thus $a_{i_0} = i_0$, $b_{i_0} > i_0$. By shifting the indices, we get the complementary sequences A', B' defined by $a'_n = a_{n+i_0-1} - i_0 + 1$, $b'_n = b_{n+i_0-1} - i_0 + 1$ for $n \geq 0$. The difference

$$c'_n = b'_n - a'_n = b_{n+i_0-1} - a_{n+i_0-1} = k\lfloor \theta(n + i_0 - 1) \rfloor$$

is approximately linear, so by Theorem 4.3, A', B' are approximately linear. Definition 2.3 then implies that also the original A, B are approximately linear.

5. Efficient algorithm for computing sequences

5.1. ***Preparation.*** We deal with complementary sequences A, B, and assume that c is approximately linear with known approximating bounds K_1, K_2, and known linearity rate $\gamma = \lim c_n/n$. Then A, B are also approximately linear by Theorem 4.3, and as we demonstrate below, we can compute some integer approximating bounds u_1, u_2, v_1, v_2, and compute $\alpha = \lim a_n/n$, $\beta = \lim b_n/n$ by the formula $1 \pm \gamma/2 + \sqrt{1 + (\gamma/2)^2}$.

Now, given a pair of integers (x, y), we would like to check whether $(x, y) = (a_n, b_n)$ for some n. If indeed $x = a_n$, then by the approximate linearity we must have $(x - u_2)/\alpha \leq n \leq (x - u_1)/\alpha$. So we need to check whether $x = a_n$ for only a fixed number of n's (at most $(u_2 - u_1)/\alpha$ such n's). If $x = a_n$ for some n, we are just left to check if $y = b_n$ in order solve the decision problem. Notice that if we know $x = a_n$, then checking whether $y = b_n$ can be done by checking whether $y - x = c_n$, thus it depends on the complexity of computing c_n.

So, we need to find an efficient method for computing the elements of a_n given n. The naïve recursive method for computing the elements of a_n, b_n is very slow and inefficient: its complexity is $O(n)$, whereas the input size is of complexity $O(\log n)$ in succinct representation. Hence the naïve method is exponential in the input size.

5.2. ***Idea.*** The algorithm computes $a_m, a_{m+1}, \ldots, a_n$ for given m, n, by operating as follows. If for some r, s we can compute $b_r, b_{r+1} \ldots, b_s$, where $b_r < a_m$ and $a_n < b_s$. Then since A, B are complementary sequences, we can compute $a_m, \ldots, a_n$ by complementing the interval $[b_r, b_s]$ with elements of $\{a_i\}$. Notice, however, that the first element of A in the interval $[b_r, b_s]$ will not necessarily be a_m, but rather a_{b_r-r+1} by Theorem 3.2. The last element of A in the interval will be a_{b_s-s}. So since we know the start and end indices, namely b_r-r+1 and b_s-s, we can immediately locate the desired $a_m, \ldots, a_n$ inside $a_{b_r-r+1}, \ldots, a_{b_s-s}$.

The problem that remains is to find r, s, which are easy to compute, and that satisfy $b_r < a_m$, $a_n < b_s$. Moreover, $b_r, \ldots, b_s$ should be efficiently computable. First we deal with finding such r. By Theorem 3.2 we can take any $r \le a_m - m$, and it will satisfy $b_r < a_m$. But we have not yet computed a_m, so how can we find $r \le a_m - m$?

By the approximate linearity of $\{a_i\}$ we know that $m(\alpha-1)+u_1 \le a_m - m$, so we can take $r = \lfloor m(\alpha-1)+u_1 \rfloor$ to satisfy $r \le m(\alpha-1)+u_1 \le a_m - m$, hence $b_r < a_m$. By similar reasoning, for $s = \lceil n(\alpha-1)+u_2 \rceil + 1$ we have $a_n - n + 1 \le n(\alpha-1)+u_2+1 \le s$, hence $a_n < b_s$ (Theorem 3.2).

So we have easy formulas for r, s, which satisfy $b_r < a_m$, $a_n < b_s$. But how can we easily compute $b_r, \ldots, b_s$?

Notice that it's enough to know $a_r, a_{r+1}, \ldots, a_s$, and then we can compute directly $b_i = a_i + c_i$. In order to compute $a_r, \ldots, a_s$, we simply recursively call the main part of the algorithm, with r, s instead of m, n, and we're done.

We now turn to formulate the algorithm, afterwards we show how it works on a specific example, and lastly we analyze its complexity.

Remark. Though the algorithm is written here recursively, it is easy to implement it only with loops, and without recursion.

5.3. ***The algorithm.***

Algorithm 1: ComputeSequenceA(m, n)

Input: Two indices m, n (integers).
Output: The elements $a_m, a_{m+1}, \ldots, a_n$.

```
If m ≤ 0 then
    For j = 0, ..., n do
```

```
3:        a_j ← mex{a_i, b_i : 0 ≤ i < j}.
4:        b_j ← a_j + c_j.
5:     End for
6:  Else {m > 0}
7:     Calculate r ← ⌊m(α − 1) + u_1⌋ and s ← ⌈n(α − 1) + u_2⌉ + 1.
8:     a_r, ..., a_s ← ComputeSequenceA(r, s).
9:     For j = r, ..., s do
10:       b_j ← a_j + c_j.
11:    End for
12:    Calculate a_{b_r−r+1}, ..., a_{b_s−s} as the complemented elements of b_r, ..., b_s
       in the interval [b_r, b_s].
13: End if
14: Return a_m, ..., a_n.
```

5.4. ***Example.*** We demonstrate the following case: A, B are complementary sequences, and $b_n = a_n + \lfloor 4n/3 \rfloor$; i.e., $c_n = \lfloor 4n/3 \rfloor$. Then $K_1 \le c_n - n\gamma \le K_2$ holds for $K_1 = -1$, $K_2 = 0$, $\gamma = \frac{4}{3}$. In the proof of Theorem 4.3 we take $r = 2$, $s = 1$ to satisfy $r \ge (K_2 - 2K_1)/\gamma = \frac{3}{2}$ and $s \ge (2K_2 - K_1)/\gamma = \frac{3}{4}$ (also $s \ge r - 1$). Then we have $c_{n+m-2} \le c_n + c_m \le c_{n+m+1}$. Note that usually this general method for finding r, s does not give the best (smallest) r, s possible, so one might try to look for better r, s if one wanted to; we don't.

Now, by Theorem 3.5, $a_{n+m-2} - 1 \le a_n + a_m + 2 - 1 \le a_{n+m+1}$, equivalently $a_{n+m} \le a_n + a_{m+2} + 0 \le a_{n+m+3} + 1$. Then in the proof of Lemma 3.6 we get that $a_{n+m} + h_1 \le a_n + a_m \le a_{n+m} + h_2$ holds for $h_1 = -a_4$, $h_2 = a_0 + a_5 + 1$. So we compute the first element of A, B by the definition:

n	0	1	2	3	4	5	6	7	...
c_n	0	1	2	4	5	6	8	9	...
a_n	0	1	3	4	6	7	9	10	...
b_n	0	2	5	8	11	13	17	19	...

Thus $h_1 = -6$, $h_2 = 0+7+1 = 8$, and Fekete's corollary yields $-6 \le a_n - n\alpha \le 8$, where $\alpha = 1 - \gamma/2 + \sqrt{1 + (\gamma/2)^2} = \frac{1}{3}(1 + \sqrt{13})$.

We begin with $m = n = 1000$. By lines 7, 8, the parameters m, n called in the recursion steps are

Recursive step #	0	1	2	3	4	5	6	7
m	1000	529	277	142	69	30	10	−1
n	1000	545	301	171	101	64	44	33

The computation of the first elements a_n, b_n begins at the recursion stop (line 2).

a_0	a_1	a_2	$\cdots$	a_{31}	a_{32}	a_{33}
0	1	3	$\cdots$	48	49	50

Computing $b_i = a_i + c_i$ (line 4) gives

b_0	a_1	b_1	a_2	a_3	$\cdots$	b_{32}	a_{60}	a_{61}	b_{33}
0		2			$\cdots$	91			94

The indices of $a_{518}, \ldots, a_{562}$ are derived from Theorem 3.2; that is, a_{b_0-0+1} and $a_{b_{33}-33}$ are the first and last elements of A inside the interval $[b_0, b_{33}]$.

By complementing the interval $[b_0, b_{33}]$ (line 12) we get $a_1, a_2, \ldots, a_{61}$. From these we extract the desired elements ($m = 10, n = 44$):

a_{10}	a_{11}	b_6	a_{12}	$\cdots$	b_{23}	a_{43}	a_{44}
15	16		18	$\cdots$		66	67

Computing $b_i = a_i + c_i$ (line 9), we get

$\vdots$

b_{277}	a_{518}	b_{278}	a_{519}	a_{520}	$\cdots$	b_{300}	a_{561}	a_{562}	b_{301}
794		796			$\cdots$	860			863

Again, the indices of $a_{518}, \ldots, a_{562}$ are derived from Theorem 3.2, i.e., $a_{b_{277}-277+1}$ and $a_{b_{518}-518}$ are the first and last elements of A inside the interval $[b_{277}, b_{518}]$.

Complementing the interval $[b_{277}, b_{301}]$, and extracting desired elements gives

a_{529}	a_{530}	b_{284}	a_{531}	$\cdots$	b_{291}	a_{544}	a_{545}
812	813		815	$\cdots$		835	836

Computing $b_i = a_i + c_i$ (line 9), we get

b_{529}	a_{989}	b_{530}	$\cdots$	a_{1016}	b_{544}	a_{1017}	b_{545}
1517		1519	$\cdots$		1560		1562

Complementing the interval $[b_{529}, b_{545}]$, and extracting the wanted element, we finally obtain

a_{1000}
1535

5.5. *Complexity analysis and validity proof.* Due to the variety of multiplication algorithms, we denote by $M(k)$ the complexity of multiplication of two k-digits numbers. We also denote by $T(k)$ the complexity of computing c_k, and by $Y(k)$ the complexity of computing the multiplication αk to integer precision.

For example, we can trivially choose $M(k) = k^2$. Then, if we take $c_k = \lfloor 4k/3 \rfloor$, we get $T(k) = O(\log k)$. Since $\gamma = \lim c_n/n = 4/3$, we have $\alpha = (1+\sqrt{13})/3$. Hence computing αk forces a computation of the square root of a $\log k$-digits integer (plus some computations with lower complexity). Thus $Y(k) = M(\log k) = O(\log^2 k)$ (complexity of square root computation is the same as of multiplication).

Now, for the complexity analysis of the algorithm. Denote by m_i, n_i the parameters m, n called on the i-th step of the recursion. So we have $m_0 = n_0 = n$ (to compute a_n we call ComputeSequenceA(n, n)), and line 7 gives the recursive relation

$$m_{i+1} = \lfloor m_i(\alpha - 1) + u_1 \rfloor, \qquad n_{i+1} = \lceil n_i(\alpha - 1) + u_2 + 1 \rceil.$$

From this it's easy to see that

$$m_{i+1} \le m_i(\alpha - 1) + u_1 \quad \text{and} \quad n_{i+1} - m_{i+1} \le (n_i - m_i)(\alpha - 1) + u_2 - u_1 + 3,$$

which implies

$$m_i \le (\alpha - 1)^i \left(n - \frac{u_1}{2-\alpha} \right) + \frac{u_1}{2-\alpha},$$

$$n_i - m_i \le -(\alpha - 1)^i \frac{u_2 - u_1 + 3}{2-\alpha} + \frac{u_2 - u_1 + 3}{2-\alpha}. \tag{2}$$

Notice from this that m_i decreases at least exponentially ($1 < \alpha < 2$, since $\alpha = 1 - \gamma/2 + \sqrt{1 + (\gamma/2)^2}$), and that though the difference $n_i - m_i$ is increasing, it is bounded by a constant, $(u_2 - u_1 + 3)/(2 - \alpha)$.

Since $0 < \alpha - 1 < 1$, and $u_1 \le a_0 - 0 = 0$, from (2) we see that the recursion stopping condition of the algorithm $m \le 0$ (line 1), is fulfilled after at most $O(\log n)$ steps of the recursion. In this case, we compute the first elements of a_i directly by definition (line 2). At the stopping condition we know that $m_i \le 0$, and since also the difference $n_i - m_i$ is bounded by a constant, we see that the number of initial n_i elements computed is bounded by a constant. Hence the complexity of the stopping part of the algorithm (line 2) is constant $O(1)$.

On line 7 we have complexity $Y(n)$ (m_i and n_i are at most n). The complexity of line 9 is just $T(n)$, since the loop index runs from m_i to n_i, and we saw that $n_i - m_i$ is bounded by a constant. We also get from this that in line 12 the size of the interval $[b_{m_i}, b_{n_i}]$ is bounded by a constant (since b_i is approximately linear), hence this line's complexity is $O(1)$.

In conclusion, the algorithm has complexity $O((Y(n)+T(n))\log n)$, which, depending on this complexity, is generally much more efficient then $O(n)$. For the example in Section 5.4, it is $O(\log^3 n)$. The arguments in Sections 5.1, 5.2 and 5.5 jointly also establish the validity of the algorithm.

6. Conclusion

We have shown that the unwieldy mex function can be harnessed to behave polynomially for a large set of interesting games, such as those given in [Hegarty and Larsson 2006; Duchêne and Gravier 2009], for which no polynomial winning strategy was known.

It is interesting that the most time consuming part of the algorithm is not the recursion (with complexity $O(\log n)$), but the calculations on lines 7 and 9–10, of complexity $O(Y(n)+T(n))$.

Another surprising aspect is that even though it seems at first that the m_i, n_i are growing apart from each other as the recursion progresses, which would affect both the time complexity and the space complexity, this effect vanishes eventually, since the difference $n_i - m_i$ is bounded by a constant, as shown by (2).

On a fundamental level, we attribute the efficiency of the algorithm to three properties of the sequences A, B:

- The two sequences partition the positive integers. This enabled use of Theorem 3.2 to compute the $\{a_n\}$ elements as the complement of a $\{b_n\}$-interval.
- The sequence $\{a_n\}$ has a simple approximation by another efficiently computable sequence, namely $s_n = n\alpha$. The sequence $S(n) = s_n - n$, is decreasing fast when composed onto itself (in our case $S(n) \approx (\alpha-1)n$, so $S^{(k)}(n) \approx (\alpha-1)^k n$ decreases exponentially).
- The recursive relation between a_n and b_n is simple. That is, knowing a_n, it is easy to compute b_n by the formula $b_n = a_n + c_n$.

One direction of further research is to extend the results to sequences to which the idea of the algorithm can be applied. For example, for $c_n = \zeta\lfloor n\theta \rfloor$, where $\zeta, \theta \in \mathbb{R}_{>0}$. This requires a generalization of the mex function, since the sequences are not integers anymore. It should not be hard to extend our results to multiple sequences that split $\mathbb{Z}_{\geq 1}$.

Acknowledgments

We wish to thank Eric Duchêne for being one of the triggers of this research by a question in his 2006 PhD thesis. We owe special thanks to Peter Hegarty and

Urban Larsson for a few short but very useful discussions. We also thank Dorit Ron for running high-precision numerical experiments on sequences during the initial phase of this work.

References

[Boshernitzan and Fraenkel 1981] M. Boshernitzan and A. S. Fraenkel, "Nonhomogeneous spectra of numbers", *Discrete Math.* **34**:3 (1981), 325–327.

[Demaine 2001] E. D. Demaine, "Playing games with algorithms: algorithmic combinatorial game theory", pp. 18–32 in *Mathematical foundations of computer science*, edited by A. Sgall, Jiříand Pultr and P. Kolman, Lecture Notes in Comput. Sci. **2136**, Springer, Berlin, 2001.

[Demaine and Hearn 2009] E. Demaine and R. Hearn, "Playing games with algorithms: algorithmic combinatorial game theory", pp. 3–56 in *Games of no chance 3*, Math. Sci. Res. Inst. Publ. **56**, Cambridge, 2009.

[Duchêne and Gravier 2009] E. Duchêne and S. Gravier, "Geometrical extensions of Wythoff's game", *Discrete Math.* **309**:11 (2009), 3595–3608.

[Duchêne and Rigo 2010] E. Duchêne and M. Rigo, "Invariant games", *Theoret. Comput. Sci.* **411**:34-36 (2010), 3169–3180.

[Fekete 1923] M. Fekete, "Über die Verteilung der Wurzeln bei gewissen algebraischen Gleichungen mit ganzzahligen Koeffizienten", *Math. Zeitschrift* **17**:1 (1923), 228–249.

[Fraenkel 1982] A. S. Fraenkel, "How to beat your Wythoff games' opponent on three fronts", *Amer. Math. Monthly* **89**:6 (1982), 353–361.

[Fraenkel 1998] A. S. Fraenkel, "Heap games, numeration systems and sequences", *Ann. Comb.* **2**:3 (1998), 197–210.

[Fraenkel 2000] A. S. Fraenkel, "Recent results and questions in combinatorial game complexities", *Theoret. Comput. Sci.* **249**:2 (2000), 265–288.

[Fraenkel 2004a] A. S. Fraenkel, "Complexity, appeal and challenges of combinatorial games", *Theoret. Comput. Sci.* **313**:3 (2004), 393–415.

[Fraenkel 2004b] A. S. Fraenkel, "New games related to old and new sequences", *Integers* **4** (2004), G6, 18.

[Fraenkel and Krieger 2004] A. S. Fraenkel and D. Krieger, "The structure of complementary sets of integers: a 3-shift theorem", *Int. J. Pure Appl. Math.* **10**:1 (2004), 1–49.

[Hegarty and Larsson 2006] P. Hegarty and U. Larsson, "Permutations of the natural numbers with prescribed difference multisets", *Integers* **6** (2006), A3, 25.

[Kimberling 2007] C. Kimberling, "Complementary equations", *J. Integer Seq.* **10**:1 (2007), Article 07.1.4, 14.

[Kimberling 2008] C. Kimberling, "Complementary equations and Wythoff sequences", *J. Integer Seq.* **11**:3 (2008), Article 08.3.3, 8.

[Larsson 2009] U. Larsson, "2-pile Nim with a restricted number of move-size imitations", *Integers* **9** (2009), G04, 671–690.

[Singmaster 1981] D. Singmaster, "Almost all games are first person games", *Eureka* **41** (1981), 33–37.

[Singmaster 1982] D. Singmaster, "Almost all partizan games are first person and almost all impartial games are maximal", *J. Combin. Inform. System Sci.* **7**:4 (1982), 270–274.

[Smith and Stănică 2002] F. Smith and P. Stănică, "Comply/constrain games or games with a Muller twist", *Integers* **2** (2002), G3, 10.

[Sun 2005] X. Sun, "Wythoff's sequence and N-heap Wythoff's conjectures", *Discrete Math.* **300**:1-3 (2005), 180–195.

[Sun and Zeilberger 2004] X. Sun and D. Zeilberger, "On Fraenkel's N-heap Wythoff's conjectures", *Ann. Comb.* **8**:2 (2004), 225–238.

[Wythoff 1906] W. A. Wythoff, "A modification of the game of Nim", *Nieuw Arch. Wisk.* **7** (1906), 199–202.

aviezri.fraenkel@weizmann.ac.il *Department of Computer Science and Applied Mathematics, Weizmann Institute of Science, 76100 Rehovot, Israel*

udi@udipeled.co.il *Department of Computer Science and Applied Mathematics, Weizmann Institute of Science, Rehovot 76100, Israel*

Games of No Chance 4
MSRI Publications
Volume **63**, 2015

The Rat game and the Mouse game

AVIEZRI S. FRAENKEL

We define three new take-away games, the Rat game, the Mouse game and the Fat Rat game. Three winning strategies are given for the Rat game and outlined for the Mouse and Fat Rat games. The efficiencies of the strategies are determined. Whereas the winning strategies of nontrivial take-away games are based on irrational numbers, our games are based on rational numbers. Another motivation stems from a problem in combinatorial number theory.

1. Description of the game

The Rat game is played on 3 piles of tokens by 2 players who play alternately. Positions in the game are denoted throughout in the form (x, y, z), with $0 \le x \le y \le z$, and moves in the form $(x, y, z) \to (u, v, w)$, where of course also $0 \le u \le v \le w$ (see below). The player first unable to move — because the position is $(0, 0, 0)$ — loses; the opponent wins. There are 3 types of moves:

(I) Take any positive number of tokens from up to 2 piles.

(II) Take $\ell > 0$ from the x pile, $k > 0$ from the y pile, and an arbitrary positive number from the z pile, subject to the constraint $|k - \ell| < a$, where

$$a = \begin{cases} 1 & \text{if } y - x \not\equiv 0 \pmod 7, \\ 2 & \text{if } y - x \equiv 0 \pmod 7. \end{cases}$$

(III) Take $\ell > 0$ from the x pile, $k > 0$ from the z pile, and an arbitrary positive number from the y pile, subject to the constraint $|k - \ell| < b$, where $b = 3$ if $w = u$; otherwise,

$$b = \begin{cases} 5 & \text{if } w - u \not\equiv 4 \pmod 7, \\ 6 & \text{if } w - u \equiv 4 \pmod 7. \end{cases}$$

In a move of type (II) we permit the permutation $x \to v$, $y \to w$, $z \to u$ (so $\ell = x - v$, $k = y - w$), in addition to $x \to u$, $y \to v$, $z \to w$ ($\ell = x - u$, $k = y - v$). No other permutations are allowed for (II), and none (except $x \to u$, $y \to v$, $z \to w$) for (III). For (I), any rearrangement is possible. When we write $(x, y, z) \to (u, v, w)$, we always mean $x \to u$, $y \to v$, $z \to w$.

Note that in (II), the congruence conditions depend only on 2 of the piles moved *from*: the smallest and the intermediate; whereas in (III) they depend only on 2 of the piles moved *to*: the smallest and the largest. The case $w = u$ in (III) is an initial condition, to accommodate the end position $(0, 0, 0)$.

Examples. Given the position $p_1 = (1, 2, 4)$. If player I takes one of the piles in its entirety, player II wins with a type (I) move to $\Phi := (0, 0, 0)$. If player I moves $p_1 \to (1, 2, t)$, $t \in \{1, 2, 3\}$, player II wins with a type (III) move to Φ. If player I moves $p_1 \to (1, 1, 4)$, player II wins with a type (II) move to Φ. It's straightforward to see that if player I makes a move of type (I), or (II) or (III), then player II can win by moving to Φ. Consider now the position $p_2 = (3, 6, 10)$. Then player I can make a type (III) move to p_1. Indeed, $(10-4)-(3-1) = 4 < 5 = b$.

What's the motivation for inventing and analyzing this game? Why are the move rules complicated? What's the connection to rats? Where's the Mouse game? How about the **Fat** *Rat game?*

2. Two characterizations of the P-positions

Let $T \subsetneqq \mathbb{Z}_{\geq 0}$. Define the mex operator by $\text{mex}(T) = \min(\mathbb{Z}_{\geq 0} \setminus T) =$ smallest nonnegative integer not in T. Recall that the set of P-positions of a game is the set of positions for which the *P*revious (second) player can win, and the set of N-positions is the set of positions for which the *N*ext (first) player can win [WW; Albert et al. 2007]. We begin with a recursive characterization of the P-positions of the Rat game.

Theorem 1. *The P-positions of the Rat game are given by*

$$R = \bigcup_{n=0}^{\infty} \{(a_n, b_n, c_n)\},$$

where $(a_0, b_0, c_0) = (0, 0, 0)$, *and for* $n \geq 1$,

$$a_n = \text{mex}\{a_i, b_i, c_i : 0 \leq i < n\},$$
$$b_n = a_n + \lfloor (7n-2)/4 \rfloor,$$
$$c_n = b_n + \lfloor (7n-3)/2 \rfloor.$$

The first few triples of R are displayed in Table 1.

We now turn to an explicit characterization of the P-positions.

Define an infinite set of triples $S_n := (A_n, B_n, C_n)$ as follows. $S_0 = (0, 0, 0)$, and for $n \in \mathbb{Z}_{\geq 1}$,

$$A_n = \lfloor 7n/4 \rfloor, \quad B_n = \lfloor 7n/2 \rfloor - 1, \quad C_n = 7n - 3. \tag{1}$$

Put $S = \bigcup_{n=0}^{\infty} S_n$.

n	a_n	b_n	c_n
0	0	0	0
1	1	2	4
2	3	6	11
3	5	9	18
4	7	13	25
5	8	16	32
6	10	20	39
7	12	23	46
8	14	27	53
9	15	30	60
10	17	34	67
11	19	37	74
12	21	41	81
13	22	44	88
14	24	48	95
15	26	51	102

Table 1. The first few P-positions of the Rat game.

Theorem 2. *The collection S constitutes the set of P-positions of the Rat game, so $S = R$.*

In Sections 4–6 we prove Theorems 1 and 2. Various extensions are given in Sections 7–8, efficiencies of the winning strategies are discussed in Section 9, and we wrap up in Section 10.

3. Preliminaries

For proving Theorem 2, which is proved first, we begin by collecting a few properties of the set S.

Lemma 1. *Let $p, q \in \mathbb{Z}_{\geq 1}$, with $p > q$, $s \in \mathbb{Z}$. Then:*

(i) *For every $t \in \mathbb{Z}$, the q values $\lfloor (pn+s)/q \rfloor$, $n \in \{t+1, \dots, t+q\}$ are distinct* (mod p).

(ii) *For every $k \in \mathbb{Z}$, $\lfloor (p(n+kq)+s)/q \rfloor = \lfloor (pn+s)/q \rfloor + kp$.*

Proof. (i) Let $n_1, n_2 \in \{t+1, \dots, t+q\}$, $n_1 \neq n_2$, say $n_2 > n_1$. Then

$$0 < (p/q) - 1 < \lfloor (pn_2+s)/q \rfloor - \lfloor (pn_1+s)/q \rfloor < (p(q-1)/q) + 1 < p,$$

as claimed.

(ii) Obvious. □

It follows that for $n \in \{1, \ldots, q\}$, $\lfloor (pn+s)/q \rfloor$ contains distinct residues $r_1 < \cdots < r_q \pmod{p}$; for $n \in \{q+1, \ldots, 2q\}$ it contains $p+r_1, \ldots, p+r_q$; for $n \in \{kq+1, \ldots, (k+1)q\}$ it contains $kp+r_1, \ldots, kp+r_q$.

Let

$$A = \bigcup_{i=1}^{\infty} A_i, \quad B = \bigcup_{i=1}^{\infty} B_i, \quad C = \bigcup_{i=1}^{\infty} C_i.$$

Lemma 2. (i) *Each of the sequences A_i, B_i, C_i is strictly increasing.*

(ii) *The sets A, B, C partition $\mathbb{Z}_{\geq 1}$.*

Proof. (i) Follows directly from the definition of the 3 sequences.

(ii) Note that $\left(\bigcup_{n=1}^{4} A_n\right) \cup \left(\bigcup_{n=1}^{2} B_n\right) \cup C_1 = \{1, \ldots, 7\}$. The result now follows from Lemma 1(i). □

For $n \in \mathbb{Z}_{\geq 0}$, let

$$d_n = B_n - A_n, \quad \delta_n = C_n - B_n, \quad \Delta_n = C_n - A_n.$$

Lemma 3. *For $n \in \mathbb{Z}_{\geq 1}$,*

$$d_n = \lfloor (7n-2)/4 \rfloor, \quad \delta_n = \lfloor (7n-3)/2 \rfloor, \quad \Delta_n = d_n + \delta_n.$$

Proof. The assertion for d_n is seen to hold for $n = 1, 2, 3, 4$. Therefore it holds for all $n \in \mathbb{Z}_{\geq 1}$ by Lemma 1(i). Similarly, the assertion about δ_n is seen to hold for $n = 1, 2$, therefore it holds for all $n \in \mathbb{Z}_{\geq 1}$. Finally,

$$\Delta_n = C_n - A_n = (C_n - B_n) + (B_n - A_n) = \delta_n + d_n. \qquad \square$$

Table 2 depicts the differences d_n, δ_n, Δ_n together with the P-positions. It also illustrates the next few lemmas.

Lemma 4. *For $n \in \mathbb{Z}_{\geq 1}$,*

$$d_n = \begin{cases} A_n & \text{if } n \equiv 1, \text{ or } 2 \pmod 4, \\ A_n - 1 & \text{if } n \equiv 0, \text{ or } 3 \pmod 4, \end{cases} \qquad \delta_n = \begin{cases} B_n - 1 & \text{if } n \text{ is even}, \\ B_n & \text{if } n \text{is odd}. \end{cases}$$

Proof. The first one follows from Lemma 3 and an easy computation. Similarly for the second. □

Lemma 5. (i) *Each of the sequences d_n, δ_n, Δ_n is strictly increasing.*

(ii) *For $n \in \mathbb{Z}_{\geq 1}$, $d_n < \delta_n < \Delta_n$.*

(iii) *For $n \in \mathbb{Z}_{\geq 1}$, the sequences d_n and δ_n are disjoint. In fact,*

$$d_n \equiv \{1, 3, 4, 6\} \pmod 7, \quad \delta_n \equiv \{2, 5\} \pmod 7,$$

n	A_n	d_n	B_n	δ_n	C_n	Δ_n
0	0	0	0	0	0	0
1	1	1	2	2	4	3
2	3	3	6	5	11	8
3	5	4	9	9	18	13
4	7	6	13	12	25	18
5	8	8	16	16	32	24
6	10	10	20	19	39	29
7	12	11	23	23	46	34
8	14	13	27	26	53	39
9	15	15	30	30	60	45
10	17	17	34	33	67	50
11	19	18	37	37	74	55
12	21	20	41	40	81	60
13	22	22	44	44	88	66
14	24	24	48	47	95	71
15	26	25	51	51	102	76

Table 2. P-positions of the Rat game with their differences.

and each of the residues (*mod* 7) *of* d_n *and* δ_n *are assumed for infinitely many* n. *Also,*

$$d_n > \delta_0 = 0, \quad \delta_n > d_0 = 0, \quad \textit{and} \quad \Delta_n \equiv \{3, 1, 6, 4\} \pmod 7.$$

(iv) $\bigcup_{n=1}^{\infty}(d_n \cup \delta_n) = \mathbb{Z}_{\geq 1} \setminus \bigcup_{i=1}^{\infty}\{7i\}$.

Proof. (i) and (ii) follow from Lemma 3.

(iii) By inspection, this holds for $n = 1, 2, 3, 4$. It follows in general from Lemma 1(i).

(iv) From (iii) we see that $\left(\bigcup_{n=1}^{4} d_n\right) \cup \left(\bigcup_{n=1}^{2} \delta_n\right) = \{1, 2, 3, 4, 5, 6\}$. The result now follows from Lemma 1(ii) by induction on k. □

Lemma 6. *For* $n \in \mathbb{Z}_{\geq 1}$, $d_{n+1} - d_n \in \{1, 2\}$, $\delta_{n+1} - \delta_n \in \{3, 4\}$,

$$\Delta_{n+1} - \Delta_n = \begin{cases} 5 & \textit{if } \Delta_n \equiv 1,\ 3 \textit{ or } 6 \pmod 7, \\ 6 & \textit{if } \Delta_n \equiv 4 \pmod 7, \end{cases}$$

and $\Delta_1 - \Delta_0 = 3$.

Proof. Table 2 shows that the statements hold for $n = 1, 2, 3, 4$. Their general truth then follows from Lemma 1. □

4. Proof of Theorem 2

It suffices to show:

(A) *Every move from* $(A_n, B_n, C_n) \in S$ *results in a position outside* S.

(B) *For every position* $(x, y, z) \notin S$, *there exists a move resulting in a position in* S.

(A) Since A_n, B_n, C_n are each strictly increasing, a move of type (I) from $(A_n, B_n, C_n) \in S$, $n \in \mathbb{Z}_{\geq 1}$, clearly results in a position not in S. Suppose that there is a move $S \to S$ of type (II) or (III). Such a move can have one of the following two forms:

(i) $(A_n, B_n, C_n) \to (A_i, B_i, C_i)$, or

(ii) $A_n \to B_i$, $B_n \to C_i$, $C_n \to A_i$.

In both cases, $i < n$. These moves have to satisfy the following conditions. Either:

(i1) $|(B_n - B_i) - (A_n - A_i)| < a$. Now $|(B_n - B_i) - (A_n - A_i)| = |d_n - d_i| \geq 1$ by Lemma 5(i). By Lemma 5(iv), $y - x = d_n \not\equiv 0 \pmod 7$. Hence by the rule of a move of type (II), $a = 1$, a contradiction. Or:

(i2) $|(C_n - C_i) - (A_n - A_i)| < b$. Now

$$|(C_n - C_i) - (A_n - A_i)| = |\Delta_n - \Delta_i| \geq |\Delta_n - \Delta_{n-1}| = b,$$

a contradiction to a move of type (III).

(ii) The constraint is $|(B_n - C_i) - (A_n - B_i)| < a$. Now

$$|(B_n - C_i) - (A_n - B_i)| = |d_n - \delta_i| \geq 1$$

by Lemma 5(i). As in case (i1) we have $a = 1$, a contradiction.

(B) Let $(x, y, z) \notin S$ with $0 < x \leq y \leq z$. (If $x = 0$ there is a type (I) move to $(0, 0, 0)$.) Throughout we use the notation:

$$d = y - x, \qquad D = z - x.$$

Since the sets A, B, C partition $\mathbb{Z}_{\geq 1}$, there exists $n \in \mathbb{Z}_{\geq 1}$, such that either

(i) $x = C_n$, or

(ii) $x = B_n$, or

(iii) $x = A_n$.

Note that since $x > 0$, we have $n > 0$, so by Lemma 3,

$$0 < A_n < B_n < C_n.$$

(i) $x = C_n$. Then do a type (I) move, $y \to A_n$, $z \to B_n$.

(ii) $x = B_n$. If $z \geq C_n$, make a type (I) move, $z \to C_n$, $y \to A_n$. So we may assume $z < C_n$. We have $A_n < B_n = x \leq y \leq z < C_n$. Then make a move of type (III): $(x, y, z) \to (A_i, B_i, C_i)$, where i is the largest index such that $\Delta_i \leq D$. This move is legal:

(a) $\Delta_i \leq D = z - x = z - B_n < z - A_n < C_n - A_n = \Delta_n$. Hence $i < n$.

(b) $x = B_n > A_n > A_i$; $y \geq x = B_n > B_i$ since $i < n$;

$$z = x + D \geq A_n + \Delta_i > A_i + \Delta_i = C_i.$$

(c) The move has to satisfy $|(z - C_i) - (x - A_i)| < b$. Indeed,

$$|(z - C_i) - (x - A_i)| = |D - \Delta_i| = D - \Delta_i < \Delta_{i+1} - \Delta_i = b,$$

by Lemma 5(iii) and Lemma 6.

(iii) $x = A_n$. If $y \geq B_n$ and $z \geq C_n$ (at least one of these inequalities is necessarily strict), then make a move of type (I), $y \to B_n$, $z \to C_n$. Below we consider the remaining 3 subcases:

(α) $y \geq B_n$, $z < C_n$,

(β) $y < B_n$, $z \geq C_n$,

(γ) $y < B_n$, $z < C_n$.

(α) $y \geq B_n$, $z < C_n$. Then $A_n = x < B_n \leq y \leq z < C_n$. We make a move of type (III) as in case (ii) above: $(x, y, z) \to (A_i, B_i, C_i)$, where i is the largest index such that $\Delta_i \leq D$. This move is legal:

(a) $\Delta_i \leq D = z - x = z - A_n < C_n - A_n = \Delta_n$. Hence $i < n$.
(b) $x = A_n > A_i$; $y \geq B_n > B_i$; $z = x + D \geq x + \Delta_i = A_n + \Delta_i > A_i + \Delta_i = C_i$.
(c) $|(z - C_i) - (x - A_i)| = |D - \Delta_i| < b$, as in case (ii)(c) above.

(β) $y < B_n$, $z \geq C_n$. We have $A_n = x \leq y < B_n < C_n \leq z$.

($\beta 1$) We first consider the case $d \not\equiv 0 \pmod 7$. Then $d \in \{d_i, \delta_i\}$ for some $i \in \mathbb{Z}_{\geq 0}$ by Lemma 5(iv). Since $d = y - A_n < B_n - A_n = d_n < \delta_n$, we have $i < n$.

($\beta 11$) Assume $d = d_i$. Then move $(x, y, z) \to (A_i, B_i, C_i)$. This is indeed a move of type (II):

(a) $x = A_n > A_i$; $y = A_n + d = A_n + d_i > A_i + d_i = B_i$; $z \geq C_n > C_i$.
(b) $|(y - B_i) - (x - A_i)| = |d - d_i| = 0 < a$.

($\beta 12$) Assume $d = \delta_i$. Then move $x \to B_i$, $y \to C_i$, $z \to A_i$. It's a move of type (II):

(a) Clearly $z \geq C_n > A_i$. It remains to show that $x > B_i$ and $y > C_i$. By Lemma 4, $A_n \geq d_n = B_n - A_n > y - x = d = \delta_i \geq B_i - 1$. Thus $A_n \geq B_i$. By Lemma 2 we then actually have $A_n > B_i$. Therefore also $y = x + d = A_n + \delta_i > B_i + \delta_i = C_i$.

(b) $|(y - C_i) - (x - B_i)| = |d - \delta_i| = 0 < 1 = a$.

(β2) We now consider the case $d \equiv 0 \pmod 7$. Then Lemma 5(iii) implies $d - 1 = d_i$ for some $i \in \mathbb{Z}_{\geq 0}$. Then move $(x, y, z) \to (A_i, B_i, C_i)$. This is a move of type (II):

(a) $d_i = d - 1 = y - x - 1 < B_n - A_n - 1 = d_n - 1 < d_n$. Hence $i < n$.

(b) $x = A_n > A_i$; $y = A_n + d = A_n + d_i + 1 > A_i + d_i = B_i$; $z \geq C_n > C_i$.

(c) $|(y - B_i) - (x - A_i)| = |d - d_i| = 1 < 2 = a$.

(γ) $y < B_n$, $z < C_n$.

(γ1) We first assume $d \not\equiv 0 \pmod 7$. By Lemma 5 there exists $i \in \mathbb{Z}_{\geq 0}$ such that $d \in \{d_i, \delta_i\}$. Now $d = y - A_n < B_n - A_n = d_n < \delta_n$. Hence $i < n$.

(γ11) We consider first the case $d = d_i$, and try a move of type (II): $(x, y, z) \to (A_i, B_i, C_i)$. Note that $x = A_n > A_i$ and

$$y = x + d = A_n + d_i > A_i + d_i = B_i.$$

If $z \geq C_i$, this is a legitimate move of type (II) (or one of type (I) if $z = C_i$). Indeed, $|(y - B_i) - (x - A_i)| = |d - d_i| = 0 < 1 = a$. So suppose that $z < C_i$. We then do a move of type (III). Move $(x, y, z) \to (A_j, B_j, C_j)$, where j is the largest index such that $\Delta_j \leq D$. This move is legal:

(a) $\Delta_j \leq D = z - A_n < C_n - A_n = \Delta_n$. Hence $j < n$.

(b) $x = A_n > A_j$. Since $z < C_i$ we have

$$\Delta_j \leq D = z - x < C_i - A_n < C_i - A_i = \Delta_i.$$

Hence $j < i$. We showed above that $y > B_i$. Hence $y > B_j$. Also $z = x + D = A_n + D \geq A_n + \Delta_j > A_j + \Delta_j = C_j$.

(c) $|(z - C_j) - (x - A_j)| = |D - \Delta_j| = D - \Delta_j < \Delta_{j+1} - \Delta_j = b$.

(γ12) We now deal with the case $d = \delta_i$. Then move $x \to B_i$, $y \to C_i$, $z \to A_i$. Recall from (γ1) that $i < n$. It's a move of type (II):

(a) Clearly $z \geq y \geq x = A_n > A_i$. It remains to show that $x > B_i$ and $y > C_i$. As in case (β12), Lemma 4 implies $x = A_n \geq d_n > d = \delta_i \geq B_i - 1$. Thus $A_n \geq B_i$. By Lemma 2 we have actually $A_n > B_i$. Therefore $y = x + d = A_n + \delta_i > B_i + \delta_i = C_i$.

(b) $|(y - C_i) - (x - B_i)| = |d - \delta_i| = 0 < 1 = a$.

(γ2) We now take up the remaining case, $d \equiv 0 \pmod 7$. Then Lemma 5(iii) implies $d - 1 = d_i$ for some $i \in \mathbb{Z}_{\geq 0}$. Then try a move of type (II) $(x, y, z) \to (A_i, B_i, C_i)$. We have:

(a) $d_i = d - 1 = y - A_n - 1 < B_n - A_n - 1 = d_n - 1 < d_n$. Hence $i < n$.

(b) $x = A_n > A_i$; $y = A_n + d = A_n + d_i + 1 > A_i + d_i = B_i$. If $z \geq C_i$ we indeed made a legitimate move of type (II) (or one of type (I) if $z = C_i$).

(c) $|(y - B_i) - (x - A_i)| = |d - d_i| = 1 < 2 = a$.

So assume $z < C_i$. We then do a move of type (III):

$$(x, y, z) \to (A_j, B_j, C_j),$$

where j is the largest index such that $\Delta_j \leq D$. The legality of this move is established in precisely the same way as for the type (III) move of case (γ11) above. □

Remark. Regarding case (B)(iii) of the proof, i.e., $x = A_n$, a move of type (III), as in case (ii), is not always possible. Example: $(x, y, z) = (17, 28, 66)$. Then $D = z - x = 49$, so Table 1 shows that a (II)(b) move would have to be to $(15, 30, 60)$. This, however, is impossible, since $y = 28 < 30$.

5. Proof of Theorem 1

For proving Theorem 1, it evidently suffices to prove the following result.

Theorem 3. *For all $n \in \mathbb{Z}_{\geq 0}$, $a_n = A_n$, $b_n = B_n$, $c_n = C_n$. In other words, the set of triples $R = \bigcup_{n=0}^{\infty}\{(a_n, b_n, c_n)\}$, defined recursively, constitutes the set of P-positions of the Rat game.*

Proof. We note that $(a_0, b_0, c_0) = (A_0, B_0, C_0) = (0, 0, 0)$. Suppose we have shown already that $(a_n, b_n, c_n) = (A_n, B_n, C_n)$ for all $n < N$ $(N \geq 1)$. Recall from Lemma 2(ii) that the sets A, B, C partition $\mathbb{Z}_{\geq 1}$, and clearly $A_n < B_n < C_n$ for all $n \in \mathbb{Z}_{\geq 1}$. Therefore $A_N = \text{mex}\{A_i, B_i, C_i : 0 \leq i < N\}$. Otherwise A_N would never be attained in the complementary sets A, B, C. Thus $A_N = a_N$. Now $B_n - A_n = \lfloor (7n - 2)/4 \rfloor$, and $C_n - B_n = \lfloor (7n - 3)/2 \rfloor$ for all $n \in \mathbb{Z}_{\geq 1}$ (Lemma 3), the same as in the recursive definition of the triples (a_n, b_n, c_n). Hence also $B_N = b_N$, and $C_N = c_N$. □

6. A numeration systems for the Rat game

Let α be a rational or irrational number satisfying $1 < \alpha < 2$. Denote its *simple continued fraction* expansion by $\alpha = [1, a_1, \ldots]$, $a_i \in \mathbb{Z}_{\geq 1}$ for all i. This expansion is unique if α is irrational. If $\alpha = [1, a_1, \ldots, a_n]$ is rational, there are 2 expansions, since for $a_n > 1$, $a_n = (a_n - 1) + \frac{1}{1}$. In the latter case we assume, for our purposes here, that $\alpha = [1, a_1, \dots, a_{2n}]$. The *convergents* $p_k/q_k = [1, a_1, \dots, a_k]$ of α ($k \le 2n$ if α is rational) are defined recursively in the form

$$p_{-1} = 1, \quad p_0 = 1, \quad p_n = a_n p_{n-1} + p_{n-2} \quad (n \ge 1),$$
$$q_{-1} = 0, \quad q_0 = 1, \quad q_n = a_n q_{n-1} + q_{n-2} \quad (n \ge 1).$$

For properties of simple continued fractions see [Hardy and Wright 2008, Chapter 10] or [Fraenkel 1982, Section 4]. It is well-known (see the previous reference) that every positive integer N has a unique representation in the form

$$N = \sum_{i=0}^{m} s_i p_i, \quad 0 \le s_i \le a_{i+1}, \; s_{i+1} = a_{i+2} \Longrightarrow s_i = 0 \quad (i \ge 0),$$

and also in the form

$$N = \sum_{i=0}^{m} t_i q_i, \quad 0 \le t_0 < a_1, \; 0 \le t_i \le a_{i+1}, \; t_i = a_{i+1} \Longrightarrow t_{i-1} = 0 \quad (i \ge 1).$$

Remarks.

- The *p-representation* of any positive integer is its representation $\sum_{i=0}^{m} s_i p_i$ in the p-numeration system. Analogously for the *q-representation*.
- If $\alpha = [1, a_1, \dots, a_{2n}]$ is rational, we may assume that there is an arbitrarily large partial quotient a_{2n+1}, so the digits s_{2n} and t_{2n} can be arbitrarily large. This permits to represent every positive number N in the numeration systems with only finitely many p_i, q_i.
- Notice that $7/4 = [1, 1, 3]$, $p_0 = 1$, $p_1 = 2$, $p_2 = 7$; $q_0 = 1$, $q_1 = 1$, $q_2 = 4$. Further, $0 \le s_0 \le 1$, $0 \le s_1 \le 3$, $s_2 \ge 0$. Since $q_1 = 1$, we have $t_0 = 0$ for the q-representation of every $N \in \mathbb{Z}_{\ge 1}$. The q- and p-numeration systems for this example are portrayed in Table 3.

Theorem 4. *The set A is identical to the set of numbers whose p-representation ends in an even number of* 0s. *The set B is identical to the set of numbers ending in* 10 *or* 30, *and the set C is identical to the set of numbers ending in* 20.

Proof. Every term in the set A must end in an even number of 0s in the numeration system, as was shown in [Fraenkel 1982, Section 4]. (There the results were proved for the continued fraction of an irrational number, but the same holds for rational numbers.) Therefore every term in B and C must end in an odd number of 0s. Now 4 is the smallest positive number in C, it has representation 20, and every subsequent number in C is larger than its predecessor by 7. Hence all numbers in C have representations of the form $t20$, $t \in \mathbb{Z}_{\ge 0}$. Since A, B, C are complementary, the representations of all numbers in B must end in 10 or 30. □

p-numeration			q-numeration			
7	2	1	4	1	1	n
		1		1	0	1
	1	0		2	0	2
	1	1		3	0	3
	2	0	1	0	0	4
	2	1	1	1	0	5
	3	0	1	2	0	6
1	0	0	1	3	0	7
1	0	1	2	0	0	8
1	1	0	2	1	0	9
1	1	1	2	2	0	10
1	2	0	2	3	0	11
1	2	1	3	0	0	12
1	3	0	3	1	0	13
2	0	0	3	2	0	14
2	0	1	3	3	0	15
2	1	0	4	0	0	16

Table 3. The q- and p-numeration systems for the Rat game.

Theorem 5. *Let $n \in \mathbb{Z}_{\geq 1}$, and let its digits in the q-numeration system be $t_2t_1t_0$. Then*

$$t_1p_1 + t_2p_2 = \begin{cases} A_n & \text{if } t_1 = 0, \\ A_n + 1 & \text{if } t_1 > 0. \end{cases}$$

Proof. The proof is similar to one given in [Fraenkel 1982, Section 4], and is therefore omitted. □

Remark. Theorem 5 states that to compute the p-representation of A_n, it suffices to compute the q-representation of n, and then *interpret* it in the p-system (i.e., replace q_i by p_i).

7. The Mouse game

The Mouse game is played on 2 piles of tokens by 2 players who play alternately. Analogously to the Rat game, positions are denoted in the form (x, y), with $0 \leq x \leq y$, and moves in the form $(x, y) \to (u, v)$, where of course also $0 \leq u \leq v$. The player first unable to move — because the position is $(0, 0)$ — loses; the opponent wins. There are 2 types of moves:

(I) Take any positive number of tokens from a single pile.

(II) Take $\ell > 0$ from one of the piles, $k > 0$ from the other, subject to the constraint $|k - \ell| < a$, where

$$a = \begin{cases} 1 & \text{if } y - x \not\equiv 0 \pmod 3, \\ 2 & \text{if } y - x \equiv 0 \pmod 3. \end{cases}$$

We then have:

Theorem 6. *The P-positions of the Mouse game are given by* $(0, 0)$, *and for* $n \geq 1$, *we have*

$$A_n = \text{mex}\{A_i, B_i : 0 \leq i < n\},$$
$$B_n = A_n + \lfloor (3n-1)/2 \rfloor.$$

The following is an explicit description of the P-positions.

Theorem 7. *The P-positions of the Mouse game are given by* $(0, 0)$, *and for* $n \geq 1$, $A_n = \lfloor 3n/2 \rfloor$, $B_n = 3n - 1$.

We omit the proofs, since they are analogous to and simplified versions of those of Theorems 1 and 2. The first few P-positions (A_n, B_n) together with their differences $d_n = B_n - A_n$ are shown in Table 4.

We leave it to the reader to characterize the P-positions of the Mouse game in terms of an appropriate numeration system.

n	A_n	d_n	B_n
0	0	0	0
1	1	1	2
2	3	2	5
3	4	4	8
4	6	5	11
5	7	7	14
6	9	8	17
7	10	10	20
8	12	11	23
9	13	13	26
10	15	14	29
11	16	16	32
12	18	17	35
13	19	19	38
14	21	20	41
15	22	22	44

Table 4. P-positions of the Mouse game with their differences d_n.

8. The Fat Rat game

The *Fat Rat game* is the case of the Rat game played on an arbitrary number of $m \in \mathbb{Z}_{\geq 2}$ piles. The games for $m \in \{2, 3\}$ were analyzed in the previous sections. The P-positions for $m = 4$ are given in Table 5. This follows the general rule of

$$A_n^k = \lfloor (2^m - 1)n/2^{m-k} \rfloor - 2^{k-1} + 1, \quad k = 1, \ldots, m, \quad n \geq 1.$$

It is not hard to see that for $m = 4$ the differences $d_n^i := A_n^{i+1} - A_n^i$ are

$$d_n^1 = \left\lfloor \frac{15n - 4}{8} \right\rfloor, \quad d_n^2 = \left\lfloor \frac{15n - 6}{4} \right\rfloor, \quad d_n^3 = \left\lfloor \frac{15n - 7}{2} \right\rfloor,$$

and

$$\bigcup_{n=1}^{\infty} \{d_n^1, d_n^2, d_n^3, \{15n\}\} = \mathbb{Z}_{\geq 1}.$$

We note that the numeration system for the case $m = 4$ is based on the continued fraction $15/8 = [1, 1, 7]$, and for the Fat Rat game,

$$\frac{2^m - 1}{2^{m-1}} = [1, 1, 2^{m-1} - 1].$$

n	A_n^1	d_n^1	A_n^2	d_n^2	A_n^3	d_n^3	A_n^4	Δ_n
1	1	1	2	2	4	4	8	7
2	3	3	6	6	12	11	23	20
3	5	5	10	9	19	19	38	33
4	7	7	14	13	27	26	53	46
5	9	8	17	17	34	34	68	59
6	11	10	21	21	42	41	83	72
7	13	12	25	24	49	49	98	85
8	15	14	29	28	57	56	113	98
9	16	16	32	32	64	64	128	112
10	18	18	36	36	72	71	143	125
11	20	20	40	39	79	79	158	138
12	22	22	44	43	87	86	173	151
13	24	23	47	47	94	94	188	164
14	26	25	51	51	102	101	203	177
15	28	27	55	54	109	109	218	190
16	30	29	59	58	117	116	233	203

Table 5. P-positions with their differences.

So we have the P-positions. But what are the game rules? Even for the case $m = 4$, there are a priori various possibilities to be checked out. It appears that 4 types of moves are required. Perhaps the case $m = 4$ will point to the game rules for general m. It appears that the transition from 3 to 4 piles is a stumbling block in a number of games. This seems to be the case, for example, for the 3-pile Tribonacci game, based on the *Tribonacci word* [Duchêne and Rigo 2008]. The P-positions of the 3-pile *Raleigh* game [Fraenkel 2007] are, for all $n \in \mathbb{Z}_{\geq 0}$,

$$A_n = \lfloor \lfloor n\varphi \rfloor \varphi \rfloor, \quad B_n = \lfloor n\varphi^2 \rfloor, \quad C_n = \lfloor \lfloor n\varphi^2 \rfloor \varphi \rfloor,$$

where $\varphi = (1 + \sqrt{5})/2$ (golden section). A natural generalization to $m > 3$ piles may also be nontrivial.

Sometimes already the transition from 2 to 3 piles looks difficult. Wythoff's game [1906; Coxeter 1953; Fraenkel 1982; Landman 2002; Duchêne and Gravier 2009] is played on 2 piles. A natural generalization to $m > 2$ piles was suggested in [Fraenkel 1996], and 2 conjectures about the asymptotic structure of their P-positions were given. Their latest form appears in [Guy and Nowakowski 2009]. Some progress on the conjectures was achieved. See [Fraenkel and Krieger 2004; X. Sun and Zeilberger 2004; X. Sun 2005]. For example, the case $m = 3$ was solved, but it is considerably more complicated than $m = 2$.

9. Complexity

We gave three winning strategies for the Rat game: recursive (Theorem 1), algebraic (Theorem 2), and arithmetic (Theorem 4). Given an arbitrary game position $(x, y) \in \mathbb{Z}^2$ of input size $O(\log x + \log y)$, what's the computational complexity of deciding whether or not (x, y) is a P-position? We indicate briefly that all three strategies are efficient.

Theorem 8. *All three winning strategies for the Rat game are polynomial-time.*

Proof. A sequence of positive integers $\{a_n\}_{n \geq 0}$ is *approximately linear*, if there exist constants $\alpha, u_1, u_2 \in \mathbb{R}$ such that $u_1 \leq a_n - n\alpha \leq u_2$ for all $n \in \mathbb{Z}_{\geq 1}$. In [Fraenkel and Peled 2015] it is shown that for approximately linear sequences, the mex function can be computed in polynomial time. Moreover, the sequences $\{a_n\}$ and $\{b_n\}$ are both approximately linear if and only if their difference $\{b_n - a_n\}$ is (Theorem 4 there). Now by Theorem 1, $b_n - a_n = \lfloor (7n - 2)/4 \rfloor$ which is clearly approximately linear, and so is $c_n - b_n = \lfloor (7n - 3)/2 \rfloor$. This implies that the recursive strategy is polynomial. For the algebraic strategy this follows from the discussion following Theorem 2 in [Fraenkel 1982], and for the arithmetic one it follows from the end of that paper. □

It is not hard to see that the same result holds for the Mouse game and the Fat Rat game.

10. Epilogue

In addition to the quest for the analysis of multipile take-away games, there is another motivation for inventing and analyzing the Rat game. The analysis of most games on piles of tokens is associated with irrational numbers. Thus, the generalized Wythoff game $W(a)$ depends on $\alpha = (2 - a + \sqrt{a^2+4})/2$ and $\beta = \alpha + a$, where a is an integer parameter [Fraenkel 1982]. Also games on more piles often depend on irrational numbers, such as the multipile Wythoff game. This is also the case for the Raleigh game.

Here we were interested in investigating whether there is a game whose strategy depends on distinct *rational* numbers. Since the sequences of P-positions of a game split $\mathbb{Z}_{\geq 1}$, this leads naturally to a question that has been solved for the integers, solved for the irrationals, but is wide open for the rationals! This fact may explain, in part, why the move rules for the Rat game are more complicated than those where the strategy depends on irrationals.

Since the game we constructed here depends on rational numbers, it is appropriately dubbed the *Rat* game. The *Mouse* game is a small Rat game with only 2 piles. This is as far as etymology is concerned.

The *m-Fat Rat* game is played on an arbitrary finite number of piles. Let $0 < \alpha_1 \leq \cdots \leq \alpha_m$, $\gamma_1, \ldots, \gamma_m$ be reals, and suppose that the $m \geq 2$ *Beatty sequences*

$$\lfloor n\alpha_i + \gamma_i \rfloor, \quad i = 1, \ldots, m, \tag{2}$$

split $\mathbb{Z}_{\geq 1}$. If the *moduli* α_i are all integers (so $\lfloor n\alpha_i + \gamma_i \rfloor = n\alpha_i + \lfloor \gamma_i \rfloor$) and $m \geq 2$, then $\alpha_m = \alpha_{m-1}$. A short "proof from the Book", due to Mirsky, D. Newman, Davenport and Rado, involving a generating function and a primitive (complex) root of unity proves this. See [Erdős 1952]. First elementary proofs were given in [Berger, Felzenbaum and Fraenkel 1986a; Simpson 1986; Lewis 1996]. The essence of the elementary proof was expounded in [Zeilberger 2001]. For finite splittings of the integers with irrational moduli there is the well-known result that if α, β are positive irrationals satisfying

$$\alpha^{-1} + \beta^{-1} = 1, \tag{3}$$

then the sequences $\lfloor n\alpha \rfloor$ and $\lfloor n\beta \rfloor$ ($n = 1, 2, \ldots$) split $\mathbb{Z}_{\geq 1}$. For a "proof from the Book", see [Fraenkel 1982]. Thus also $\lfloor 2n\alpha \rfloor$, $\lfloor (2n-1)\alpha \rfloor$, $\lfloor n\beta \rfloor$, where $n = 1, 2, \ldots$, is a splitting. This is a simple example of the *composition* of the integer splitting $2n$, $2n-1$ with an irrational splitting. In fact, the composition of one or two integer moduli splittings with one or both of any irrational moduli α, β respectively satisfying (3) is also a splitting.

So far we have seen that any splitting of the positive integers by m sequences that are multiples of moduli, contains at least two sequences with identical moduli

($m \geq 2$ for integer moduli, $m \geq 3$ for irrational moduli). But a counterexample for the case when the α_k are rational was constructed in [Fraenkel 1973]:

$$\left\lfloor \frac{2^m - 1}{2^{m-k}} n \right\rfloor - 2^{k-1} + 1, \quad k = 1, \ldots, m, \quad n \geq 1,$$

which splits the positive integers for every $m \in \mathbb{Z}_{\geq 2}$. In fact, the following was conjectured there:

Conjecture 1. If $0 < \alpha_1 < \cdots < \alpha_m$ are any real numbers and $m \geq 3$, then the system (2) splits $\mathbb{Z}_{\geq 1}$ if and only if

$$\alpha_k = \frac{2^m - 1}{2^{m-k}} \quad \text{for } k = 1, \ldots, m.$$

In other words, the only disjoint covering system with *distinct* moduli is conjectured to have this form. Some special cases were proved in [Fraenkel 1973]. See also [Erdős and Graham 1980, Section 1] and [Berger, Felzenbaum and Fraenkel 1986b]. The most substantial progress towards settling the conjecture was made by Graham [1973], who showed that if the moduli are irrational and $m \geq 3$, then again $\alpha_i = \alpha_j$ for some $i \neq j$. He did this by proving that any finite irrational splitting of the integers by at least 3 sequences is a composition as stated above, so the result follows from the integer case. However, the remaining rational case appears to be rather stubborn.

Morikawa [1982/83; 1985b] (and in a number of additional papers) investigated splittings of $\mathbb{Z}_{\geq 1}$ by rational Beatty sequences and proved the conjecture for $m = 3$. This case was proved independently in [Tijdeman 1996]. Morikawa [1985a] gave necessary and sufficient conditions for two rational Beatty sequences to be disjoint. Simpson [2004] simplified this proof and dubbed it "Japanese remainder theorem" in honor of Morikawa. He also gave there a generating function method for the splitting of $\mathbb{Z}_{\geq 1}$ by rational Beatty sequences, similar to that of Mirsky, D. Newman, Davenport and Rado for expressing the splitting of $\mathbb{Z}_{\geq 1}$ by integer Beatty sequences. Simpson [1991] proved that the conjecture holds if $\alpha_1 \leq 3/2$. Altman, Gaujal and Hordijk [Altman et al. 2000] proved it for $m = 4$, using the notion of balanced words. Using this method, Tijdeman [2000a; 2000b] established it for $m = 5$ and 6. Using the same method, Barát and Varjú [2003] extended it to $m = 7$. The conjecture was generalized by Graham and O'Bryant [2005] to exact t-fold coverings, and proved several special cases of it using Fourier methods. For further developments see [Vuillon 2003; Paquin and Vuillon 2007].

It is of some interest to note that the conjecture has applications in the theory of scheduling and just-in-time sequencing. See, e.g., [Kubiak 2003; Brauner and Jost 2008; Brauner and Crama 2004]. These applications, in turn, precipitated proofs of special cases of the conjecture.

But despite all these really nice results, the conjecture is still open. Ron Graham summed it all up in a slight rephrasing of Piet Hein's saying:

A problem worthy of attack,
proves its worth by fighting back.

Summarizing, the 3-fold motivation for this work: (i) a games approach might help to settle this conjecture, and (ii) find a take-away game whose winning strategy depends on *rational* numbers, and (iii) find another analyzable nonNim take-away game played on more than 2 piles.

We remark finally that it may be of interest to characterize infinite disjoint covering systems with rational moduli. For the case of integer moduli (arithmetic sequences), see [Barát and Varjú 2005; Z.-W. Sun 2005].

Acknowledgment

I wish to thank the two anonymous referees, whose remarks eliminated some of the typos (zillions still remain), and made the presentation more reader-friendly.

References

[Albert et al. 2007] M. Albert, R. J. Nowakowski, and D. Wolfe, *Lessons in play: An introduction to combinatorial game theory*, A K Peters, Wellesley, MA, 2007.

[Altman et al. 2000] E. Altman, B. Gaujal, and A. Hordijk, "Balanced sequences and optimal routing", *J. ACM* **47**:4 (2000), 752–775.

[Barát and Varjú 2003] J. Barát and P. P. Varjú, "Partitioning the positive integers to seven Beatty sequences", *Indag. Math. (N.S.)* **14**:2 (2003), 149–161.

[Barát and Varjú 2005] J. Barát and P. P. Varjú, "A contribution to infinite disjoint covering systems", *J. Théor. Nombres Bordeaux* **17**:1 (2005), 51–55.

[Berger, Felzenbaum and Fraenkel 1986a] M. A. Berger, A. Felzenbaum, and A. S. Fraenkel, "A nonanalytic proof of the Newman–Znám result for disjoint covering systems", *Combinatorica* **6**:3 (1986), 235–243.

[Berger, Felzenbaum and Fraenkel 1986b] M. A. Berger, A. Felzenbaum, and A. S. Fraenkel, "Disjoint covering systems of rational Beatty sequences", *J. Combin. Theory Ser. A* **42**:1 (1986), 150–153.

[Brauner and Crama 2004] N. Brauner and Y. Crama, "The maximum deviation just-in-time scheduling problem", *Discrete Appl. Math.* **134**:1-3 (2004), 25–50.

[Brauner and Jost 2008] N. Brauner and V. Jost, "Small deviations, JIT sequencing and symmetric case of Fraenkel's conjecture", *Discrete Math.* **308**:11 (2008), 2319–2324.

[Coxeter 1953] H. S. M. Coxeter, "The golden section, phyllotaxis, and Wythoff's game", *Scripta Math.* **19** (1953), 135–143.

[Duchêne and Gravier 2009] E. Duchêne and S. Gravier, "Geometrical extensions of Wythoff's game", *Discrete Math.* **309**:11 (2009), 3595–3608.

[Duchêne and Rigo 2008] E. Duchêne and M. Rigo, "A morphic approach to combinatorial games: the Tribonacci case", *Theor. Inform. Appl.* **42**:2 (2008), 375–393.

[Erdős 1952] P. Erdős, "On a problem concerning congruence systems", *Mat. Lapok* **3** (1952), 122–128. In Hungarian; English summary.

[Erdős and Graham 1980] P. Erdős and R. L. Graham, *Old and new problems and results in combinatorial number theory*, vol. 28, Monographies de L'Enseignement Mathématique [Monographs of L'Enseignement Mathématique], Université de Genève, L'Enseignement Mathématique, Geneva, 1980.

[Fraenkel 1973] A. S. Fraenkel, "Complementing and exactly covering sequences", *J. Combinatorial Theory Ser. A* **14** (1973), 8–20.

[Fraenkel 1982] A. S. Fraenkel, "How to beat your Wythoff games' opponent on three fronts", *Amer. Math. Monthly* **89**:6 (1982), 353–361.

[Fraenkel 1996] A. S. Fraenkel, "Scenic trails ascending from sea-level Nim to alpine chess", pp. 13–42 in *Games of no chance* (Berkeley, CA, 1994), edited by R. Nowakowski, Math. Sci. Res. Inst. Publ. **29**, Cambridge Univ. Press, 1996.

[Fraenkel 2007] A. S. Fraenkel, "The Raleigh game", pp. 199–208 in *Combinatorial number theory*, edited by B. Landman et al., de Gruyter, Berlin, 2007. Appeared also in *Integers* **7(20)**, #A13, http://www.integers-ejcnt.org/vol7(2).html.

[Fraenkel and Krieger 2004] A. S. Fraenkel and D. Krieger, "The structure of complementary sets of integers: A 3-shift theorem", *Int. J. Pure Appl. Math.* **10**:1 (2004), 1–49.

[Fraenkel and Peled 2015] A. S. Fraenkel and U. Peled, "Harnessing the unwieldy MEX function", pp. 77–94 in *Games of No Chance 4*, Mathematical Sciences Research Institute Publications **63**, Cambridge University Press, New York, 2015.

[Graham 1973] R. L. Graham, "Covering the positive integers by disjoint sets of the form $\{[n\alpha + \beta] : n = 1, 2, \dots\}$", *J. Combinatorial Theory Ser. A* **15** (1973), 354–358.

[Graham and O'Bryant 2005] R. Graham and K. O'Bryant, "A discrete Fourier kernel and Fraenkel's tiling conjecture", *Acta Arith.* **118**:3 (2005), 283–304.

[Guy and Nowakowski 2009] R. K. Guy and R. J. Nowakowski, "Unsolved problems in combinatorial games", pp. 135–150 in *Games of no chance 3*, edited by M. Albert and R. Nowakowski, Math. Sci. Res. Inst. Publ. **56**, Cambridge, 2009.

[Hardy and Wright 2008] G. H. Hardy and E. M. Wright, *An introduction to the theory of numbers*, Sixth ed., Oxford University Press, 2008. Revised by D. R. Heath-Brown and J. H. Silverman.

[Kubiak 2003] W. Kubiak, "On small deviations conjecture", *Bull. Pol. Acad. Sci., Tech. Sci.* **51**:2 (2003), 189–203.

[Landman 2002] H. A. Landman, "A simple FSM-based proof of the additive periodicity of the Sprague–Grundy function of Wythoff's game", pp. 383–386 in *More games of no chance* (Berkeley, CA, 2000), edited by R. Nowakowski, Math. Sci. Res. Inst. Publ. **42**, Cambridge Univ. Press, 2002.

[Lewis 1996] E. Lewis, "Infinite covering systems of congruences which don't exist", *Proc. Amer. Math. Soc.* **124**:2 (1996), 355–360.

[Morikawa 1982/83] R. Morikawa, "On eventually covering families generated by the bracket function", *Bull. Fac. Liberal Arts Nagasaki Univ.* **23**:1 (1982/83), 17–22.

[Morikawa 1985a] R. Morikawa, "Disjoint sequences generated by the bracket function", *Bull. Fac. Liberal Arts Nagasaki Univ.* **26**:1 (1985), 1–13.

[Morikawa 1985b] R. Morikawa, "On eventually covering families generated by the bracket function, IV", *Bull. Fac. Liberal Arts Nagasaki Univ.* **25**:2 (1985), 1–8.

[Paquin and Vuillon 2007] G. Paquin and L. Vuillon, "A characterization of balanced episturmian sequences", *Electron. J. Combin.* **14**:1 (2007), R33, 12.

[Simpson 1986] R. J. Simpson, "Exact coverings of the integers by arithmetic progressions", *Discrete Math.* **59**:1-2 (1986), 181–190.

[Simpson 1991] R. J. Simpson, "Disjoint covering systems of rational Beatty sequences", *Discrete Math.* **92**:1-3 (1991), 361–369.

[Simpson 2004] J. Simpson, "Disjoint Beatty sequences", *Integers* **4** (2004), A12, 10.

[X. Sun 2005] X. Sun, "Wythoff's sequence and N-heap Wythoff's conjectures", *Discrete Math.* **300**:1-3 (2005), 180–195.

[X. Sun and Zeilberger 2004] X. Sun and D. Zeilberger, "On Fraenkel's N-heap Wythoff's conjectures", *Ann. Comb.* **8**:2 (2004), 225–238.

[Z.-W. Sun 2005] Z.-W. Sun, "On the range of a covering function", *J. Number Theory* **111**:1 (2005), 190–196.

[Tijdeman 1996] R. Tijdeman, "On complementary triples of Sturmian bisequences", *Indag. Math. (N.S.)* **7**:3 (1996), 419–424.

[Tijdeman 2000a] R. Tijdeman, "Exact covers of balanced sequences and Fraenkel's conjecture", pp. 467–483 in *Algebraic number theory and Diophantine analysis* (Graz, 1998), edited by F. Halter-Koch and R. F. Tichy, de Gruyter, Berlin, 2000.

[Tijdeman 2000b] R. Tijdeman, "Fraenkel's conjecture for six sequences", *Discrete Math.* **222**:1-3 (2000), 223–234.

[Vuillon 2003] L. Vuillon, "Balanced words", *Bull. Belg. Math. Soc. Simon Stevin* **10**:suppl. (2003), 787–805.

[WW] E. R. Berlekamp, J. H. Conway, and R. K. Guy, *Winning ways for your mathematical plays*, 2nd ed., A K Peters, Wellesley, MA, vol. 1 (2001), vols. 2, 3 (2003), vol. 4 (2004).

[Wythoff 1906] W. A. Wythoff, "A modification of the game of Nim", *Nieuw Arch. Wisk.* **7** (1906), 199–202.

[Zeilberger 2001] D. Zeilberger, "How Berger, Felzenbaum and Fraenkel revolutionized covering systems the same way that George Boole revolutionized logic", *Electron. J. Combin.* **8**:2 (2001), A1, 10.

aviezri.fraenkel@weizmann.ac.il *Department of Computer Science and Applied Mathematics, Weizmann Institute of Science, 76100 Rehovot, Israel*

Games of No Chance 4
MSRI Publications
Volume **63**, 2015

A ruler regularity in hexadecimal games

J.P. GROSSMAN AND RICHARD J. NOWAKOWSKI

An important problem in the theory of impartial games is to determine the regularities of their nim-sequences. Subtraction games have periodic nim-sequences and those of octal games are conjectured to be periodic, but the possible regularities of the nim-sequence of a hexadecimal game are unknown. Periodic and arithmetic periodic nim-sequences have been discovered but other patterns also exist. We present an infinite set of hexadecimal games, based on the game **0.2048**, that exhibit a regularity — ruler regularity — not yet reported or codified.

1. Introduction

A taking-and-breaking game [Albert et al. 2007; Berlekamp et al. 2001] is an impartial combinatorial game, played with heaps of beans on a table. A move for either player consists of choosing a heap, removing a certain number of beans from the heap, and then possibly splitting the remainder into several heaps; the winner is the player making the last move. For example, both Grundy's Game (choose a heap and split it into two unequal heaps) and Couples-Are-Forever (choose a heap with at least three beans and split it into two) are taking-and-breaking games with very simple rules, however neither has been solved.

We present an overview of the required theory of *impartial* games. The reader can consult the references above for a more in-depth grounding in the theory of, and for more details about, subtraction and octal games.

The *followers* of a game are all those positions which can be reached in one move. The *minimum excluded value* of a set S is the least nonnegative integer which is not included in S and is denoted $\operatorname{mex}(S)$. The *nim-value* of an impartial game G, denoted by $\mathcal{G}(G)$, is given by $\mathcal{G}(G) = \operatorname{mex}\{\mathcal{G}(H) \mid H \text{ is a follower of } G\}$. The values in the set $\{\mathcal{G}(H) \mid H \text{ is a follower of } G\}$ are called *excluded values* for $\mathcal{G}(G)$. An impartial game G is a previous player win (i.e., the next player has no good move) if and only if $\mathcal{G}(G) = 0$. The *exclusive or* (XOR), or *nim-sum*, of two nonnegative integers a, b, written $a \oplus b$, consists of adding their binary

Partially supported by grants from NSERC.
Keywords: hexadecimal game, ruler regular, nim-sequence, periods.

representations with no carrying; for example, $3 \oplus 6 = (11 \oplus 110)_{\text{base } 2} = 101_{\text{base } 2} = 5$. The *disjunctive sum* of games G and H, written $G + H$, is the game where a player chooses either of the two and plays in that game; the nim-value of a disjunctive sum satisfies $\mathcal{G}(G+H) = \mathcal{G}(G) \oplus \mathcal{G}(H)$. Taking-and-breaking games are examples of disjunctive games — choose one heap and play in it. To know how to play these game well, it suffices to know what the nim-values are for individual heaps. For a given game G, let $\mathcal{G}(i)$ be the nim-value of G played with a heap of size i. We define the nim-sequence for a taking-and-breaking game to be the sequence $\mathcal{G}(0), \mathcal{G}(1), \mathcal{G}(2), \ldots$. A nim sequence is *periodic* if there exist N and p such that $\mathcal{G}(n+p) = \mathcal{G}(n)$ for all $n \geq N$. It is *arithmetic-periodic* if there exist N, p and s such that $\mathcal{G}(n+p) = \mathcal{G}(n) + s$ for all $n \geq N$, where s is called the *saltus*.

In a hexadecimal game, after removing beans from a heap the remainder can be split into at most three heaps. The rules for a hexadecimal game are described by a hexadecimal code $\mathbf{0.}d_1 d_2 \ldots d_u$ where $0 \leq d_i \leq 15$. This is an extension of the octal code used in [Berlekamp et al. 2001]. We use the letters A, B, C, D, E, F for the numbers 10 through 15 respectively. If $d_i = 0$ then a player cannot take i beans away from a heap. If $d_i = a_3 2^3 + a_2 2^2 + a_1 2^1 + a_0 2^0$ where a_i is 0 or 1, a player can remove i beans from the heap provided he leaves the remainder in exactly j heaps for some j with $a_j = 1$. If $0 \leq d_i \leq 7$ for all i then this is called an *octal* game. This restriction allows a heap to be split into no more than 2 heaps. A subtraction game has $d_i = 0$ or 3; that is, a player can remove beans but cannot split the heap.

The nim-sequence of a subtraction game with a finite subtraction set is periodic [Albert et al. 2007; Berlekamp et al. 2001]. In [Althöfer and Bültermann 1995], it is shown that the sequence of nim-values of games with small subtraction sets can have long periods. It is conjectured that all octal games also have periodic nim-sequence in unsolved problem 2 of [Guy and Nowakowski 2002]. This appears to be hard. For example, in the game **0.106** Flammenkamp computed the period length as 328226140474 and preperiod length as 465384263797. See [Flammenkamp 2012] for this and other searches for long periods, also see [Berlekamp et al. 2001; Caines et al. 1999; Gangolli and Plambeck 1989]. It is interesting to note that while octal games cannot have arithmetic-periodic nim-sequences [Austin 1976], if a single pass move is allowed then not only do arithmetic-period nim-sequences occur, but also nim-sequences composed of a periodic subsequence and an arithmetic-periodic subsequence occur called *sapp* sequences [Horrocks and Nowakowski 2003].

For hexadecimal games, there is an even richer selection of behaviours [Howse and Nowakowski 2004]):

- *Periodic*: **0.B**, period 2 with no preperiod.

- *Arithmetic-periodic*: **0.137F** has nim-sequence 0, 1, 1, 2, 2, 3, 3, ..., where $\mathcal{G}(2m-1) = \mathcal{G}(2m) = m$ and $\mathcal{G}(2m+1) = \mathcal{G}(2m)+1 = m+1$ for $m \geq 1$. In this case, the saltus is 1 and the period length is 2.
- *Sapp* regularity: **0.205200C** has the nim-sequence consisting of
  (a) periodic subsequences $\mathcal{G}(40k+19) = 6$ and $\mathcal{G}(40k+39) = 14$, and
  (b) arithmetic-periodic subsequences
$$\mathcal{G}(40k+j) = \mathcal{G}(40(k-1)+j)+16, \quad j \neq 19, 39,\ k \geq 0,$$
  with preperiod length of 4, period length of 40.
- In [Howse and Nowakowski 2004], it is noted that **0.123456789** exhibits another type of regularity. Starting with $n = 0$, the first fifteen nim-values are 0, 1, 0, 2, 2, 1, 1, 3, 2, 4, 4, 5, 5, 6, 4, and thereafter
$$\mathcal{G}(2m-1) = \mathcal{G}(2m) = m-1,$$
  except $\mathcal{G}(2^k+6) = 2^k - 1$. The nim-sequence is essentially arithmetic-periodic, but with an infinite number of exceptional values that occur in a geometric fashion.

In this paper, we report another new regularity: *ruler-regularity*. This concept is taken from the markings of a foot ruler. (See [Berlekamp et al. 2003, page 470, Figure 7], *"The $\mathcal{G}$-values for the* RULER *game"*, except for us the ruler needs to be rotated by 45 degrees and then flipped!)

Definition 1. A sequence is *ruler regular* if there are positive integers N, p, s, r and a finite set of integers K such that $\mathcal{G}(n+p) = \mathcal{G}(n)+s$ for all $n \geq N$ except
$$n = r((q+1)2^{m+1} + 2^m + 1) + k, \quad k \in K, q, m \geq 0,$$
when
$$\mathcal{G}(r((q+1)2^{m+1} + 2^m + 1) + k) = \mathcal{G}(r(q2^{m+1} + 2^m + 1) + k) + 2^{m+3}.$$

This is illustrated in Figure 1 for **0.2048**, where $r = 13$. New "lines" start at $(m, q) = (1, 0), (2, 0), \ldots$, and are indicated in the top graph. A ruler-regular nim-sequence appears arithmetic-periodic regardless of how far we extend the sequence, but before the sequence goes far enough to ensure arithmetic-periodicity (approximately $3e + 8ps^3$, where e is the size of the largest heap which is not in the period; see [Howse and Nowakowski 2004, Theorem 4]), a new term arises which essentially doubles the length of the apparent period. In this paper we show that **0.20...048**, where there are an odd number of 0s in the hexadecimal code, are ruler-regular. These are not the only such games, for instance **0.21317809532**, **0.31711188** (Figure 2, top), **0.321432132900903213**,

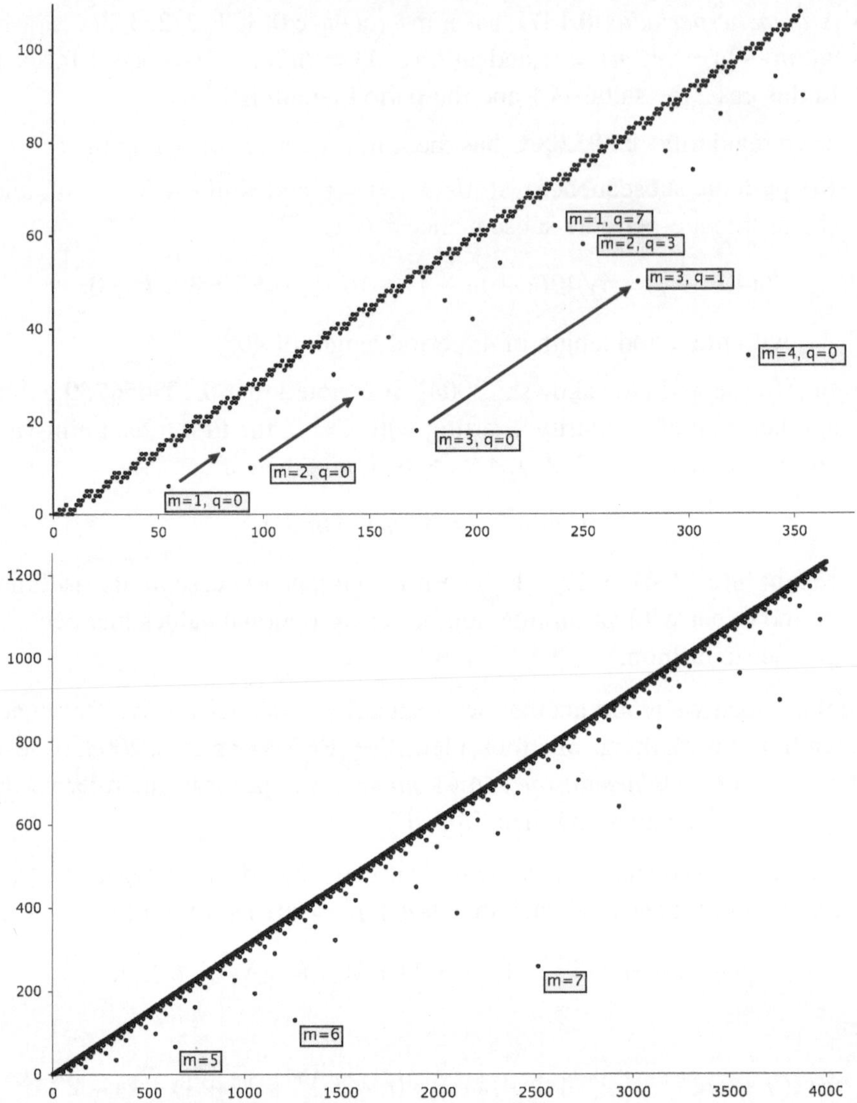

Figure 1. The first 350 values (top) and the first 4000 (bottom) of the $\mathcal{G}$-sequence of **0.2048**.

and **0.404008** (Figure 2, bottom) all appear to be ruler-regular with $|K| = 1$. (Actually, **0.404008** isn't precisely ruler-regular; the sharp-eyed reader will have noticed the same behaviour as in **0.123456789** happening along the first diagonal underneath the main diagonal, in tandem with the point starting the new "line".) The game **0.9138B835B** has $|K| = 3$; see Figure 3.

All of the previously known regularities — periodic, arithmetic-periodic and sapp — have the property that only a finite number of nim-values have to be

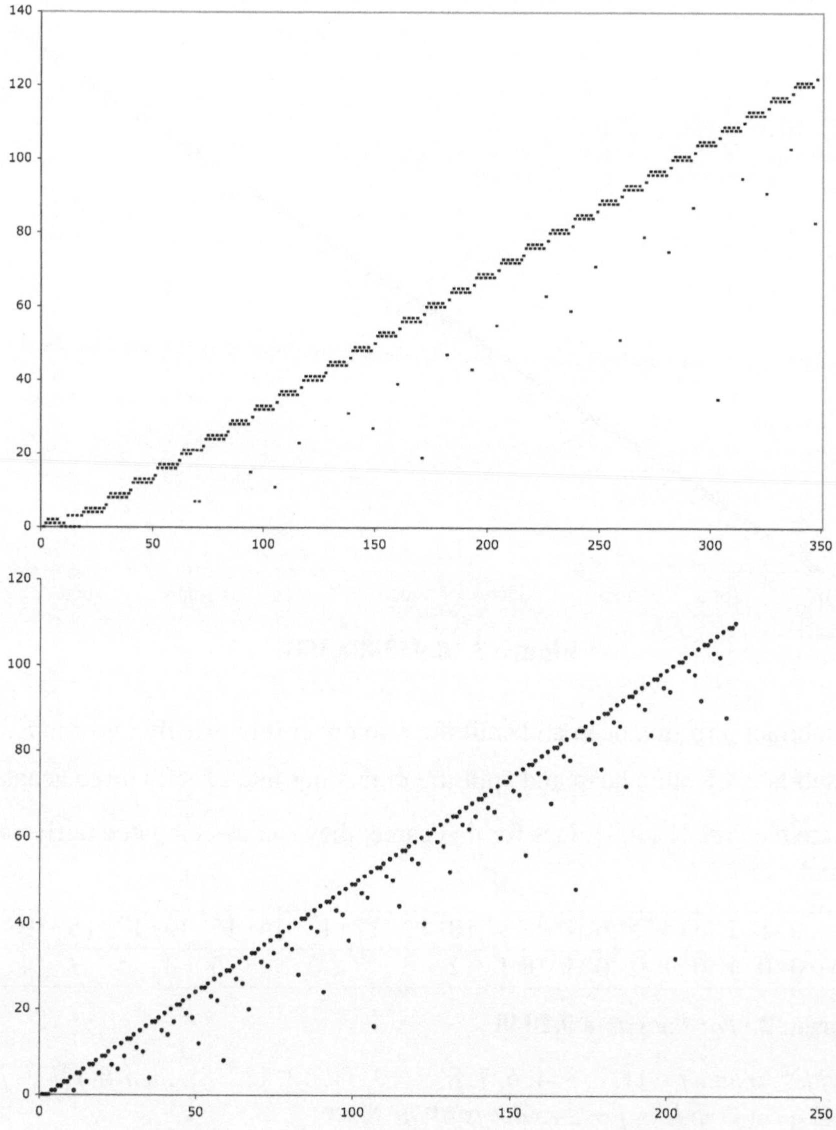

Figure 2. Top: **0.31711188**. Bottom: **0.404008**.

calculated to identify the type of regularity. However, as yet we have no similar mechanism to check for ruler-regularity.

2. The hexadecimal game 0.2048

In the hexadecimal game **0.2048**, the legal moves are to:

- subtract 1 from a heap of size ≥ 2;

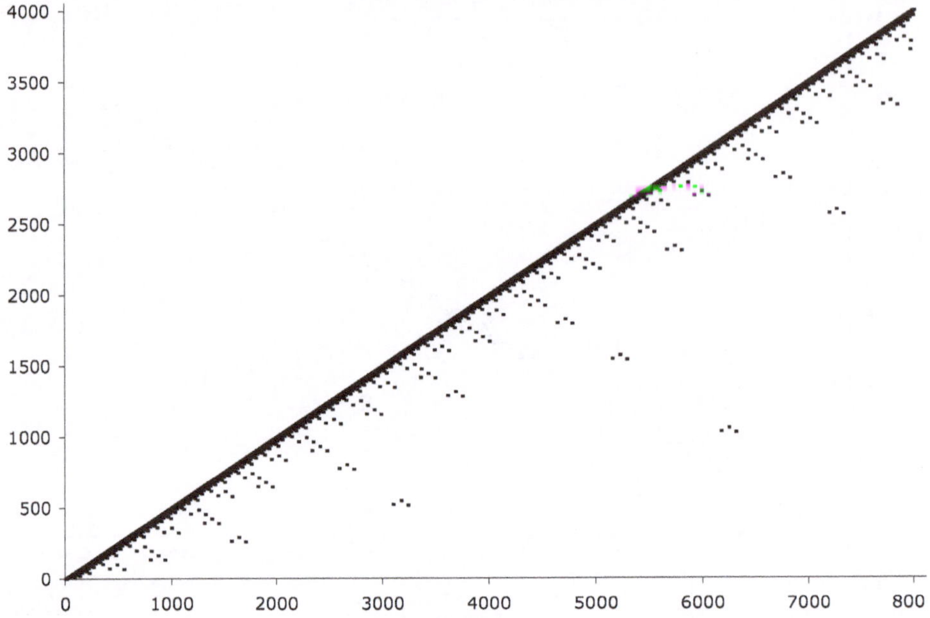

Figure 3. 0.9138B835B.

- subtract 3 from a heap and split the remainder into exactly two heaps;
- subtract 4 from a heap and split the remainder into exactly three heaps.

Here are the first 21 nim-values for this game; they can be computed fairly easily by hand.

n	0	1	2	3	4	5	6	7	8	9	10	11	12	13	14	15	16	17	18	19	20
$\mathcal{G}(n)$	0	0	1	0	1	2	0	1	0	1	2	3	2	3	4	5	4	5	3	4	5

Theorem 2. *For the game* **0.2048**:

(a) *If $k \geq 0$ and $j \in \{1, 2, 3, 4, 6, 7, 8, 9, 10, 11, 12, 13, 18\}$, then $\mathcal{G}(13k + j) = 4k + \mathcal{G}(j)$ unless $j = 2$ and k is of the form*

$$(q+1)2^{m+1} + 2^m + 1, \quad (q, m \geq 0).$$

(b) *For $q, m \geq 0$, $\mathcal{G}\big(13((q+1)2^{m+1} + 2^m + 1) + 2\big) = 2^{m+3}q + 2^{m+2} + 2$.*

Part (b) of the theorem describes the exceptional values which form the ruler pattern. Equivalently stated, if

$$k = (q+1)2^{m+1} + 2^m + 1$$

then

$$\mathcal{G}(13k + 2) = 4(k - 2^{m+1} - 1) + 2.$$

m	$q=0$	$q=1$	$q=2$	$q=3$	$q=4$	Differences
0	(54,6)	(80,14)	(106,22)	(132,30)	(158,38)	(13·2,8)
1	(93,10)	(145,26)	(197,42)	(249,58)	(301,74)	(13·4,16)
2	(171,18)	(275,50)	(379,82)	(483,114)	(587,146)	(13·8,32)
3	(327,34)	(535,98)	(743,162)	(951,226)	(1159,290)	(13·16,64)
4	(639,66)	(1055,194)	(1471,322)	(1887,450)	(2303,578)	(13·32,128)
5	(1263,130)	(2095,386)	(2927,642)	(3759,898)	(4591,1154)	(13·64,256)
6	(2511,258)	(4175,770)	(5839,1282)	(7503,1794)	(9167,2306)	(13·128,512)

Table 1. The "lines" $m = 0, 1, \ldots, 6$.

These exception values form lines, each of slope $\frac{4}{13}$, indexed by $m \geq 0$ starting at $(39 \times 2^m + 15, 2^{m+2} + 2)$, and the other points on the "line" are given by $(39 \times 2^m + 15, 2^{m+2} + 2) + 13(2^{m+1}, 2^{m+3})$. This is illustrated in Table 1. The annotations "$m =$" in Figure 1 indicate the start of the m-th line. In the top half of the figure, the four points $m = 4$, $q = 0$ through $m = 1$, $q = 7$ lie on a "rule" of negative slope. For a given $k \geq 0$, starting with the point in the bottom right position of this line, the coordinates of the k-th new rule are given by Theorem 2(b) with

$$(m, q) = (k, 0),\ (k-1, 1),\ (k-2, 3),\ \ldots,\ (k-i, 2^i - 1),\ \ldots,\ (0, 2^k - 1).$$

The difference between consecutive points is $(-13 \cdot 2^{m-1}, 2^{m+1})$ which gives a slope of $-\frac{4}{13}$. The other set of interesting points are the ones given by $(m, 0)$, $m = 1, 2, \ldots$. These also fall on a line and this has slope $\frac{4}{3 \cdot 13}$.

Before we embark on the proof of Theorem 2, let us take a moment to discuss the statement of the theorem and make a few simple observations that will be of use later on. First, the set $J = \{1, 2, 3, 4, 6, 7, 8, 9, 10, 11, 12, 13, 18\}$ is a modified set of remainders (mod 13) chosen so that for $j \in J$ we have $0 \leq \mathcal{G}(j) \leq 3$. These are the values of $\mathcal{G}(j)$ for $j \in J$:

j	1	2	3	4	6	7	8	9	10	11	12	13	18
$\mathcal{G}(j)$	0	1	0	1	0	1	0	1	2	3	2	3	3

Our proof of Theorem 2 is by induction, where we divide into cases based on j, the residue modulo 13. For each $j \in J$, we must show that $\mathcal{G}(13k + j)$ is as described by the theorem. In order to show that $\mathcal{G}(n) = g$, it is necessary to (i) prove that for all $g' < g$ there is a move from a heap of size n to a position with nim-value g', and (ii) prove that there is no move from a heap of size n to a position with nim-value g. Note that (a) holds trivially with $k = 0$, so in all that follows, let $k > 0$ and $j \in J$, and suppose inductively that the statement of Theorem 2 holds for heaps smaller than $13k + j$. We begin with a lemma:

Lemma 3. *For all $1 \leq a \leq k$ there is a move from a heap of size $13k + j$ to a position with nim-value $4(k-a)+2$ unless $k = (q+1)2^{m+1} + 2^m + 1$, $j = 2$ and $a = 2^{m+1} + 1$, in which case there is no move from a heap of size $13k + j$ to a position with nim-value $4(k-a)+2$.*

Proof. We first consider the nonexceptional cases where $j \neq 2$ or $a \neq 2^{m+1} + 1$ or $k \neq (q+1)2^{m+1} + 2^m + 1$, and show that there is a move from a heap of size $13k+j$ to a position with nim-value $4(k-a)+2$. We can subtract 4 and split the remainder into three heaps of size r, r and $13k + j - 4 - 2r$. The nim-value of this position is simply $\mathcal{G}(13k + j - 4 - 2r)$, so it certainly suffices to find $r \geq 1$ such that $\mathcal{G}(13k + j - 4 - 2r) = 4(k-a) + 2$. In other words, it suffices to find $n \leq 13k + j - 6$ such that $13k + j - n \equiv 0 \pmod 2$ and $\mathcal{G}(n) = 4(k-a)+2$. For $a = k$, either $n = 5$ or $n = 10$ will satisfy these conditions unless $13k + j = 14$, but for this case we simply subtract 3 and divide into heaps of 10 and 1. For $a < k$ we consider two cases:

Case 1: $j + a \equiv 0 \pmod 2$. Take $n = 13(k-a) + 10$. Then by induction, $\mathcal{G}(n) = 4(k-a) + 2$, and also $13k + j - n \equiv j + a \equiv 0 \pmod 2$, so we are done if $13(k-a) + 10 \leq 13k + j - 6$, which is equivalent to $16 \leq 13a + j$. This is certainly the case if $a > 1$, so we need only to consider $a = 1$ in which case the inequality becomes $j \geq 3$. This in turn is true unless $j = 1$ or $j = 2$, but $j \neq 2$ because we are considering the case $j + a \equiv 0 \pmod 2$. Thus, it remains to show that there is a move from a heap of size $13k + 1$ to a position with nim-value $4(k-1) + 2$. We can accomplish this by subtracting 3 and splitting the remainder into two heaps of size 1 and $13k - 3$; by induction $\mathcal{G}(1) \oplus \mathcal{G}(13k - 3) = 0 \oplus \mathcal{G}(13(k-1) + 10) = 4(k-1) + 2$.

Case 2: $j + a \equiv 1 \pmod 2$. Let m be the number of trailing zeros in the binary representation of $k - a$. Theorem 2(b) states that $4(k-a)+2$ appears as an exceptional value $\mathcal{G}(n)$ for $n = 13(k - a + 2^{m+1} + 1) + 2$. If either $a > 2^{m+1} + 1$ or $a = 2^{m+1} + 1$ and $j > 2$ then $n = 13(k - a + 2^{m+1} + 1) + 2 < 13k + j$, so by induction we can assume that $\mathcal{G}(n) = 4(k-a)+2$. Then there is some q such that $4(k-a)+2 = 2^{m+3}q + 2^{m+2} + 2$, and so $13k + j - n \equiv j + a - 1 \equiv 0 \pmod 2$, so we're done if $n \leq 13k + j - 6$, that is, $8 \leq 13(a - 2^{m+1} - 1) + j$. This in turn is true if either $a > 2^{m+1} + 1$ or $a = 2^{m+1} + 1$ and $j \geq 8$; note that these conditions on (a, j) subsume the previous ones. Thus, we may assume from this point forward that $a \leq 2^{m+1} + 1$ and either $a < 2^{m+1} + 1$ or $j < 8$. In particular, for $j \geq 8$ we may assume that $a < 2^{m+1} + 1$. Note that by definition of m, if $d < 2^m$ then the 1s in the binary representations of $k - a$ and d do not overlap, so $d \oplus (k-a) = d + (k-a)$, which we rewrite as

$$k - a = d \oplus (k - a + d). \qquad (1)$$

We proceed by considering three cases.

Case 2.1: $a \equiv 0 \pmod 2$, $j \equiv 1 \pmod 2$. Since $a \geq 1$ we in fact have $a \geq 2$. Let $d = (a-2)/2$. Then $d \leq (2^{m+1}-1)/2 < 2^m$ and (1) becomes

$$k - a = d \oplus (k - 2 - d). \tag{2}$$

Our strategy is now to subtract 4 and split the remaining $13k + j - 4$ into exactly three heaps of sizes $13(k-2-d)+j_1$, $13d+j_2$ and j_3 where $j_1+j_2+j_3=22+j$, $\mathcal{G}(j_1) \oplus \mathcal{G}(j_2) \oplus \mathcal{G}(j_3) = 2$, $j_3 \in J \cup \{5\}$ and $j_1, j_2 \in J$. Using induction, (2) and the fact that $0 \leq \mathcal{G}(j_i) \leq 3$, the nim-value of this position is

$$\begin{aligned}(4(k-2-d) + \mathcal{G}(j_1)) \oplus (4d + \mathcal{G}(j_2)) \oplus \mathcal{G}(j_3) \\ = 4(d \oplus (k-2-d)) + \mathcal{G}(j_1) \oplus \mathcal{G}(j_2) \oplus \mathcal{G}(j_3) = 4(k-a)+2.\end{aligned}$$

Since $j \equiv 1 \pmod 2$, we only need to consider $j \in \{1, 3, 7, 9, 11, 13\}$. We can verify that the j_i in the following table satisfy the given conditions for each choice of j:

$22+j$	23	25	29	31	33	35
j_1	9	10	12	13	7	8
j_2	9	10	12	13	8	9
j_3	5	5	5	5	18	18

Case 2.2a: $a \equiv 1 \pmod 2$, $j \equiv 0 \pmod 2$, $a < 2^{m+1}+1$, $j \geq 8$ *or* $j = 2$. Let $d = (a-1)/2 < 2^{m+1}/2 = 2^m$. Then (1) becomes

$$k - a = d \oplus (k - 1 - d). \tag{3}$$

The conditions on j imply that $j \in \{2, 8, 10, 12, 18\}$ and hence

$$(j+4)/2 \in \{3, 6, 7, 8, 11\} \subset J.$$

We can therefore subtract 4 and split the remaining $13k + j - 4$ into three heaps of size $13(k-1-d) + (j+4)/2$, $13d + (j+4)/2$ and 5. Using induction and (3), we again find that the nim-value of this position is $4(k-a)+2$.

Case 2.2b: $a \equiv 1 \pmod 2$, $a \leq 2^{m+1}+1$, $j = 4, 6$. We dispense with the case $a = 1$ by observing that we can subtract 4 from $13k+4$ and split the remaining $13k$ into heaps of size 5, 4 and $13(k-1)+4$; we can subtract 3 from $13k+6$ and split the remaining $13k+3$ into heaps of size 6 and $13(k-1)+10$. By induction, the nim-value of each resulting position is $4(k-1)+2$. Now assume that $a \geq 3$, and let $d = (a-3)/2 \leq (2^{m+1}-2)/2 < 2^m$. Then (1) becomes

$$k - a = d \oplus (k - 3 - d). \tag{4}$$

Our strategy is to again subtract 4 and split the remaining $13k+j-4$ into exactly three heaps of size $13(k-3-d)+j_1$, $13d+j_2$ and j_3 where $j_1+j_2+j_3=35+j$, $\mathcal{G}(j_1)\oplus\mathcal{G}(j_2)\oplus\mathcal{G}(j_3)=2$, $j_3\in J\cup\{5\}$ and $j_1, j_2\in J$. Using induction, (4) and the fact that $0\le\mathcal{G}(j_i)\le 3$, we again find that the nim-value of this position is $4(k-a)+2$. We can verify that for $j=4$ we can take $(j_1, j_2, j_3)=(10, 11, 18)$ and for $j=6$ we can take $(j_1, j_2, j_3)=(11, 12, 18)$.

The only remaining values of j and a to consider are $j=2$ and $a=2^{m+1}+1$; in all other cases we have shown that there is a move from a heap of size $13k+j$ to a position with nim-value $4(k-a)+2$. Recall that m is the number of trailing zeros in the binary representation of $k-a$, thus $k-a$ is of the form $q\cdot 2^{m+1}+2^m$ for some $q\ge 0$, that is $k=(k-a)+a=(q+1)\cdot 2^{m+1}+2^m+1$. This is exactly the case excepted by the statement of the lemma, so all that remains to be shown is that in this case there is no move from a heap of size $13k+j$ to a position with nim-value $4(k-a)+2$. First, observe that subtracting 1 leaves a heap which by induction has nim-value $4k$. Next, suppose we subtract 4 and divide the remaining $13k-2$ into exactly 3 heaps. We can express the size of these heaps as $13k_1+j_1$, $13k_2+j_2$, and $13k_3+j_3$, where for $i=1, 2, 3$ either $j_i\in J$ or $j_i=5$ and $k_i=0$. By induction, the nim-value of the resulting position is:

$$(4k_1+\mathcal{G}(j_1))\oplus(4k_2+\mathcal{G}(j_2))\oplus(4k_3+\mathcal{G}(j_3))$$
$$=4(k_1\oplus k_2\oplus k_3)+\mathcal{G}(j_1)\oplus\mathcal{G}(j_2)\oplus\mathcal{G}(j_3).$$

If this is equal to $4(k-a)+2$ then the following three equations must hold:

$$(13k_1+j_1)+(13k_2+j_2)+(13k_3+j_3)=13k-2, \tag{5}$$
$$k_1\oplus k_2\oplus k_3=k-a, \tag{6}$$
$$\mathcal{G}(j_1)\oplus\mathcal{G}(j_2)\oplus\mathcal{G}(j_3)=2. \tag{7}$$

From (5) we have $j_1+j_2+j_3\equiv -2 \pmod{13}$. On the other hand, since $j_i\le 18$ we have $j_1+j_2+j_3\le 54$, thus $j_1+j_2+j_3\in\{11, 24, 37, 50\}$. There are no choices of $j_1, j_2, j_3\in J$ which sum to 50. If $j_1+j_2+j_3=37$ then they can't all be ≤ 13, so one of them is 18 and the other two sum to 19, but we can verify by hand that for all such choices we have $\mathcal{G}(j_1)\oplus\mathcal{G}(j_2)\oplus\mathcal{G}(j_3)=0$. If $j_1+j_2+j_3=24$ then from (5) $k_1+k_2+k_3=k-2$. But $a=2^{m+1}+1$ is odd and $k_1\oplus k_2\oplus k_3=k-a$, so in this case $k_1+k_2+k_3$ and $k_1\oplus k_2\oplus k_3$ would have opposite parity which is impossible. Finally, if $j_1+j_2+j_3=11$ then each $j_i\le 9$, so in order to satisfy (7) one of them must be 5, say j_3, hence $k_3=0$. We are then left with two equations:

$$k_1+k_2=k-1, \tag{8}$$
$$k_1\oplus k_2=k-a. \tag{9}$$

Now $k_1 + k_2 = k_1 \oplus k_2 + 2(k_1 \& k_2)$, where & denotes bitwise Boolean AND. Equations (8) and (9) therefore give us $2(k_1\&k_2) = k_1+k_2-k_1\oplus k_2 = a-1 = 2^{m+1}$ which implies that $k_1 \& k_2 = 2^m$ and so k_1 and k_2 both have a 1 in the 2^m place of their binary representations. But $k_1 \oplus k_2 = k - a = q \cdot 2^{m+1} + 2^m$ so that exactly one of k_1 and k_2 have a 1 in the 2^m place of their binary representations, a contradiction. Hence there is no way to subtract 4 and divide the remaining $13k-2$ into three heaps such that the resulting position has nim-value $4(k-a)+2$. Finally, suppose we subtract 3 and divide the remaining $13k-1$ into exactly 2 heaps of size $13k_1+j_1$ and $13k_2+j_2$ where again either $j_i \in J$ or $j_i = 5$ and $k_i = 0$. By induction, the nim-value of the resulting position is $4(k_1 \oplus k_2) + \mathcal{G}(j_1) \oplus \mathcal{G}(j_2)$. We therefore have the following three equations, corresponding to Equations (5)–(7):

$$(13k_1 + j_1) + (13k_2 + j_2) = 13k - 1, \tag{10}$$

$$k_1 \oplus k_2 = k - a, \tag{11}$$

$$\mathcal{G}(j_1) \oplus \mathcal{G}(j_2) = 2. \tag{12}$$

From (10) we have $j_1 + j_2 \equiv -1 \pmod{13}$, but $j_i \le 18$ so that $j_1 + j_2 \le 36$ giving $j_1 + j_2 \in \{12, 25\}$. Hence from (10), $k_1 + k_2 = k - 1$ or $k_1 + k_2 = k - 2$. But we saw previously that $k_1 + k_2 = k - 1$ is inconsistent with (11) (which is the same as (9)), and $k - 2$, $k - a$ have opposite parity. Hence there is no way to subtract 3 and divide the remaining $13k - 1$ into two heaps such that the resulting position has nim-value $4(k - a) + 2$, and this concludes the proof of Lemma 3. □

Lemma 4. *For all* $a \ge 1$, *there is a move from a heap of size* $13k + j$ *to positions with nim-values* $4(k - a) + 1$, $4(k - a)$ *and* $4(k - a) - 1$.

Proof. As in the proof of the previous lemma, we can move to a position with a given nim-value if a heap of size n has the desired value where $n \le 13k + j - 6$ and $13k + j - n \equiv 0 \pmod 2$. Since $j \ge 1$, this will certainly be the case if $n \le 13(k-1)+8$ and $13k+j-n$ is even. Hence, we obtain nim-value $4(k-a)+1$ with either $n = 13(k - a) + 7$ or $n = 13(k - a) + 4$ (one or the other will always do since they have opposite parity), we obtain nim-value $4(k - a)$ with either $n = 13(k - a) + 6$ or $n = 13(k - a) + 3$, and we obtain nim-value $4(k - a) - 1$ with either $n = 13(k - a - 1) + 11$ or $n = 13(k - a - 1) + 18$. □

We now know that from a heap of size $13k + j$, we can move to a position with nim-value g for any $g \le 4k - 2$, with one exception given by Lemma 3. This in fact proves part (b) of Theorem 2, as it shows that from a heap of size

$$13((q + 1) \cdot 2^{m+1} + 2^m + 1) + 2,$$

we cannot move to a position with value $2^{m+3}q+2^{m+2}+2$, but we can move to a position with value g for all $g < 2^{m+3}q+2^{m+2}+2$, so

$$\mathcal{G}(13((q+1)\cdot 2^{m+1}+2^m+1)+2) = 2^{m+3}q+2^{m+2}+2.$$

To finish the proof of Theorem 2, we must show that for all nonexceptional heaps $13k+j$ we can move to a position with value g for $4k-1 \le g < 4k+\mathcal{G}(j)$, but we cannot move to a position with value $4k+\mathcal{G}(j)$. Table 2 accomplishes the former, giving explicit moves for each $j \in J$.

It remains to show that there is no move from a heap of size $13k+j$ to a position with nim-value $4k+\mathcal{G}(j)$. First, subtracting 1 leaves $13k+j-1$. By induction, if $j-1 \in$J, then $\mathcal{G}(13k+j-1) = 4k+\mathcal{G}(j-1)$, but $\mathcal{G}(j) \ne \mathcal{G}(j-1)$ for $j,\, j-1 \in J$; otherwise$j+12 \in J$ and

$$\mathcal{G}(13k+j-1) = \mathcal{G}(13(k-1)+j+12) = 4(k-1)+\mathcal{G}(j+12) < 4k+\ G(j).$$

This leaves us with two cases to consider: moves that split $13k+j-4$ into exactly three heaps, and moves that split $13k+j-3$ into exactly two heaps. Following the approach in the proof of Lemma 3, splitting $13k+j-4$ into heaps of size $13k_i+j_i$ with either $j_i \in J$ or $j_i = 5$ and $k_i = 0$ gives us three equations, for which we must show there is no solution:

$$(13k_1+j_1)+(13k_2+j_2)+(13k_3+j_3) = 13k+j-4, \tag{13}$$

$$k_1 \oplus k_2 \oplus k_3 = k, \tag{14}$$

$$\mathcal{G}(j_1) \oplus \mathcal{G}(j_2) \oplus \mathcal{G}(j_3) = \mathcal{G}(j). \tag{15}$$

Equation (14) tells us that $k_1+k_2+k_3 \ge k_1 \oplus k_2 \oplus k_3 = k$, so we can rewrite (13) as:

$$j-4-j_1-j_2-j_3 = 13(k_1+k_2+k_3-k) \ge 0. \tag{16}$$

For each j there are limited choices for $j_1,\, j_2,\, j_3$ satisfying both

$$j_1+j_2+j_3 \equiv j-4 \pmod{13} \quad \text{and} \quad j_1+j_2+j_3 \le j-4,$$

and we can verify by hand that none of them also satisfy (15). The same approach can be used for moves which split $13k+j-3$ into two heaps; in this case we verify by hand that there are no $j_1,\, j_2$ satisfying $j_1+j_2 \equiv j-3 \pmod{13}$, $j_1+j_2 \le j-3$, and $\mathcal{G}(j_1) \oplus \mathcal{G}(j_2) = \mathcal{G}(j)$. This completes the proof of Theorem 2.

3. The games 0.200048, 0.20000048, 0.2000000048, ...

Theorem 5. *Let G be the game* $0.2(0)^{2n+1}48$ (*i.e.*, $2n+1$ *zeroes between the* 2 *and* 4) *and let* $A = 6(n+2)+1$.

j	Move	Nim-value
1	$(13k+1)-1=(13(k-1)+13)$	$4(k-1)+3=4k-1$
2	$(13k+2)-3=(1)+(13(k-1)+11)$ $(13k+2)-1=(13k+1)$	$4(k-1)+3=4k-1$ $4k$
3	$(13k+3)-4=(3)+(5)+(13(k-1)+4)$	$4(k-1)+3=4k-1$
4	$(13k+4)-4=(1)+(1)+(13(k-1)+11)$ $(13k+4)-1=(13k+3)$	$4(k-1)+3=4k-1$ $4k$
6	$(13k+6)-1=(13(k-1)+18)$	$4(k-1)+3=4k-1$
7	$(13k+7)-3=(6)+(13(k-1)+11)$ $(13k+7)-1=(13k+6)$	$4(k-1)+3=4k-1$ $4k$
8	$(13k+8)-4=(2)+(2)+(13(k-1)+13)$	$4(k-1)+3=4k-1$
9	$(13k+9)-3=(8)+(13(k-1)+11)$ $(13k+9)-1=(13k+8)$	$4(k-1)+3=4k-1$ $4k$
10	$(13k+10)-4=(3)+(3)+(13(k-1)+13)$ $(13k+10)-3=(1)+(13k+6)$ $(13k+10)-1=(13k+9)$	$4(k-1)+3=4k-1$ $4k$ $4k+1$
11	$(13k+11)-3=(3)+(13(k-1)+18)$ $(13k+11)-4=(2)+(2)+(13k+3)$ $(13k+11)-3=(1)+(13k+7)$ $(13k+11)-1=(13k+10)$	$4k-1$ $4k$ $4k+1$ $4(k-1)+3=4k+2$
12	$(13k+12)-4=(4)+(4)+(13(k-1)+13)$ $(13k+12)-4=(1)+(1)+(13k+6)$ $(13k+12)-4=(2)+(2)+(13k+4)$	$4(k-1)+3=4k-1$ $4k$ $4k+1$
13	$(13k+13)-4=(1)+(3)+(13(k-1)+18)$ $(13k+13)-4=(3)+(3)+(13k+3)$ $(13k+13)-3=(3)+(13k+7)$ $(13k+13)-1=(13k+12)$	$4(k-1)+3=4k-1$ $4k$ $4k+1$ $4k+2$
18	$(13k+18)-4=(7)+(7)+(13(k-1)+13)$ $(13k+18)-4=(3)+(3)+(13k+8)$ $(13k+18)-4=(5)+(5)+(13k+4)$ $(13k+18)-4=(1)+(1)+(13k+12)$	$4(k-1)+3=4k-1$ $4k$ $4k+1$ $4k+2$

Table 2. Moving to a position with nim-value g, $4k-1 \le g < 4k+\mathcal{G}(j)$.

(a) *If* $k > 0$ *then* $\mathcal{G}(Ak + j) = 4k + \mathcal{G}(j)$ *unless* $j = 2$ *and* k *is of the form* $(q+1) \cdot 2^{m+1} + 2^m + 1$, $q, m \geq 0$.

(b) *For* $q, m \geq 0$, $\mathcal{G}(A((q+1) \cdot 2^{m+1} + 2^m + 1) + 2) = 2^{m+3}q + 2^{m+2} + 2$.

Proof. It is not too difficult to prove that the generic game $\mathbf{0.2(0)}^{2n+1}\mathbf{48}$ has initial nim-sequence 0, $(0, 1)^{n+2}$, 2, $(0, 1)^{n+2}$, $(2, 3)^{n+2}$. The proof is now a repetition of that for **0.2048**, with fewer special cases to be considered. We leave this to the reader. □

References

[Albert et al. 2007] M. Albert, R. J. Nowakowski, and D. Wolfe, *Lessons in play*, A. K. Peters, Natick, MA, 2007.

[Althöfer and Bültermann 1995] I. Althöfer and J. Bültermann, "Superlinear period lengths in some subtraction games", *Theoret. Comput. Sci.* **148**:1 (1995), 111–119.

[Austin 1976] R. B. Austin, *Impartial and partizan games*, Master's thesis, University of Calgary, 1976.

[Berlekamp et al. 2001] E. R. Berlekamp, J. H. Conway, and R. K. Guy, *Winning ways for your mathematical plays, I*, 2nd ed., A. K. Peters, Natick, MA, 2001.

[Berlekamp et al. 2003] E. R. Berlekamp, J. H. Conway, and R. K. Guy, *Winning ways for your mathematical plays, III*, 2nd ed., A. K. Peters, Natick, MA, 2003.

[Caines et al. 1999] I. Caines, C. Gates, R. K. Guy, and R. J. Nowakowski, "Unsolved problems: periods in taking and splitting games", *Amer. Math. Monthly* **106**:4 (1999), 359–361.

[Flammenkamp 2012] A. Flammenkamp, "Sprague-nim-values of some octal games", 2012, http://wwwhomes.uni-bielefeld.de/achim/octal.html.

[Gangolli and Plambeck 1989] A. Gangolli and T. Plambeck, "A note on periodicity in some octal games", *Internat. J. Game Theory* **18**:3 (1989), 311–320.

[Guy and Nowakowski 2002] R. K. Guy and R. J. Nowakowski, "Unsolved problems in combinatorial games", pp. 457–473 in *More games of no chance* (Berkeley, CA, 2000), edited by R. J. Nowakowski, Math. Sci. Res. Inst. Publ. **42**, Cambridge Univ. Press, 2002.

[Horrocks and Nowakowski 2003] D. G. Horrocks and R. J. Nowakowski, "Regularity in the $\mathcal{G}$-sequences of octal games with a pass", *Integers* **3** (2003), G1, 10.

[Howse and Nowakowski 2004] S. Howse and R. J. Nowakowski, "Periodicity and arithmetic-periodicity in hexadecimal games", *Theoret. Comput. Sci.* **313**:3 (2004), 463–472.

jpg@alum.mit.edu *D. E. Shaw Research, 120 West 45th Street, 39th Floor, New York, New York 10036, United States*

rjn@mathstat.dal.ca *Mathematics and Statistics, Dalhousie University, Halifax B3H 3J5, Canada*

Games of No Chance 4
MSRI Publications
Volume **63**, 2015

A handicap strategy for Hex

PHILIP HENDERSON AND RYAN B. HAYWARD

We give an $\lceil \frac{n+1}{6} \rceil$-cell handicap strategy for the game of Hex on an $n \times n$ board: the first player is guaranteed victory if she is allowed to colour $\lceil \frac{n+1}{6} \rceil$ cells on her first move. Our strategy exploits a new kind of inferior Hex cell.

1. Introduction

Hex was invented independently by Piet Hein [1942] and John Nash [1952]. The game is played by two players, Black and White, on a board with hexagonal cells. The players alternate turns, colouring any single uncoloured cell with their colour. The winner is the player who creates a path of her colour connecting her two opposing board sides. See Figure 1.

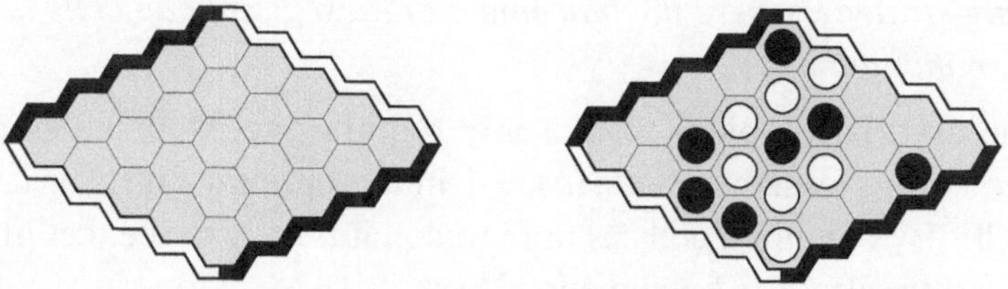

Figure 1. An empty 5×5 Hex board (left) and a completed game position (right).

Hein and Nash observed that Hex cannot end in a draw [Hein 1942; Nash 1952]: exactly one player has a winning path if all cells are coloured [Beck et al. 1969]. Also, an extra coloured cell is never disadvantageous for the player with that colour [Nash 1952]. For $n \times n$ boards, Nash showed the existence of a first-player winning strategy [1952]; however, his proof reveals nothing about the nature of such a strategy. For 8×8 and smaller boards, computer search can find all winning first moves [Hayward et al. 2004; Henderson et al. 2009]. For the 9×9 board, Yang found by human search that moving to the centre cell is a winning first move.

The authors gratefully acknowledge financial support from NSERC, Alberta Ingenuity Fund, and iCORE .

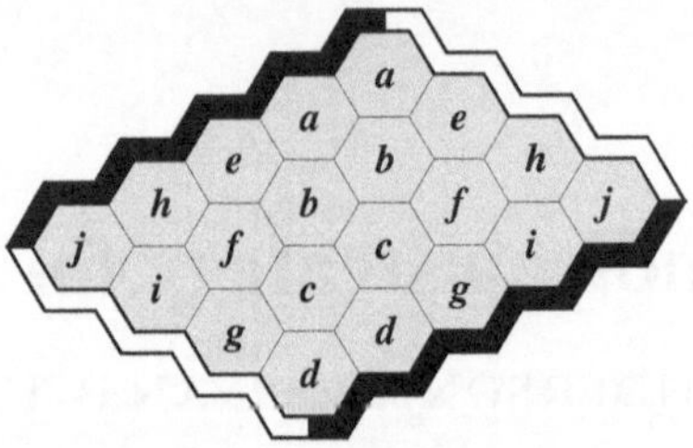

Figure 2. Shannon's pairing strategy on a 4×5 Hex board.

For $m \times n$ boards with $m \neq n$ (hereafter *irregular boards*), the player whose opposing board sides are closer together can win, even as the second player, via a pairing strategy due to Shannon [Gardner 1959]. Figure 2 shows this strategy for the 5×4 Hex board. By contrast, Reisch [1981] showed that solving arbitrary Hex positions is PSPACE-complete.

The problem of efficiently (i.e., in polynomial time) identifying a winning first move on an empty $n \times n$ Hex board has been unsolved for roughly 60 years. As a step towards solving this problem we ask the following:

For the empty $n \times n$ Hex board, let $h(n)$ represent the number of cells the first player needs to colour in order to reach a position with a known and polynomial-time winning strategy, where the placement of these handicap cells is specified. What is the minimum $h(n)$?

Solving the original problem would only require one handicap stone for all values of n, so it is possible that $h(n) = 1$ in the optimal case; for instance, if $P = PSPACE$. By colouring cells as in Figure 3, it is easy to see that $h(n) \leq \lceil \frac{n}{2} \rceil$. Before now, little else was known about $h(n)$.

In this paper, we show that $h(n) \leq \lceil \frac{n+1}{6} \rceil$. Our handicap strategy combines Shannon's $(n-1) \times n$ pairing strategy with the exploitation of a new kind of inferior cell: we colour handicap cells in the second row in a way that allows us

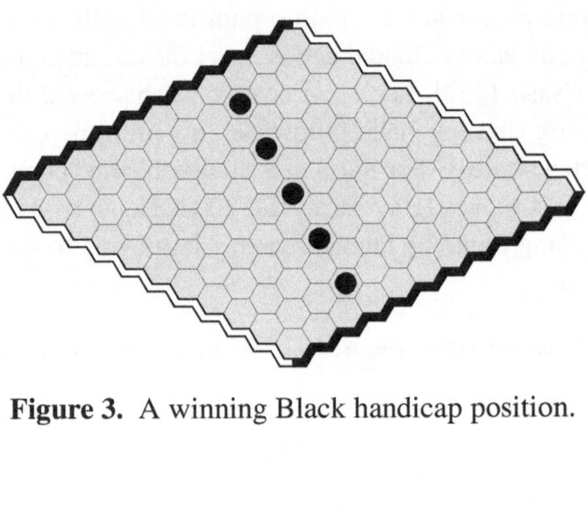

Figure 3. A winning Black handicap position.

to negate any opponent moves in the first row. The resulting $\lceil \frac{n+1}{6} \rceil$-cell handicap strategy is both explicit and efficient.

In Section 2 we review previous Hex inferior cell analysis, in Section 3 we introduce a new type of inferior cell, and in Section 4 we present our handicap strategy.

2. Inferior cell analysis

As discussed in [Berlekamp et al. 2001], inferior moves — such as dominated and reversible moves — can be pruned from consideration or bypassed; similar results have been proven in Hex. Following observations by Beck, Bleicher and Crowe [Beck et al. 1969] and Schensted and Titus [1975], Hayward and van Rijswijck defined a class of provably useless cells: with respect to a particular Hex position, an uncoloured cell c is *dead* if there does not exist some *completion* of the position (namely, a colouring of all uncoloured cells) in which changing c's colour changes the winner in the completion [Hayward and van Rijswijck 2006]. An equivalent definition is that a cell is dead if and only if it is not on any minimal winning path.

Although identifying dead cells is co-NP-complete on planar graphs [Björnsson et al. 2007], some dead cells can be recognized by matching patterns of neighbouring cells. For example, for each pattern in Figure 4 the uncoloured cell is dead [Hayward et al. 2004]. Since there is no completion in which a dead cell's colour matters, it follows that a dead cell can be assigned an arbitrary colour without changing the position's outcome (i.e., winner assuming optimal play) and that any move to a dead cell is useless.

An uncoloured cell c is *vulnerable* for player P if her opponent $\overline{P}$ has a move that makes c dead; this $\overline{P}$-move is c's *killer*. For example, in each pattern in Figure 5 the dotted cell is White-vulnerable to the shaded cell. Moves to vulnerable cells are reversible, but instead of being bypassed they can be pruned entirely from consideration; if a player has a winning strategy in a Hex position,

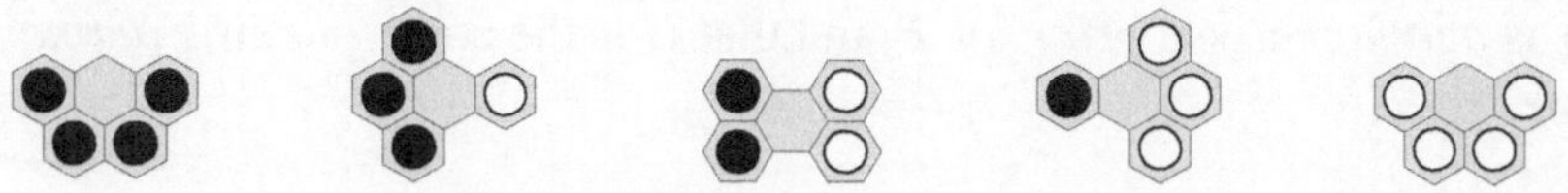

Figure 4. Dead patterns.

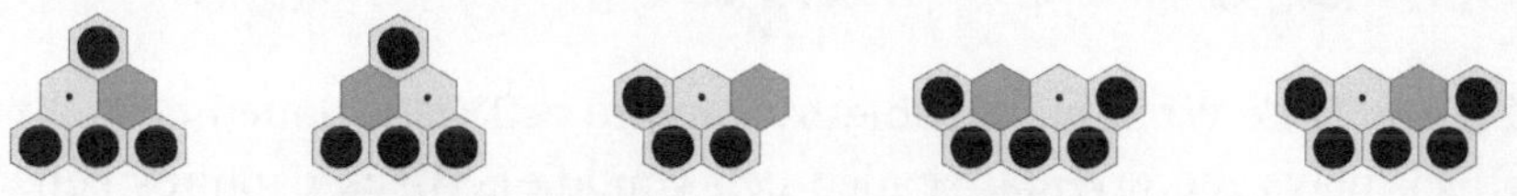

Figure 5. White-vulnerable patterns.

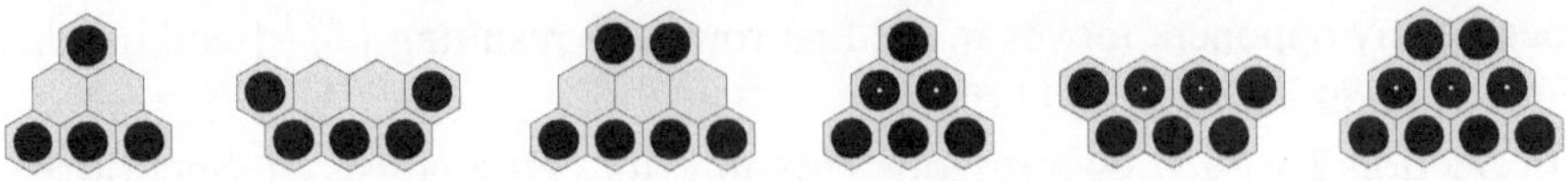

Figure 6. Black-captured patterns (left). Black-colouring a Black-captured set does not alter a position's value (right).

then they have a winning strategy that does not consider any dead or vulnerable cells [Hayward and van Rijswijck 2006].

A set S of uncoloured cells is *captured* by P if she has a second-player strategy to make all cells in S dead or her colour. P-colouring the cells of a P-captured set does not alter a position's outcome: the winning strategy in the original position is simply the combinatorial sum of the captured set strategy with the winning strategy on the filled-in board [Hayward 2003]. Figure 6 illustrates captured patterns relevant to our handicap strategy. See [Björnsson et al. 2007; Hayward 2003; Hayward et al. 2004; Hayward and van Rijswijck 2006] for more on Hex inferior cell analysis.

3. Permanently inferior cells

Before describing a new kind of inferior Hex cell, we first generalize the notion of vulnerability. A cell c is *vulnerable-by-capture* for player P if her opponent $\overline{P}$ has a move that captures a set S which when $\overline{P}$-coloured makes c dead; the $\overline{P}$-move is the *killer*, and the subsequent capturing strategy is the *killing strategy*. See Figure 7. S can be empty, so vulnerable-by-capture generalizes vulnerable. As with vulnerable cells, if player P has a winning strategy then P has a winning strategy that does not consider any P-vulnerable-by-capture cells.

We now introduce a new kind of inferior cell:

Definition 1. Given a Hex position with a set C of uncoloured cells and distinct cells $c_1, c_2 \in C$ such that c_1 is $\overline{P}$-vulnerable to c_2, and each cell in $C \setminus \{c_2\}$ is P-vulnerable-by-capture to a killing strategy using only cells in C, then we say that c_1 is *permanently inferior* for P and that C is the corresponding *permanently inferior region*.

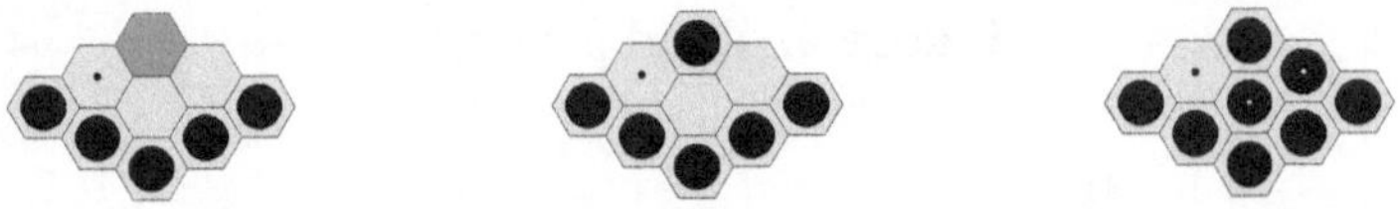

Figure 7. A White-vulnerable-by-capture cell (left, dotted cell). If Black plays the original shaded cell (middle), Black captures cells which in turn kill the original dotted cell (right).

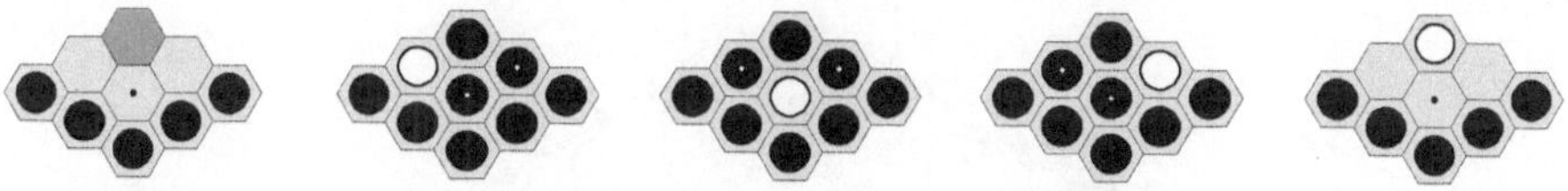

Figure 8. A White-permanently inferior pattern (left): the dotted cell is White-permanently inferior. The three unshaded cells are each White-vulnerable-by-capture to the shaded cell (middle three). Also, White-colouring the shaded cell kills the dotted cell (right).

Figure 9. Two more White-permanently inferior patterns. Black-colouring a (dotted) White-permanently inferior cell does not alter a position's value by Lemma 1.

See Figure 8 for the White-permanently inferior pattern that will be applied in the handicap strategy. Figure 9 shows the only other known White-permanently inferior patterns.

Lemma 1. *Let* c_1 *be a* P*-permanently inferior cell in a Hex position* S*, and let* T *be the position obtained from* S *by* $\overline{P}$*-colouring* c_1*. Then the outcome of* S *equals the outcome of* T*.*

Proof. Extra $\overline{P}$-coloured cells are never disadvantageous for $\overline{P}$, so if $\overline{P}$ wins S then $\overline{P}$ wins T and we are done. Suppose then that P wins S. Thus P has a winning strategy for S in which she never plays a dead or P-vulnerable-by-capture cell. Let c_2 and C be as in Definition 1. Notice that the uncoloured cells of $C \setminus \{c_2\}$ remain P-inferior in any continuation from S in which P has not coloured any cell of $C \setminus \{c_2\}$: they will be either dead or P-vulnerable-by-capture. It follows that if P ever plays in C, then her first such move is to c_2, at which point c_1 is killed unless it is already $\overline{P}$-coloured. Thus P never plays at c_1, and her winning strategy also applies to position T. □

As with captured cells, colouring a permanently inferior cell and finding a winning strategy on the filled-in board allows us to construct a winning strategy for the original position: if the opponent ever plays a vulnerable-by-capture cell in the permanently inferior region, follow the killing strategy; otherwise follow the filled-in board strategy.

4. Handicap strategy

We now describe Black's $\lceil \frac{n+1}{6} \rceil$-handicap strategy for $n \times n$ Hex, $n > 1$. The *handicap cells* are always in the second row, and located in column 4 for $n \geq 6$,

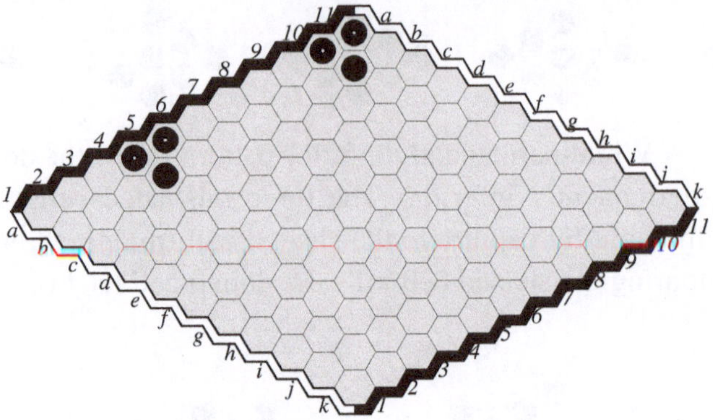

Figure 10. Handicap cell placement: handicap cells are solid black and primary cells are dotted black.

and in columns $n-1-6j$ for each j in $\{0, 1, \ldots, \max(0, \lfloor \frac{n}{6} \rfloor - 1)\}$. The *primary cells* are the first row cells adjacent to a handicap cell. See Figure 10.

Lemma 2. *Let S be an empty $n \times n$ Hex board with the addition of handicap cells as described above, and let T be the board obtained from S by filling-in the entire first row with Black cells. Then the outcome of S equals the outcome of T.*

Proof. The placement of handicap cells is such that two consecutive handicap cells are never separated by more than five uncoloured cells. Thus their respective primary cells are never separated by more than four uncoloured cells. By the captured patterns in Figure 6, it follows that all primary cells can be filled-in without altering the value of S; this completes the fill-in of all first row cells between two handicap cells that are adjacent or separated by a single uncoloured cell. See the two leftmost cases in Figure 11.

If the gap between two primary cells is of size one or two, then the first row cells in that gap are dead or captured. See the third and fourth cases in Figure 11, and the corresponding patterns in Figures 4 and 6. Thus in both these cases we can fill-in all first row cells between the two handicap cells.

If the gap between two primary cells is of size three or four, then the permanently inferior pattern in Figure 8 applies, filling-in the first row cells closest to the primary cells. See the rightmost case in Figure 11. Following this fill-in, the remaining first row gap is reduced to one or two uncoloured cells, which can once again be filled-in via dead or captured patterns as above. Similar arguments hold for any first row cells between a handicap cell and a White side. □

Theorem 1. *Black has a winning $\lceil \frac{n+1}{6} \rceil$-handicap strategy for $n \times n$ Hex.*

Proof. Placing the handicap cells as stated above requires $\lceil \frac{n+1}{6} \rceil$ cells for any $n \times n$ Hex board with $n > 1$. By Lemma 2 this initial handicap position is

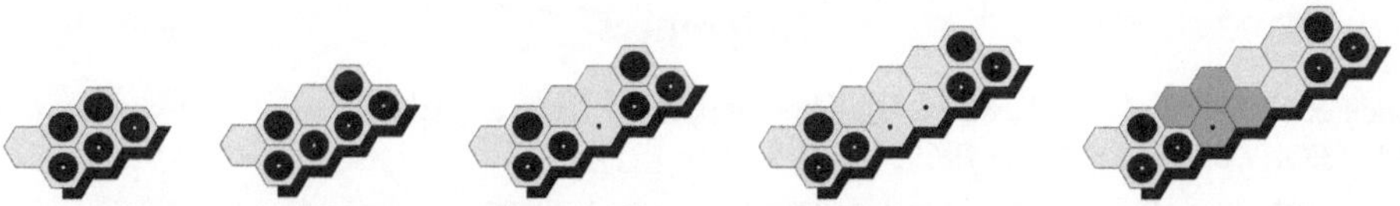

Figure 11. Strategies when the gap between two handicap cells is less than five. Primary cells (dotted black) are filled-in by captured patterns. Gaps of size one or two (unshaded dotted) are dead or captured as a consequence of capturing primary cells. Shaded regions indicate permanently inferior regions.

equivalent to an $(n-1) \times n$ board in Black's favour with $\lceil \frac{n+1}{6} \rceil$ first row Black cells on the board. Since additional Black cells cannot be detrimental to Black, it follows from Shannon's pairing strategy that this handicap position is a second-player Black win. □

Using the observations from Sections 2 and 3, we can construct a winning strategy for the original handicap position by combining the pairing strategy with the captured set and permanently inferior region strategies:

Black's handicap strategy. Colour the $\lceil \frac{n+1}{6} \rceil$ handicap cells. In response to each White move:

(1) if White colours a vulnerable-by-capture cell in a permanently inferior region (on the first two rows), then colour any killer;

(2) otherwise if White colours any Black-captured cell in the first two rows (i.e., primary cells, captured cells between primary cells, or cells in the killing strategy of a vulnerable-by-capture cell), then colour the corresponding reply in the capturing strategy;

(3) otherwise if White colours any cell with an uncoloured pair in the $(n-1) \times n$ Shannon strategy (obtained by ignoring the first row), then colour its pair;

(4) otherwise, colour any random cell.

As far as we know, this is the first two-stone handicap strategy for 11×11 Hex. The late Claude Berge, who was a Hex enthusiast [Berge 1977; 1981], would often give beginners three handicap stones on 11×11 Hex boards, suggesting that he did not expect them to find a winning strategy requiring fewer than four handicap stones. We would like to think that our result would have surprised him.

Acknowledgements

We thank the University of Alberta's Hex and GAMES group members for their feedback on this research, especially Broderick Arneson, Michael Johanson, and Morgan Kan.

References

[Beck et al. 1969] A. Beck, M. N. Bleicher, and D. W. Crowe, *Excursions into mathematics*, pp. 327–339, Worth, New York, 1969.

[Berge 1977] C. Berge, "L'art subtil du Hex", manuscript, 1977.

[Berge 1981] C. Berge, "Some remarks about a Hex problem", pp. 25–27 in *The mathematical Gardner*, edited by D. Klarner, Wadsworth, Belmont, CA, 1981.

[Berlekamp et al. 2001] E. R. Berlekamp, J. H. Conway, and R. K. Guy, *Winning ways for your mathematical plays, I*, 2nd ed., A. K. Peters, Natick, MA, 2001.

[Björnsson et al. 2007] Y. Björnsson, R. Hayward, M. Johanson, and J. van Rijswijck, "Dead cell analysis in Hex and the Shannon game", pp. 45–59 in *Graph theory in Paris*, edited by A. Bondy et al., Trends Math. **6**, Birkhäuser, Basel, 2007.

[Gardner 1959] M. Gardner, *The Scientific American book of mathematical puzzles and diversions*, Simon and Schuster, New York, 1959.

[Hayward 2003] R. B. Hayward, "A Note on Domination in Hex", University of Alberta, 2003, http://cs.ualberta.ca/~hayward/publications.html.

[Hayward and van Rijswijck 2006] R. B. Hayward and J. van Rijswijck, "Hex and combinatorics", *Discrete Math.* **306**:19-20 (2006), 2515–2528.

[Hayward et al. 2004] R. Hayward, Y. Björnsson, M. Johanson, M. Kan, N. Po, and J. van Rijswijck, "Solving 7 × 7 Hex: virtual connections and game-state reduction", pp. 261–278 in *Advances in computer games* (Graz, 2003), edited by H. J. van den Herik et al., IFIP Int. Fed. Inf. Process. **135**, Kluwer Acad. Publ., Boston, MA, 2004.

[Hein 1942] P. Hein, "Vil de laere Polygon?", article in *Politiken* newspaper, 26 December, 1942.

[Henderson et al. 2009] P. Henderson, B. Arneson, and R. Hayward, "Solving 8x8 Hex", pp. 505–510 in *IJCAI-09*, Int. Joint Conferences on Artificial Intelligence, 2009.

[Nash 1952] J. Nash, "Some games and machines for playing them", D-1164, RAND, February 1952.

[Reisch 1981] S. Reisch, "Hex ist PSPACE-vollständig", *Acta Inform.* **15**:2 (1981), 167–191.

[Schensted and Titus 1975] C. Schensted and C. Titus, *Mudcrack Y and poly-Y*, Neo Press, Peaks Island, ME, 1975.

ph@cs.ualberta.ca *Department of Computing Science, University of Alberta, Edmonton T6G 2E8, Canada*

hayward@ualberta.ca *Department of Computing Science, University of Alberta, Edmonton T6G 2E8, Canada*

Games of No Chance 4
MSRI Publications
Volume **63**, 2015

Restrictions of m-Wythoff Nim and p-complementary Beatty sequences

URBAN LARSSON

Let m be a positive integer. The game of *m-Wythoff Nim* (A. S. Fraenkel, 1982) is a well-known extension of Wythoff Nim, also known as Corner the Queen. Its set of P-positions may be represented by a pair of increasing sequences of nonnegative integers. It is well-known that these sequences are so-called *complementary homogeneous Beatty sequences*, that is they satisfy Beatty's theorem. For a positive integer p, we generalize the solution of m-Wythoff Nim to a pair of *p-complementary*— each positive integer occurs exactly p times— homogeneous Beatty sequences $a = (a_n)_{n\in\mathbb{Z}_{\geq 0}}$ and $b = (b_n)_{n\in\mathbb{Z}_{\geq 0}}$, which, for all n, satisfies $b_n - a_n = mn$. By the latter property, we show that a and b are unique among *all* pairs of nondecreasing p-complementary sequences. We prove that such pairs can be partitioned into p pairs of complementary Beatty sequences. Our main results are that $\{\{a_n, b_n\} \mid n \in \mathbb{Z}_{\geq 0}\}$ represents the solution to three new "p-restrictions" of m-Wythoff Nim — of which one has a *blocking maneuver* on the *rook-type* options. C. Kimberling has shown that the solution of Wythoff Nim satisfies the *complementary equation* $x_{x_n} = y_n - 1$. We generalize this formula to a certain "p-complementary equation" satisfied by our pair a and b. We also show that one may obtain our new pair of sequences by three so-called *Minimal EXclusive* algorithms. We conclude with an appendix by Aviezri Fraenkel.

1. Introduction and notation

The combinatorial game of *Wythoff Nim* [Wythoff 1907] is a so-called (2-player) impartial game played on two piles of tokens. As an addition to the rules of the game of Nim [Bouton 1901/02], where the players alternate in removing any finite number of tokens from precisely one of the piles (at most the whole pile), Wythoff Nim also allows removal of the same number of tokens from both piles. The player who removes the last token wins.

This game is better known as Corner the Queen, invented by R. P. Isaacs (1960), because the game can be played on a (large) Chess board with one single

Keywords: Beatty sequence, blocking maneuver, complementary sequences, congruence, impartial game, Muller twist, Wythoff Nim.

Queen. Two players move the Queen alternately but with the restriction that, for each move, the (L^1) distance to the lower left corner, position (0, 0), must decrease. (The Queen must at all times remain on the board.) The player who moves to this *final/terminal* position wins.

In this paper we follow the convention to denote our players with the *next player* (the player who is in turn to move) and the *previous player*. A *P-position* is a position from which the previous player can win (given perfect play). An *N*-position is a position from which the next player can win. Any position is either a *P*-position or an *N*-position. We denote *the solution*, the set of all *P*-positions, of an impartial game G, by $\mathcal{P} = \mathcal{P}(G)$ and the set of all *N*-positions by $\mathcal{N} = \mathcal{N}(G)$. The positive integers are denoted by $\mathbb{Z}_{>0}$ and the nonnegative integers by $\mathbb{Z}_{\geq 0}$. Let $x = (x_i)$ denote an integer sequence over some index set and let $\xi \in \mathbb{Z}$. Then, we let $x_{>\xi}$ denote $(x_i)_{i>\xi}$.

1.1. *Restrictions of m-Wythoff Nim.* Let $m \in \mathbb{Z}_{>0}$. We next turn to a certain m-extension of Wythoff Nim, studied by A. S. Fraenkel [1982] . In the game of *m-Wythoff Nim*, or just mWN (our notation), the Queen's "bishop-type" options are extended so that $(x+i, y+j) \to (x, y)$ is legal if $|i-j| < m$, $i, j, x, y \in \mathbb{Z}_{\geq 0}$, $i > 0$ or $j > 0$. The *rook-type* options are as in Nim. Hence 1-Wythoff Nim is identical to Wythoff Nim.

In this paper we define three new *restrictions* of m-Wythoff Nim — here a rough outline:

- The first has a so-called *blocking maneuver* or *Muller twist* (see also [Holshouser and Reiter 2010; Smith and Stănică 2002] and Section 1.6 of this paper) on the rook-type options, before the next player moves, the previous player may announce some of these options as forbidden.
- The second has a certain congruence restriction on the rook-type options.
- For the third, a rectangular piece is removed from the lower left corner of the game board (including position (0, 0)).

Depending on the particular setup, in addition to (0, 0), the first two games may have a finite number of final positions of the form $\{0, x\}$[1], $x \in \mathbb{Z}_{>0}$. In the third game, there are precisely two final positions $\neq (0, 0)$.

1.2. *Beatty sequences and p-complementarity.* A (general) *Beatty sequence* is a nondecreasing integer sequence of the form

$$(\lfloor n\alpha + \gamma \rfloor), \tag{1}$$

[1]For integers $0 \leq x \leq y$ we use the notation $\{x, y\}$ whenever (x, y) and (y, x) are considered the same.

usually indexed over $\mathbb{Z}$, $\mathbb{Z}_{\geq 0}$ or $\mathbb{Z}_{>0}$, where α is a positive irrational and $\gamma \in \mathbb{R}$. S. Beatty is probably most known for a (re)discovery[2] [1926] of (the statement of) the following theorem: If α and β are positive real numbers such that $\frac{1}{\alpha} + \frac{1}{\beta} = 1$ then $(\lfloor n\alpha \rfloor)_{\mathbb{Z}_{>0}}$ and $(\lfloor n\beta \rfloor)_{\mathbb{Z}_{>0}}$ partition $\mathbb{Z}_{>0}$ if and only if they are Beatty sequences. This was proved in [Beatty et al. 1927] (see also [Fraenkel 1982]). A pair of sequences that partition $\mathbb{Z}_{>0}$ ($\mathbb{Z}_{\geq 0}$, $\mathbb{Z}$) is usually called *complementary* (see [Fraenkel 1969; 1973; Kimberling 2007; 2008]). Let us generalize this notion.

Definition 1. Let $p \in \mathbb{Z}_{>0}$ and Q, R, $S \subset \mathbb{Z}$. Two sequences $(x_i)_{i \in Q}$ and $(y_i)_{i \in R}$ of nonnegative integers are *p-complementary* (on S), if, for each $n \in S$,

$$\#\{i \in Q \mid x_i = n\} + \#\{i \in R \mid y_i = n\} = p.$$

Remark 1. Since the main topics in this paper are three games mostly played on $\mathbb{Z}_{\geq 0} \times \mathbb{Z}_{\geq 0}$, we often find it convenient to use $S = \mathbb{Z}_{\geq 0}$ in Definition 1. Also, for our purposes, it will be convenient use $\mathbb{Z}_{\geq 0}$ or $\mathbb{Z}_{>0}$ as the index sets R and Q. In the Appendix Aviezri Fraenkel provides an alternative approach.

We study the Beatty sequences $a = (a_n)_{n \in \mathbb{Z}_{\geq 0}}$ and $b = (b_n)_{n \in \mathbb{Z}_{\geq 0}}$, where for all $n \in \mathbb{Z}_{\geq 0}$,

$$a_n = a_n^{m,p} = \left\lfloor \frac{n\phi_{(mp)}}{p} \right\rfloor \tag{2}$$

and

$$b_n = b_n^{m,p} = \left\lfloor \frac{n(\phi_{(mp)} + mp)}{p} \right\rfloor, \tag{3}$$

and where

$$\phi_k = \frac{2 - k + \sqrt{k^2 + 4}}{2}. \tag{4}$$

We show that a and $b_{>0}$ are p-complementary.

W. A. Wythoff [1907] proved that the solution of Wythoff Nim is given by $\{\{a_n^{1,1}, b_n^{1,1}\} \mid n \in \mathbb{Z}_{\geq 0}\}$. Then in [Fraenkel 1982] it was shown that the solution of m-Wythoff Nim is

$$\{\{a_n^{m,1}, b_n^{m,1}\} \mid n \in \mathbb{Z}_{\geq 0}\}.$$

1.3. *Recurrence.* Let X be a strict subset of the nonnegative integers. Then the *Minimal EXclusive*, for short mex, of X is defined as usual (see [Conway 2001]):

$$\operatorname{mex} X := \min(\mathbb{Z}_{\geq 0} \setminus X).$$

[2]This theorem was in fact discovered by J. W. Rayleigh, see [Rayleigh 1894; O'Bryant 2003].

For $n \in \mathbb{Z}_{\geq 0}$ put

$$x_n = \operatorname{mex}\big\{\{x_i, y_i \mid i \in \{0, 1, \dots, n-1\}\big\} \quad \text{and} \quad y_n = x_n + mn. \tag{5}$$

With notation as in (5), it was proved in [Fraenkel 1982] that $(x_n) = (a_n^{m,1})$ and $(y_n) = (b_n^{m,1})$. The mex-algorithm in (5) gives an exponential time solution to mWN whereas the Beatty-pair in (2) and (3) give a polynomial time ditto. (For interesting discussions on complexity issues for combinatorial games, see for example [Fraenkel 2004; Fraenkel and Peled 2009].) We show that one may, for general m and p, obtain a and b by three mex-algorithms, which in various ways generalize (5).

It is well-known that the solution of Wythoff Nim satisfies the *complementary equation* (see for example [Kimberling 1995; 2007; 2008])

$$x_{x_n} = y_n - 1.$$

For arbitrary positive integers m and p, we generalize this formula to a "p-complementary equation"

$$x_{\varphi_n} = y_n - 1, \tag{6}$$

where $\varphi_n := py_n - n = \big(x_n + (mp-1)y_n\big)/m$, and show that a solution is given by $x = a$ and $y = b$.

1.4. ***I. G. Connell's restriction of Wythoff Nim.*** In the literature there is another generalization of Wythoff Nim that is of special interest to us. Let $p \in \mathbb{Z}_{>0}$. I. G. Connell [1959a] studied the restriction of Wythoff Nim, where the rook-type options are restricted to jumps of precise multiples of p. This game we call Wythoff modulo-p Nim and denote with $\mathrm{WN}^{(p)}$. Hence Wythoff modulo-1 Nim equals Wythoff Nim.

From [Connell 1959a] one derives that $\mathcal{P}(\mathrm{WN}^{(p)}) = \big\{\{a_n^{1,p}, b_n^{1,p}\} \mid n \in \mathbb{Z}_{\geq 0}\big\}$.

Remark 2. In Connell's presentation, for the proof of the above formulas, he uses p pairs of complementary sequences of integers (in analogy with the discovery of a new formulation of Beatty's theorem in [Skolem 1957]). We have indicated this pattern of P-positions with different shades in Figure 1. In fact, the black positions, starting with (0, 0) are P-positions of 3-Wythoff Nim. More generally, it is immediate by (2) and (3) that, for all p and n, $a_n^{p,1} = a_{pn}^{1,p}$ and $b_n^{p,1} = b_{pn}^{1,p}$.

Remark 3. Fraenkel and I. Borosh [1973] study yet another variation of both m-Wythoff Nim and Wythoff modulo-p Nim which includes a (different from ours) Beatty-type characterization of the P-positions.

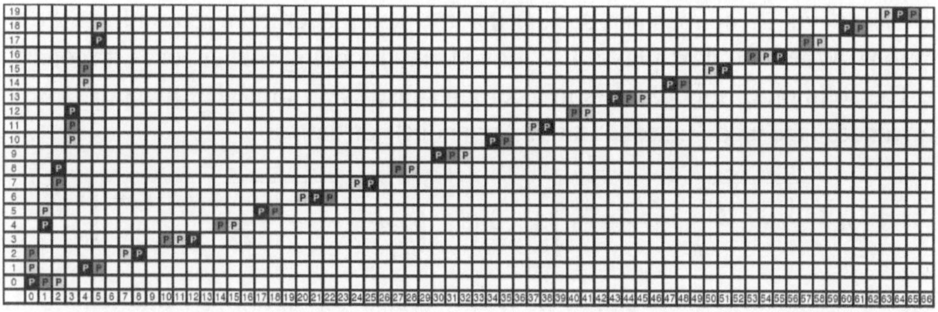

Figure 1. The P-positions of $\text{WN}^{(3)}$ are the positions nearest the origin such that there are precisely three positions in each row and column and one position in each NE-SW-diagonal. The black positions represent the first few P-positions of 3-Wythoff Nim, namely the positions nearest the origin such that there is precisely one position in each row and one position in every third NE-SW diagonal. Positions with lighter shades represent the solutions of games in our third variation.

1.5. ***Exposition.*** In Section 2, given fixed game constants m, $p \in \mathbb{Z}_{>0}$, we define our games, exemplify them and state our main theorem. Roughly: For each of our games, a position is P if and only if it is of the form $\{a_n^{m,p}, b_n^{m,p}\}$, with a and b as in (2) and (3) respectively. In Section 3 we generalize Beatty's theorem to pairs of p-complementary sequences, establish that such sequences can be partitioned into p pairs of complementary Beatty sequences and at last prove some arithmetic properties of a and b — most important of which is that a and $b_{>0}$ are p-complementary and, for all $n \in \mathbb{Z}_{\geq 0}$, satisfy $b_n = a_n + mn$. Then, in Section 4, we prove that the latter two properties make a and b unique among all pairs of nondecreasing sequences. Section 5 is devoted to Equation (6) and three mex-algorithms. In Section 6 we prove our game theory results (stated in Section 2) and in Section 7 a few questions are posed. At last there is an appendix, provided by Aviezri Fraenkel.

Let us, before we move on to our games, give some more background to the so-called *blocking maneuver* in the context of Wythoff Nim.

1.6. ***A bishop-type blocking variation of m-Wythoff Nim.*** Let m, $p \in \mathbb{Z}_{>0}$. In [Hegarty and Larsson 2006] we gave an exponential time solution to a variation of m-Wythoff Nim with a 'bishop-type' blocking maneuver, denoted by p-Blocking m-Wythoff Nim (and with (m, p)-Wythoff Nim in [Larsson 2009]).

The rules are as in m-Wythoff Nim, except that before the next player moves, the previous player is allowed to block off (at most) $p-1$ bishop-type — note, not

m-bishop-type — options and declare that the next player must refrain from these options. When the next player has moved, any blocking maneuver is forgotten.

The solution of this game is in a certain sense "very close" to pairs of Beatty sequences (see also [Larsson 2009, Appendix]) of the form

$$\left(\left\lfloor n\frac{\sqrt{m^2+4p^2}+2p-m}{2p}\right\rfloor\right) \quad \text{and} \quad \left(\left\lfloor n\frac{\sqrt{m^2+4p^2}+2p+m}{2p}\right\rfloor\right).$$

But we explain why there can be no Beatty-type solution to this game for $p > 1$. However, in [Larsson 2009], for the cases $p \mid m$, we give a certain "Beatty-type" characterization. For these kind of questions, see also [Boshernitzan and Fraenkel 1984]. However, a recent discovery, in [Hadad 2008; Fraenkel and Peled 2009], provides a polynomial time algorithm for the solution of (m, p)-Wythoff Nim (for any combination of m and p).

An interesting connection to 4-Blocking 2-Wythoff Nim is presented in [Duchêne and Gravier 2009], where the authors give an explicit bijection of solutions to a variation of Wythoff's original game, where a player's bishop-type move is restricted to jumps by multiples of a predetermined positive integer.

For another variation, we defined in [Larsson 2009] the rules of a so-called move-size dynamic variation of two-pile Nim, (m, p)-Imitation Nim, for which the P-positions, treated as starting positions, are identical to the P-positions of (m, p)-Wythoff Nim.

This discovery of a "dual" game to (m, p)-Wythoff Nim has in its turn motivated the study of "dual" constructions of the rook-type blocking maneuver in this paper.

2. Three games

This section is devoted to defining and exemplifying our game rules and stating our main results. We begin by introducing some (nonstandard) notation whereby we decompose the Queen's moves into *rook-type* and *bishop-type* ditto.

Definition 2. Fix $m, p \in \mathbb{Z}_{>0}$ and an $l \in \{0, 1, \ldots, m\}$.

(i) An (l, p)*-rook* moves as in Nim, but the length of a move must be $ip+j > 0$ positions for some $i \in \mathbb{Z}_{\geq 0}$ and $j \in \{0, 1, \ldots, l-1\}$ (we denote a $(0, p)$-rook by a p-rook and a (p, p)-rook simply by a *rook*);

(ii) A m*-bishop* may move $0 \leq i < m$ rook-type positions and then any number of, say $j \geq 0$, bishop-type positions (a *bishop* moves as in Chess), all in one and the same move, provided $i + j > 0$ and the L^1-distance to $(0, 0)$ decreases.

2.1. *Game definitions.* As is clear from Definition 2 the rook-type options intersect the m-bishop-type options precisely when $m > 1$. For example, $(2,3) \to (1,3)$ is both a 2-bishop-type and a rook-type option. We will make use of this fact when defining the blocking maneuver. Therefore, let us introduce some new terminology.

Definition 3. Fix $m \in \mathbb{Z}_{>0}$. A rook-type option, which is not of the form of the m-bishop as in Definition 2(ii), is a *roob-type*[3] option.

Hence, for $m = 2$, a *roob* may move $(2,3) \to (2,1)$, but not $(2,3) \to (2,2)$ (both are rook-type options). Let us define our games.

Definition 4. Fix $m, p \in \mathbb{Z}_{>0}$.

(i) The game of *m-Wythoff p-Blocking Nim*, or $m\text{WN}^p$, is a restriction of m-Wythoff Nim with a roob-type blocking maneuver. The Queen moves as in m-Wythoff Nim (that is, as the m-bishop or the rook), but with one exception: Before the next player moves, the previous player may *block off* (at most) $p - 1$ of the next player's roob-type options. Any blocked option is unavailable for the next player. As usual, each blocking maneuver is particular to a specific move; that is, when the next player has moved, any blocking maneuver is forgotten and has no further impact on the game. (For $p = 1$ this game equals m-Wythoff Nim.)

(ii) Fix an integer $0 \le l < p$. In the game of *m-Wythoff Modulo-p l-Nim*, or $m\text{WN}^{(l,p)}$, the Queen moves as the m-bishop or the (l, p)-rook. For $l = 0$ we denote this game by *m-Wythoff Modulo-p Nim* or $m\text{WN}^{(p)}$. (In case $l = p$ the game is simply m-Wythoff Nim.)

(iiia) Fix an integer $0 \le l < p$. In the game of *l-Shifted $m \times p$-Wythoff Nim*, or $m \times p\text{WN}_l$, the Queen moves as in (mp)-Wythoff Nim (that is, as the (mp)-bishop or the rook), except that, if $l > 0$, it is not allowed to move to a position of the form (i, j), where $0 \le i < ml$ and $0 \le j < m(p - l)$. Hence, for this case, the terminal positions are $(ml, 0)$ and $(0, m(p - l))$.[4] On the other hand $m \times p\text{WN}_0$ is identical to (mp)-Wythoff Nim.

(iiib) The game of $m \times p$-Wythoff Nim, $m \times p\text{WN}$: Before the first player moves, the second player may decide the parameter l as in (iiia). Once the parameter

[3] Think of "roob" as "ROOk minus m-Bishop", or maybe "ROOk Blocking".

[4] One might want to think of the game board as if a rectangle with circumference $2mp$ is cut out from its lower left corner. By symmetry, there are $\lceil p/2 \rceil$ rectangle shapes but (given a starting position) p distinct game boards; see also (3b). Of course, if cutting out a corner of the nice game board does not appeal to the players, one might just as well define all positions inside the rectangle as N. Notice the close relation of these games to the misère version of m-Wythoff Nim studied in [Fraenkel 1984].

l is fixed, it remains the same until the game has terminated, so that for the remainder of the game, the rules are as in $m \times p\mathrm{WN}_l$.

2.2. *Examples.* Let us illustrate some of our games, where our players are *Alice* and *Bob* — Alice makes the first move (and Bob makes the first blocking maneuver in case the game has a Muller twist).

Example 1. Suppose the starting position is (0, 2) and the game is $2\mathrm{WN}^2$. Then the only bishop-type move is $(0, 2) \to (0, 1)$. There is precisely one roob-type option, namely (0, 0). Since this is a terminal position Bob will block it off from Alice's options, so that Alice has to move to (0, 1). The move $(0, 1) \to (0, 0)$ cannot be blocked off for the same reason, so Bob wins. If $y \geq 3$ there is always a move $(0, y) \to (0, x)$, where $x = 0$ or 2. This is because the previous player may block off at most one option. Altogether, this gives that $\{0, y\}$ is P if and only if $y = 0$ or 2.

Example 2. Suppose the starting position is (0, 2) and the game is $2\mathrm{WN}^{(2)}$. Alice can move to (0, 0), since $0 \equiv 2 \pmod 2$, so (0, 2) is N. On the other hand, the position (0, 3) is P since the only options are (0, 2) and (0, 1). (The latter is N since the 2-bishop can move $(0, 1) \to (0, 0)$.)

Example 3. Suppose the starting position is (0, 2) and the game is $2\mathrm{WN}^{(2,4)}$. Alice cannot move to a P-position since $2 - 0 \not\equiv 0, 1 \pmod 4$ and the 2-bishop's move is restricted to (0, 1), which is N. Hence (0, 2) is P. More generally, $(0, y)$ is N for all $y \geq 3$, since $(0, y) \to (0, 0)$ is legal if $0 < y \equiv 0, 1 \pmod 4$ and $(0, y) \to (0, 2)$ is legal if $2 < y \equiv 2, 3 \pmod 4$.

Example 4. Suppose the starting position is (0, 4) and the game is $2\mathrm{WN}^3$. Then the only bishop-type move is $(0, 4) \to (0, 3)$, so that the roob-type options are (0, 0), (0, 1), (0, 2). Bob may block off 2 of these positions, say (0, 0), (0, 2). Then if Alice moves to (0, 1) she will loose (since she may not block off (0,0)), so suppose rather that she moves to (0, 3). Than she may not block off (0, 2) so Bob moves $(0, 3) \to (0, 2)$ and blocks off (0, 0). Hence (0, 4) is a P-position.

Example 5. Suppose the starting position is (0, 4) and the game is $2\mathrm{WN}^{(3)}$. Alice cannot move to (0, 0) or (0, 2). But $(0, 1) \to (0, 0)$ is a 2-bishop-type option and $(0, 3) \to (0, 0)$ is a 3-rook-type option. This shows that (0, 4) is a P-position.

Notice that, in comparison to Examples 4 and 5, the P-positions in the Examples 1 and 2 are distinct in spite the identical game constants ($m = p = 2$). On the other hand, the P-positions in Examples 1 and 3 coincide.

Example 6. Suppose the game is $2 \times 3\mathrm{WN}_1$, then the terminal positions are (2, 0) and (0, 4). On the other hand, for the game $2 \times 3\mathrm{WN}_2$, the positions (0, 2)

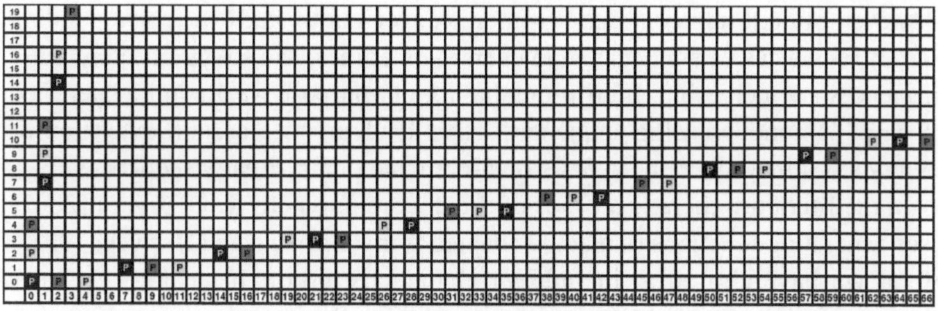

Figure 2. P-positions of $2\text{WN}^{(3)}$, 2WN^3, $2\text{WN}^{2,6}$ and $2 \times 3\text{WN}$— the positions nearest the origin such that there are precisely three positions in each row and column and one position in every second NE-SW-diagonal. The palest colored squares represent P-positions of $2 \times 3\text{WN}_1$. They are of the form (a_{3n+1}, b_{3n+1}) or (b_{3n+2}, a_{3n+2}). The darkest squares, $\left(\{a_{3i}^{2,3}, b_{3i}^{2,3}\}\right)$, represent the solution of 6WN.

and (4, 0) are terminal. Suppose now that the starting position of $2 \times 3\text{WN}_2$ is (1, 9). Then Alice wins by moving to (0, 4). If the starting position is the same, but the game is $2 \times 3\text{WN}_1$, then Alice cannot move to (0, 2) and hence Bob wins. Similarly, if the starting position of $2 \times 3\text{WN}_0$ is (1, 7) Alice may not move to (0, 0) and hence Bob wins.

2.3. ***Game theory results.*** We may now state our main results. We prove them in Section 6, since our proofs depend on some arithmetic results presented in Section 3, 4 and 5.

Theorem 2.1 (main theorem). *Fix* $m, p \in \mathbb{Z}_{>0}$ *and let* a *and* b *be as in* (2) *and* (3). *Then*

(i) $\mathcal{P}(m\text{WN}^p) = \{\{a_i, b_i\} \mid i \in \mathbb{Z}_{\geq 0}\}$;

(ii) (a) $\mathcal{P}(m\text{WN}^{(p)}) = \{\{a_i, b_i\} \mid i \in \mathbb{Z}_{\geq 0}\}$ *if and only if* $\gcd(m, p) = 1$;
(b) $\mathcal{P}(m\text{WN}^{(m,mp)}) = \{\{a_i, b_i\} \mid i \in \mathbb{Z}_{\geq 0}\}$;

(iii) (a) $\mathcal{P}(m \times p\text{WN}_l) = \{(a_{ip+l}, b_{ip+l}) \mid i \in \mathbb{Z}_{\geq 0}\} \cup \{(b_{ip-l}, a_{ip-l}) \mid i \in \mathbb{Z}_{>0}\}$;
(b) $\mathcal{P}(m \times p\text{WN}) = \{\{a_i, b_i\} \mid i \in \mathbb{Z}_{\geq 0}\}$.

By this result it is clear that, in terms of game complexity, the solution of each of our games is polynomial.

3. More on p-complementary Beatty sequences

As we have seen, it is customary to represent the solution of "a removal game on two heaps of tokens" as a sequence of (ordered) pairs of nonnegative integers.

However, often it turns out that it is more convenient to study the corresponding pair of sequences of nonnegative integers. Sometimes, as in Wythoff Nim, these sequences are increasing. It turns out that for our purpose we are more interested in pairs of nondecreasing sequences. This leads us to a certain extension of Beatty's original theorem, to pairs of p-complementary Beatty sequences.

In the literature there is a proof of this theorem, where K. O'Bryant [2002] uses generating functions (a method adapted from [Borwein and Borwein 1993]). Here, we have chosen to include an elementary proof, in analogy to ideas presented in [Beatty et al. 1927; Fraenkel 1982]. (See also the Appendix.)

Theorem 3.1 (O'Bryant). *Let $p \in \mathbb{Z}_{>0}$ and let $0 < \alpha < \beta$ be real numbers such that*

$$\frac{1}{\alpha} + \frac{1}{\beta} = p. \tag{7}$$

Then $(\lfloor i\alpha \rfloor)_{i \in \mathbb{Z}_{\geq 0}}$ and $(\lfloor i\beta \rfloor)_{i \in \mathbb{Z}_{>0}}$ are p-complementary on $\mathbb{Z}_{\geq 0}$ if and only if α, β are irrational.

Proof. It suffices to establish that exactly p members of the set

$$S = \{0, \alpha, \beta, 2\alpha, 2\beta, \ldots\}$$

is in the interval $[n, n+1)$ for each $n \in \mathbb{Z}_{\geq 0}$. But for any fixed n we have

$$\begin{aligned} \#(S \cap [0, n]) &= \#(\{0, \alpha, 2\alpha, \ldots\} \cap [0, n]) + \#(\{\beta, 2\beta, \ldots\} \cap [1, n]) \\ &= \lfloor n/\alpha \rfloor + 1 + \lfloor n/\beta \rfloor. \end{aligned}$$

But α and β are irrational if and only if, for all n,

$$\begin{aligned} np - 1 = n/\alpha + n/\beta - 1 &< \lfloor n/\alpha \rfloor + 1 + \lfloor n/\beta \rfloor \\ &< n/\alpha + n/\beta + 1 = np + 1. \end{aligned}$$

This gives $\lfloor n/\alpha \rfloor + 1 + \lfloor n/\beta \rfloor = np$. Going from n to $n+1$ gives the result. □

The following result establishes that a pair of p-complementary homogeneous Beatty sequences can always be partitioned into p complementary pairs of Beatty sequences. For the proof we use a generalization of Beatty's theorem in [Skolem 1957; Fraenkel 1969; O'Bryant 2003].

Proposition 3.2. *Let $2 \leq p \in \mathbb{Z}$. Suppose that $(x_i) = (\lfloor \alpha i \rfloor)_{i \in \mathbb{Z}_{\geq 0}}$ and $(y_i) = (\lfloor \beta i \rfloor)_{i \in \mathbb{Z}_{>0}}$ are p-complementary homogeneous Beatty sequences with $0 < \alpha < \beta$. Then, for any fixed integer $0 \leq l < p$, the pair of sequences*

$$(x_{pi+l})_{i \in \mathbb{Z}_{\geq 0}} \text{ and } (y_{pi-l})_{i \in \mathbb{Z}_{>0}}$$

is complementary on $\mathbb{Z}_{\geq 0}$.

Proof. Since (x_i) and (y_i) are nondecreasing and p-complementary, we get that both (x_{pi+l}) and (y_{pi-l}) are increasing. Then, by $x_l = \min\{x_{pi+l}\}$ and $y_{p-l} = \min\{y_{pi-l}\}$, again, p-complementarity gives that

$$\max\{x_l, y_{p-l}\} > \min\{x_l, y_{p-l}\} = 0$$

and so, we may conclude that the integer 0 occurs precisely once together in (x_{pi+l}) and (y_{pi-l}). Let us adapt to the terminology in [O'Bryant 2003].

Case $x_l = 0$: Then $l < 1/\alpha$. For all $n \in \mathbb{Z}_{>0}$, we set

$$\left\lfloor \frac{n-\gamma'}{\gamma} \right\rfloor = x_{pn+l} \quad \text{and} \quad \left\lfloor \frac{n-\eta'}{\eta} \right\rfloor = y_{pn-l}.$$

This gives $\gamma = 1/p\alpha,\ \gamma' = -l/p,\ \eta = 1/p\beta,\ \eta' = l/p$. Then according to Fraenkel's partition theorem in [O'Bryant 2003, p. 5], since α is irrational, (x_{pn+l}) and (y_{pn-l}) are complementary on $\mathbb{Z}_{>0}$ if and only if

(i) $0 < \gamma < 1$,

(ii) $\gamma + \eta = 1$,

(iii) $0 \le \gamma + \gamma' \le 1$,

(iv) $\gamma' + \eta' = 0$ and $k\gamma + \gamma' \notin \mathbb{Z}$ for $2 \le k \in \mathbb{Z}$.

These four items are easy to verify.

(i) This follows since (7) together with $0 < \alpha < \beta$ is equivalent to $1/p < \alpha < 2/p$.

(ii) This is immediate by (7).

(iii) We have that $0 \le \gamma + \gamma' \le 1$ if and only if $0 \le 1/p\alpha - l/p \le 1$ if and only if $l \le 1/\alpha \le p + l$.

(iv) We have that $\eta' + \gamma' = l/p - l/p = 0$. Since γ is irrational and γ' is rational the latter part holds as well.

Case $y_{p-l} = 0$: Then $p - l < 1/\beta = p - 1/\alpha$, so that $l > 1/\alpha$. For all $n \in \mathbb{Z}_{>0}$, we set

$$\left\lfloor \frac{n-\gamma'}{\gamma} \right\rfloor = x_{p(n-1)+l} \quad \text{and} \quad \left\lfloor \frac{n-\eta'}{\eta} \right\rfloor = y_{p(n+1)-l}.$$

This gives

$$\gamma = \frac{1}{p\alpha}, \quad \gamma' = \frac{p-l}{p}, \quad \eta = \frac{1}{p\beta}, \quad \eta' = \frac{l-p}{p}.$$

Then items (i), (ii) and (iv) are treated in analogy with the first case. For (iii), since $\alpha l > 1$, we get

$$0 \le \gamma + \gamma' = \frac{1}{p\alpha} + \frac{p-l}{p} = 1 + \frac{1-\alpha l}{\alpha p} < 1. \qquad \square$$

We will now focus on the properties of the sequences a and b. The next result is central to the rest of the paper.

Lemma 3.3. *Fix $m, p \in \mathbb{Z}_{>0}$ and let a and b be as in* (2) *and* (3) *respectively. Then for each $n \in \mathbb{Z}_{\ge 0}$ we have that*

(i) *a and b are p-complementary;*

(ii) *$b_n - a_n = mn$;*

(iii) *if $p = 1$, then*
   (a) *$a_{n+1} - a_n = 1$ and $b_{n+1} - b_n = m + 1$, or*
   (b) *$a_{n+1} - a_n = 2$ and $b_{n+1} - b_n = m + 2$;*

(iv) *if $p > 1$, then*
   (a) *$a_{n+1} - a_n = 0$ and $b_{n+1} - b_n = m$, or*
   (b) *$a_{n+1} - a_n = 1$ and $b_{n+1} - b_n = m + 1$.*

Proof. Since ϕ_x is irrational and $1/\phi_x + 1/(\phi_x + x) = 1$, case (i) is immediate from Theorem 3.1.

For case (ii) put $\nu = \nu(m, p) = \phi_{mp}/p + m/2$ and observe that

$$b_n - a_n = \left\lfloor n\left(\nu + \frac{m}{2}\right)\right\rfloor - \left\lfloor n\left(\nu - \frac{m}{2}\right)\right\rfloor.$$

If mn is even, we are done, so suppose that $mn - 1 = 2k$, $k \in \mathbb{Z}_{\ge 0}$. Then

$$b_n - a_n = \left\lfloor n\nu + \frac{1}{2}\right\rfloor - \left\lfloor n\nu - \frac{1}{2}\right\rfloor + 2k = 1 + 2k = mn.$$

For case (iii), by [Fraenkel 1982], we are done. In case $p > 1$, by the triangle inequality, we get

$$0 < \frac{\phi_{mp}}{p} = \frac{1}{p} - \frac{m}{2} + \sqrt{\frac{m^2}{4} + \frac{1}{p^2}} < \frac{1}{p} + \frac{1}{p} \le 1,$$

the last step since $p \ge 2$. Thus we may estimate

$$a_{n+1} - a_n = \left\lfloor \frac{(n+1)\phi_{mp}}{p} \right\rfloor - \left\lfloor \frac{n\phi_{mp}}{p} \right\rfloor \in \{0, 1\}.$$

Then by (ii) we have

$$b_{n+1} - b_n = a_{n+1} + m(n+1) - a_n - mn = a_{n+1} - a_n + m,$$

so that (iv) holds. $\square$

4. A unique pair of p-complementary Beatty sequences

For fixed p and m we now present a certain uniqueness property for our pair of p-complementary Beatty sequences (in case $p = 1$ see also [Hegarty and Larsson 2006] for extensive generalizations).

Theorem 4.1. *Fix $m, p \in \mathbb{Z}_{>0}$. Suppose $x = (x_i)_{i \in \mathbb{Z}_{\geq 0}}$ and $y = (y_i)_{i \in \mathbb{Z}_{\geq 0}}$ are nondecreasing sequences of nonnegative integers. Then the following two items are equivalent,*

(i) *x and $y_{>0}$ are p-complementary and, for all n, $y_n - x_n = mn$;*

(ii) *for all n, $x_n = a_n^{m,p}$ and $y_n = b_n^{m,p}$.*

Proof. By Lemma 3.3 it is clear that (ii) implies (i). Hence, it suffices to prove the other direction.

It is given that $x_0 = y_0 = a_0 = b_0 = 0$. Since x is nondecreasing the condition $y_n - x_n = mn$ implies that y is increasing. Suppose that Lemma 3.3(iv) holds for a fixed $n \geq 0$, but with a exchanged for x and b exchanged for y. Then, since x and $y_{>0}$ are p-complementary and $y_{n+1} > x_{n+1}$, we must have $x_{n+1} - x_n = 0$ if

$$\#\{i \mid x_i = x_n \text{ or } y_{i+1} = x_n, 0 \leq i \leq n\} < p,$$

and $x_{n+1} - x_n = 1$ if

$$\#\{i \mid x_i = x_n \text{ or } y_{i+1} = x_n, 0 \leq i \leq n\} = p.$$

But, by Lemma 3.3, this also holds for the sequence (a_i). In conclusion, $y_{n+1} = x_{n+1} + m(n+1) = a_{n+1} + m(n+1) = b_{n+1}$ gives the result. □

5. Recurrence results

In Proposition 5.2 in this section, we generalize the mex-algorithm in (5). Each game family in Definition 4 has motivated the study of a "corresponding" mex-algorithm. The numberings in Definition 4 and Proposition 5.2 respectively are in accordance with this correspondence.

But first we explain why the sequences a and b satisfy the "p-complementary equation" in (6).

Theorem 5.1. *Fix $m, p \in \mathbb{Z}_{>0}$ and let a and b be as in* (2) *and* (3). *For each $n \in \mathbb{Z}_{\geq 0}$, define*

$$\varphi_n = \varphi_n(m, p) := \frac{a_n + (mp - 1)b_n}{m}.$$

Then, for each $n \in \mathbb{Z}_{>0}$, φ_n is the greatest integer such that

$$b_n - 1 = a_{\varphi_n}. \tag{8}$$

Proof. Notice that, for all n,

$$\varphi_n = \frac{a_n + (mp-1)b_n}{m} = \frac{mpb_n - mn}{m} = pb_n - n, \tag{9}$$

so that

$$\varphi_{n+1} - \varphi_n = pb_{n+1} - (n+1) - (pb_n - n) = p(b_{n+1} - b_n) - 1. \tag{10}$$

The proof is by induction. For the base case, notice that $b_1 = m$, $a_1 = 0$ and $\varphi_1 = (mp-1)$. The only representative from b in the interval $[0, p-1]$ is $b_0 = 0$ (which we by definition do not count). Hence, by $a_0 = 0$ and p-complementarity, we get that

$$a_{\varphi_1} = a_{mp-1} = m - 1 = b_1 - 1$$

and

$$a_{\varphi_1+1} = a_{mp} = m = b_1.$$

Suppose that (8) holds for all $i \le n$. Then we need to show that $b_{n+1} - 1 = a_{\varphi_{n+1}}$ and $b_{n+1} = a_{\varphi_{n+1}+1}$.

If $a_{\varphi_{n+1}} - a_{\varphi_n} = b_{n+1} - b_n$, by $b_n - 1 = a_{\varphi_n}$ and $b_n = a_{\varphi_n+1}$, we are done, so assume that either

(A) $a_{\varphi_{n+1}} - a_{\varphi_n} < b_{n+1} - b_n$, or

(B) $a_{\varphi_{n+1}} - a_{\varphi_n} > b_{n+1} - b_n$.

Again, by p-complementarity, the total number of elements from a and b in the interval

$$I_n := (a_{\varphi_n}, a_{\varphi_{n+1}}] = (a_{\varphi_n}, a_{\varphi_n + p(b_{n+1}-b_n)-1)}]$$

is $R_n := p(a_{\varphi_{n+1}} - a_{\varphi_n})$, and where the equality is by (10). By assumption, $a_{\varphi_n+1} \in I_n$ so that we have at least $p(b_{n+1} - b_n) - 1$ representatives from a in I_n. But also $b_n = a_{\varphi_n} + 1 \in I_n$ so that altogether we have at least $p(b_{n+1} - b_n)$ representatives in I_n. Hence

$$p(b_{n+1} - b_n) \le R_n = p(a_{\varphi_{n+1}} - a_{\varphi_n}),$$

which rules out case (A).

Notice that case (B) implies that b_{n+1} lies in I_n so that $a_{\varphi_n+1} = b_n < b_{n+1} \le a_{\varphi_{n+1}}$. Since both b_n and b_{n+1} lie in I_n, the total number of representatives in I_n is

$$p(a_{\varphi_{n+1}} - a_{\varphi_n}) \le 2 + \varphi_{n+1} - \varphi_n = p(b_{n+1} - b_n) + 1. \tag{11}$$

In case $p > 1$, since a and b are integer sequences, we are done, so suppose $p = 1$. Then, in fact, by complementarity, we must have $a_{\varphi_{n+1}} < b_n < b_{n+1} < a_{\varphi_{n+1}}$, contradicting (11). □

Remark 4. For arbitrary $m > 0$ and $p = 1$ it is well-known that a and b solve $x_{y_n} = x_n + y_n$. This complementary equation is studied in for example [Connell 1959b; Fraenkel and Kimberling 1994; Kimberling 2007]. However, we have not been able to find any references for the complementary equation $y_n - 1 = x_{y_n - n}$. By (9), for the cases $p = 1$, this equation is also resolved by a and b.

A *multiset* (or a sequence) X may be represented as (another) sequence of nonnegative integers $\xi = (\xi^i)_{i \in \mathbb{Z}_{\geq 0}}$, where, for each $i \in \mathbb{Z}_{\geq 0}$, $\xi^i = \xi^i(X)$ counts the number of occurrences of i in X. For a positive integer p, let $\operatorname{mex}^p \xi$ denote the least nonnegative integer $i \in X$ such that $\xi^i < p$.

Proposition 5.2. *Let $m, p \in \mathbb{Z}_{>0}$. Then the definitions of the sequences x and y in* (i), (ii) *and* (iii) *are equivalent. In fact, for each $n \in \mathbb{Z}_{\geq 0}$, we have that $x_n = a_n^{m,p}$ and $y_n = b_n^{m,p}$.*

(i) *For $n \geq 0$,*

$$x_n = \operatorname{mex}^p \xi_n,$$

where ξ_n is the multiset, where for each $i \in \mathbb{Z}_{\geq 0}$,

$$\xi_n^i = \#\{j \mid i = x_j \text{ or } i = y_j, 0 \leq j < n\},$$
$$y_n = x_n + mn.$$

(ii) *For $n \geq 0$,*

$$x_n = \operatorname{mex}\{\nu_i^n, \mu_i^n \mid 0 \leq i < n\}, \text{ where}$$
$$\nu_i^n = x_i \text{ if } n \equiv i \pmod{p}, \text{ else } \nu_i^n = \infty,$$
$$\mu_i^n = y_i \text{ if } n \equiv -i \pmod{p}, \text{ else } \mu_i^n = \infty,$$
$$y_n = x_n + mn.$$

(iii) *For $n \geq 0$,*

$$x_{pn} = \operatorname{mex}\{x_{pi}, y_{pi} \mid 0 \leq i < n\},$$
$$y_{pn} = x_{pn} + mpn,$$

and for each integer $0 < l < p$,

$$x_{pn+l} = \operatorname{mex}\{x_{pi+l}, y_{p(i+1)-l} \mid 0 \leq i < n\},$$
$$y_{pn+l} = x_{pn+l} + m(pn + l).$$

Proof. For $p = 1$ it is a straightforward task to check that each recurrence is equivalent to (5). Hence, let $p > 1$. Observe that in (i), by definition, x and y are nondecreasing, p-complementary and, for all n,

$$y_n = x_n + mn. \tag{12}$$

Hence, for this case, Theorem 4.1 gives the result.

Let us now study the definitions of x and y in (ii). For $z \in \mathbb{Z}$, let $\overline{z}$ denote the congruence class of z modulo p. Here, it is not immediately clear that the sequences are nondecreasing. Neither is it obvious that they are p-complementary. But, at least we have that, for each $n \in \mathbb{Z}_{\geq 0}$, (12) holds.

Hence, if (ii) fails (by $a_0 = x_0$) there has to exist a least index $n' \in \mathbb{Z}_{>0}$ such that $a_{n'} \neq x_{n'}$. But notice that $0 \leq n < p$ implies $\nu_i^n = \mu_i^n = \infty$, for all $0 \leq i < n$, which in its turn implies $a_n = x_n = 0$. This gives $n' \geq p$. We have two cases to consider:

(a) $r := x_{n'} < a_{n'}$. By Theorem 5.1 there are two cases to consider.

Case 1: There is an $i \geq 0$ such that $\varphi(i) + p - 1 < n'$ and

$$y_i = x_{\varphi(i)+1} = x_{\varphi(i)+2} = \cdots = x_{\varphi(i)+p-1} = r.$$

But then, by

$$\{\,\overline{-i},\ \overline{-i+1},\ \ldots,\ \overline{-i+p-1}\,\} = \{\,\overline{0},\ \overline{1},\ \ldots,\ \overline{p-1}\,\} \tag{13}$$

and

$$\varphi_n = pb_n - n \equiv -n \pmod{p}, \tag{14}$$

there is a $j \in \{i, \varphi(i)+1, \ldots, \varphi(i)+p-1\}$ such that either $n' \equiv j \pmod{p}$ and $j \in \{\varphi(i)+1, \ldots \varphi(i)+p-1\}$ which implies $\nu_j^{n'} = r$, or $n' \equiv -j \pmod{p}$ and $j = i$ which implies $\mu_j^{n'} = r$. In either case the choice of $x_{n'} = r$ contradicts the definition of mex.

Case 2: There is an $i \geq 0$ such that $i + p - 1 < n'$ and

$$r = x_i = x_{i+1} = x_{i+2} = \cdots = x_{i+p-1}.$$

This case is similar but simpler, since for this case we rather use that

$$\{\,\overline{i},\ \overline{i+1},\ \ldots,\ \overline{i+p-1}\,\} = \{\,\overline{0},\ \overline{1},\ \ldots,\ \overline{p-1}\,\}. \tag{15}$$

(b) $r := a_{n'} < x_{n'}$: Then our mex-algorithm has refused r as the choice for $x_{n'}$. But then there must be an index $0 \leq j < n'$ such that either $\nu_j^{n'} = r$ or $\mu_j^{n'} = r$. Hence, we get to consider two cases.

Case 1: $\overline{j} = \overline{n'}$ and $r = x_j$. On the one hand, there is a $k \in \mathbb{Z}_{>0}$ such that $kp + j = n'$ On the other hand, there is a greatest $k' \in \mathbb{Z}_{>0}$ such that $a_{n'-k'} = a_{n'-k'+1} = \cdots = a_{n'}$ and by p-complementarity $0 \leq k' < p$. But then, since $n' - k' > n' - kp = j$, we get $a_j < r = x_j$, which contradicts the minimality of n'.

Case 2: $\overline{-j} = \overline{n'}$ and $r = y_j$. Then, by Theorem 5.1, $\varphi_j + 1$ is the least index such that $a_{\varphi_j+1} = a_{n'}$. Then, by minimality of n', $a_j = x_j$ gives $b_j = y_j$ so that $a_{n'} = b_j$. For this case, p-complementarity gives $n' - (\varphi_j + 1) + 1 \le p - 1$. Then $0 < k' := n' - \varphi_j < p$ and so

$$\overline{-j + k'} = \overline{\varphi(j) + k'} = \overline{n'} = \overline{-j},$$

which is nonsense.

For (iii), suppose that there is a least index $n' \ge p$ such that $a_{n'} \ne x_{n'}$. (The case $n' < p$ may be ruled out in analogy with (ii).) Then, there exist unique integers, $0 < t$ and $0 \le l < p$, such that $tp + l = n'$.

Suppose that $r := a_{n'} < x_{n'}$. Since the mex-algorithm did not choose $x_{n'} = r$, there must be an index $0 \le t' < t$ such that either $x_{t'p+l} = r$ or $y_{(t'+1)p-l} = r$. But then, by assumption, either $a_{t'p+l} = x_{t'p+l} = a_{n'}$ or $b_{(t'+1)p-l} = y_{(t'+1)p-l} = a_{n'}$. But, by Proposition 3.2 a and b are complementary, so either case is ridiculous.

Hence, assume $r := a_{n'} > x_{n'}$. Then again, by Proposition 3.2, there is an index $0 \le t' < t$ such that either $a_{t'p+l} = x_{n'}$ or $b_{(t'+1)p-l} = x_{n'}$. But, again, by minimality of n', this contradicts the mex-algorithm's choice of $x_{n'} < a_{n'}$. □

6. The games' final section

The proof of Theorem 2.1 is based on standard impartial games arguments — with occasional references to facts already established in for example Proposition 5.2. Notice that the blocking variation in (i) is "simpler" (and more elegant?) than the other, namely it depends only on results from Section 3.

Proof of Theorem 2.1. For $p = 1$, the games have identical rules. This case has been established in [Fraenkel 1982]. The case $m = 1$ has been studied in [Connell 1959a] for games of form (ii). (and implicitly for $1 \times p\mathrm{WN}_l$).

For the rest of the proof assume that $p > 1$. For each game we need to prove that, if (x, y)

(A) is of the form $\{a_i, b_i\}$, then none of its options is;

(B) is not of the form $\{a_i, b_i\}$, then it has an option of this form.

(We will need a slightly different notation for Case (iii) below.) By symmetry, we may assume that $0 \le x \le y$. Clearly, any final position satisfies (A) but not (B).

Game (i): We need to prove that $\mathcal{P}(m\mathrm{WN}^p) = \{\{a_i, b_i\} \mid i \in \mathbb{Z}_{\ge 0}\}$. Suppose $(x, y) = (a_i, b_i)$ for some $i \in \mathbb{Z}_{\ge 0}$. By parts (i) and (ii) of Lemma 3.3, a and $b_>$ are p-complementary and $b_i - b_j \ge m$ for all $j < i$. Then any roob-type option of the form $\{a_j, b_j\}$ may be blocked off, unless perhaps $a_j < a_i$ and $b_j = b_i$ for some $j < i$. But this is ridiculous since b is strictly increasing. By Lemma 3.3(ii) we get that, for $j < i$, $b_i - a_i \pm (b_j - a_j) \geq m$. Then an m-bishop cannot move $(x, y) \to \{a_j, b_j\}$, This proves (A).

For (B), since $p \geq 2$, we may assume $x = a_i$, for some i, but $y \neq b_i$. Then, by Lemma 3.3(iv): (*) There exists a $j < i$ such that an m-bishop can move $(x, y) \to (a_j, b_j)$ unless $y - x - (b_j - a_j) \geq m$ for all j such that $a_j \leq x$. Then, for all j such that $a_j = x$, we have that $y \geq x + m(j+1) > b_j$. But then, by Lemma 3.3(i), there are p options of (x, y) of the form $\{a_i, bi\}$. By the rule of game, they cannot all be blocked off.

Game (iia): We are going to prove that $\mathcal{P}(m\mathrm{WN}^{(p)}) = \{\{a_i, b_i\} \mid i \in \mathbb{Z}_{\geq 0}\}$ if and only if $\gcd(m, p) = 1$. Let us first explain the "only if" direction. Denote with $\gamma = \gcd(m, p)$, $p' = \frac{p}{\gamma}$ and $m' = \frac{m}{\gamma}$. Then, clearly, the positions of the form $(0, mi)$, where $0 \leq i < p'$, are P-positions of $m\mathrm{WN}^{(p)}$. Now, $(0, mp')$ is an N-position because $m'p = mp'$ implies that $(0, mp') \to (0, 0)$ is an option. But, by definition, $b_{p'} = mp'$ if and only if $p' < p$ if and only if $\gamma > 1$. Hence $\gcd(m, p) = 1$ is a necessary requirement.

For this game, the options of the m-bishop are identical to those in (i). Hence, let us analyze the p-rook.

For (A), suppose that $(x, y) = (a_i, b_i)$ for some $i \in \mathbb{Z}_{\geq 0}$ but that, for a contradiction, that a p-rook can move to $\{a_j, b_j\}$. Then, since b is strictly increasing, there is a $0 \leq j < i$, such that either $b_i \equiv b_j \pmod p$ and $a_i = a_j$, or $b_i \equiv a_j \pmod p$ and $a_i = b_j$. But then, for the first case (using the same notation as in Section 5), since

$$\overline{mj} = \overline{b_j - a_j} = \overline{b_i - a_i} = \overline{mi}$$

and $\gcd(m, p) = 1$ we must have $\overline{j} = \overline{i}$. This is ridiculous, since by p-complementarity and a nondecreasing we have $0 < i - j < p$. For the second case, by Theorem 5.1, we have that

$$\overline{-mj} = \overline{a_j - b_j} = \overline{b_i - a_i} = \overline{mi} = \overline{m(\varphi(j) + t)} = \overline{m(-j + t)},$$

for some $t \in \{1, \ldots, p-1\}$. This implies $\overline{0} = \overline{mt}$ but then again $\gcd(m, p) = 1$ gives a contradiction.

For (B), we follow the ideas in the second part of Case (i) up until (*). Then, for this game, we rather need to show that there is a $j < i$ such that $y \equiv b_j \pmod p$ and $a_j = x$ or $y \equiv a_j \pmod p$ and $b_j = x$. But this follows directly from the proof of Proposition 5.2(ii)(a).

Game (iib): We are now going to show that $\mathcal{P}(m\mathrm{WN}^{(m,mp)}) = \{\{a_i, b_i\} \mid i \in \mathbb{Z}_{\geq 0}\}$. For (A), suppose $(x, y) = (a_i, b_i)$ for some $i \in \mathbb{Z}_{\geq 0}$ but that there is a $j < i$ such that the (m, mp)-rook can move to $\{a_j, b_j\}$. Then, we have two cases:

Case 1: $b_i \equiv b_j - r \pmod{mp}$ and $a_i = a_j$, for some $r \in \{0, 1, \dots, m-1\}$. Then $b_i - a_i \equiv b_j - a_j - r \pmod{mp}$ so that $mi \equiv mj - r \pmod{mp}$ and so $m(i-j) \equiv -r \pmod{mp}$. But this forces $r = 0$ and $i - j \equiv 0 \pmod{p}$ which is impossible since Lemma 3.3 (i) and (iv) imply $i - j \in \{1, 2, \dots, p-1\}$.

Case 2: $b_i \equiv a_j - r \pmod{mp}$ and $a_i = b_j$, for some $r \in \{0, 1, \dots, m-1\}$. Then $b_i - a_i \equiv a_j - b_j - r \pmod{mp}$ so that $mi \equiv -mj - r \pmod{mp}$ and so $m(i+j) \equiv -r \pmod{mp}$. By Theorem 5.1 we have that $i = \varphi(j) + s$ for some $s \in \{1, 2, \dots, p-1\}$. Further, by (14), we have $\varphi(j) \equiv -j \pmod{p}$, so that $m(\varphi(j) + s + j) = ms \equiv -r \pmod{mp}$. Once again we have reached a contradiction.

For (B), in analogy with (*), it suffices to study the (m, mp)-rook's options where y is such that $y - x - (b_j - a_j) \geq m$ for all j such that $a_j \leq x = a_i$. Hence, we need to show that there are a j and an $r \in \{0, 1, \dots m-1\}$ such that

$$y \equiv b_j - r \pmod{mp} \quad \text{and} \quad a_j = x$$

or

$$y \equiv a_j - r \pmod{mp} \quad \text{and} \quad b_j = x.$$

Clearly, we may choose r such that $y - x + r \equiv 0 \pmod{m}$. Then, for all j, we get $ms := y - x + r \equiv \pm(b_j - a_j) \pmod{m}$. Hence, it suffices to find a specific j such that

$$j = \frac{b_j - a_j}{m} \equiv s \pmod{p} \quad \text{and} \quad a_j = x$$

or

$$-j = \frac{a_j - b_j}{m} \equiv s \pmod{p} \quad \text{and} \quad b_j = x.$$

But then, by (13) or (15), we are done.

Game (iiia): We are now going to show that $\mathcal{P}(m \times p\mathrm{WN}_l) = \{(a_{ip+l}, b_{ip+l}) \mid i \in \mathbb{Z}_{\geq 0}\} \cup \{(b_{ip-l}, a_{ip-l}) \mid i \in \mathbb{Z}_{>0}\}$. We may assume that $l > 0$. We have already seen that $(a'_i) := (a_{pi+l})_{i \geq 0}$ and $(b'_i) := (b_{p(i+1)-l})_{i \geq 0}$ are complementary. Our proof will be a straightforward extension of those in [Fraenkel 1982] (which deals with the case $l = 0$) and [Connell 1959a] (which implicitly deals with the case $m = 0$). Observe that $a'_0 = a_l = 0$ and $b'_0 = b_{p-l} = m(p-l)$.

For (A), let $(x, y) = (a'_i, b'_i)$. In case $i = 0$ (by Definition 4 (iiia)), the Queen has no options at all, so assume $i > 0$. Proposition 5.2(iii) gives that $b'_i - a'_i \pm (b'_j - a'_j) \geq mp$ for all $0 \leq j < i$. Then the mp-bishop cannot move $(x, y) \to (a'_j, b'_j)$ for any $0 \leq j < i$. Since a' and b' are complementary there is no rook-type option $(a'_i, b'_i) \to \{a'_j, b'_j\}$.

For (B), we adjust the statement (*) accordingly: Suppose $x = a_i'$. By Proposition 5.2(iii): If the mp-bishop cannot move to (a_j', b_j') for any $j < i$ we get that either $i = 0$ or $y - x - (b_j' - a_j') \geq mp$ for all $j < i$. If $i = 0$ there is a rook-type option to (a_0', b_0') (we may assume here that $y > b_0'$), so suppose $i > 0$. But then, by Proposition 5.2(iii), we get $y \geq b_j' + mp + x - a_j' \geq b_i' + a_i' - a_j' > b_i'$. Hence, for this case, the rook-type move $(x, y) \to (a_i', b_i')$ suffices. Suppose on the other hand that $x = b_i'$ with $i \geq 0$. Then, since $y \geq x = b_i' > a_i'$, the desired rook-type move is $(x, y) \to (b_i', a_i')$.

Game (iiib) It only remains to demonstrate that $\mathcal{P}(m \times p\text{WN}) = \{\{a_i, b_i\} \mid i \in \mathbb{Z}_{\geq 0}\}$. Suppose that the starting position is (a_i, b_i). Then $i = pj + l'$ for some (unique) pair $j \in \mathbb{Z}_{\geq 0}$ and $0 \leq l' < p$. The second player should choose $l = l'$. If, on the other hand, the starting position is (b_i, a_i). Then $i = pj - l'$ for some (unique) pair $j \in \mathbb{Z}_{>0}$ and $0 < l' \leq p$. The second player should choose $l = p - l'$. In either case, by (iiia), there is no option of the form (a_i', b_i').

If the starting position (x, y) is not of the form $\{a_i, b_i\}$, again, by (iiia), for any choice of $0 \leq l < p$, there is a move $(x, y) \to \{a_i', b_i'\}$ for some $i \geq 0$. □

7. Questions

Can one find a polynomial time solution of $m\text{WN}^{(l,p)}$ for some integers $l \geq 0$, $m > 0$ and $p > 0$ whenever

- $\gcd(m, p) \neq 1$ and $l = 0$, or
- $0 < l \neq m$ or $m \nmid p$?

If this turns out to be complicated, can one at least say something about its asymptotic behavior?

Denote the solution of $m\text{WN}^{(l,p)}$ with $\{\{c_i^{(l,m,p)}, d_i^{(l,m,p)}\}\}_{i \in \mathbb{Z}_{\geq 0}}$. Let us finish off with two tables of the initial P-positions of such games.

From these tables one may conclude that: The infinite arithmetic progressions of the sequences

$$(b_i^{m,p} - a_i^{m,p})_{i \in \mathbb{Z}_{\geq 0}} = (mi)_{i \in \mathbb{Z}_{\geq 0}}$$

(see also Tables 1 and 2) are not in general seen among the sequences

$$(d_i^{(l,m,p)} - c_i^{(l,m,p)})_{i \in \mathbb{Z}_{\geq 0}}.$$

We believe that the latter sequence is an arithmetic progression if and only if none of the items in our above question is satisfied. We also believe that, for arbitrary constants, $(c_i^{(l,m,p)})_{i \in \mathbb{Z}_{\geq 0}}$ and $(d_i^{(l,m,p)})_{i \in \mathbb{Z}_{>0}}$ are p-complementary. But the solution of these questions are left for some future work.

Remark 5. We may also define generalizations of $m\text{WN}^p$ and $m \times p\text{WN}_l$:

$b_n^{1,3}$	0	1	2	4	5	7	8	10	11	12	14	15	17	18	20	21	22
$a_n^{1,3}$	0	0	0	1	1	2	2	3	3	3	4	4	5	5	6	6	6
n	0	1	2	3	4	5	6	7	8	9	10	11	12	13	14	15	16

Table 1. The initial P-positions of Connell's restriction of Wythoff Nim, $\mathrm{WN}^{(3)}$.

$b_n^{2,3}$	0	2	4	7	9	11	14	16	19	21	23	26	28	31	33	35	38
$a_n^{2,3}$	0	0	0	1	1	1	2	2	3	3	3	4	4	5	5	5	6
$b_n - a_n$	0	2	4	6	8	10	12	14	16	18	20	22	24	26	28	30	32
n	0	1	2	3	4	5	6	7	8	9	10	11	12	13	14	15	16

Table 2. Some initial values of the Beatty pairs defined in (2) and (3), here $m = 2$ and $p = 3$, together with the differences of their coordinates $(= 2n)$.

$d_n^{(0,2,2)}$	0	3	6	9	12	15	19	22	25	28	31	34	37	40	43	46	49
$c_n^{(0,2,2)}$	0	0	1	1	2	2	3	4	4	5	5	6	7	7	8	8	9
$d_n - c_n$	0	3	5	8	10	13	16	18	21	23	26	28	30	33	35	38	40
n	0	1	2	3	4	5	6	7	8	9	10	11	12	13	14	15	16

Table 3. The first few P-positions of $2\mathrm{WN}^2$ together with the respective differences of their coordinates.

$d_n^{(1,2,3)}$	0	2	5	7	11	14	16	19	21	26	29	31	36	39	41	44	46
$c_n^{(1,2,3)}$	0	0	1	1	2	3	3	4	4	5	6	6	7	8	8	9	9
$d_n - c_n$	0	2	4	6	9	11	13	15	17	21	23	25	29	31	33	35	37
n	0	1	2	3	4	5	6	7	8	9	10	11	12	13	14	15	16

Table 4. The first few P-positions of $2\mathrm{WN}^{(1,3)}$. Notice that (as in Table 3) the successive differences of their coordinates are not in arithmetic progression.

Fix $l \in \mathbb{Z}_{>0}$. Let $m\mathrm{WN}_l^p$ be as $m\mathrm{WN}^p$ but where the player may only block off l*-roob-type options* (recall, non-l-bishop options). Otherwise, the Queen moves as the m-bishop or the rook. Then $m\mathrm{WN}_m^p = m\mathrm{WN}^p$. On the other hand $m\mathrm{WN}_1^p$ is the blocking variation of m-Wythoff Nim where the previous player may block off *any* $p - 1$ rook-type options.

Let $u, v \in \mathbb{Z}_{>0}$ and let $m \times p\mathrm{WN}_{u,v}$ be as $m \times p\mathrm{WN}_l$, but the removed (lower left) rectangle has base u and height v. Then for this game the final positions are $(u, 0)$ and $(0, v)$. If $l > 0$, $u = ml$ and $v = m(p-l)$ we get $m \times p\mathrm{WN}_{lm,m(p-l)} = m \times p\mathrm{WN}_l$. Some of these games are identical to misère versions of Wythoff Nim, see [Fraenkel 1984].

One may ask questions in analogy to the above for these variations. For example, we have found a minimal exclusive algorithm satisfying $\mathcal{P}(m\mathrm{WN}_1^p)$ which is related to a polynomial time construction in [Fraenkel 1998]. Is there an analog polynomial time construction for $\mathcal{P}(m\mathrm{WN}_1^p)$? Another question is if any of these further generalized games coincide via identical set of P-positions?

Appendix

The following discussion, provided by Aviezri Fraenkel, provides a complementary analysis of "p-complementarity"/"p-fold complementarity" and homogeneous Beatty sequences:

Definition. Let $p \in \mathbb{Z}_{>0}$. The multisets S, T of positive integers are 1-*upper p-fold complementary*, for short: *p-fold complementary*, if $S \cup T = p \times \mathbb{Z}_{>0}$.

If the multisets S, T satisfy Definition 1 and have irrational densities α^{-1}, β^{-1}, say $\alpha \le \beta$, then a necessary condition for p-fold complementarity is $\alpha^{-1} + \beta^{-1} = p$. Thus $a := \beta - \alpha > 0$. Then $\alpha = (2 - ap + \sqrt{a^2p^2+4})/2p$, so $p^{-1} < \alpha < 2p^{-1}$. Then $1/\beta = p - 1/\alpha$, so $\beta > 2/p$.

Let $M = \lfloor 1/\alpha \rfloor + 1$, $N = \lfloor 1/\beta \rfloor + 1$. Notice that $\alpha(M-1) < 1 < \alpha M$, $\beta(N-1) < 1 < \beta N$. From now on we let $S = \{\lfloor n\alpha \rfloor\}_{n \ge M}$, $T = \{\lfloor n\beta \rfloor\}_{n \ge N}$.

Theorem. *The multisets S, T are p-fold complementary.*

Proof. For any $k \in \mathbb{Z}_{>0}$, since α is irrational, the number of terms less than k in $S \cup T$ is

$$\begin{aligned}\lfloor k/\alpha \rfloor - (M-1) + \lfloor k/\beta \rfloor - (N-1) &= \lfloor k/\alpha \rfloor + \lfloor k(p - \alpha^{-1}) \rfloor - M - N + 2 \\ &= kp + \lfloor k/\alpha \rfloor + \lfloor -k/\alpha \rfloor - M - N + 2 \\ &= kp - M - N + 1.\end{aligned}$$

Similarly, $S \cup T$ contains $(k+1)p - M - N + 1$ terms $< k+1$. Hence there are exactly p terms $< k+1$ but not $< k$. They are the terms k with multiplicity p. $\square$

Remarks. (i) $\lfloor 1/\alpha \rfloor = p + \lfloor -1/\beta \rfloor = p - 1 - \lfloor 1/\beta \rfloor = p - N$. Hence $M = \lfloor 1/\alpha \rfloor + 1 = p - N + 1$.

(ii) Clearly $(\lfloor (M-1)\alpha \rfloor, \lfloor (N-1)\beta \rfloor) = (0, 0)$. Since $\alpha < \beta$, we have $N \le M$. Hence, for all $N \le n < M$, we have that $(\lfloor n\alpha \rfloor, \lfloor n\beta \rfloor) = (0, \lfloor n\beta \rfloor)$ where $\lfloor n\beta \rfloor > 0$. Thus there are precisely $M - N$ couples $(\lfloor n\alpha \rfloor, \lfloor n\beta \rfloor)$ with $\lfloor n\alpha \rfloor = 0$ and $\lfloor n\beta \rfloor > 0$, containing $M - N$ 0s. Thus, for $0 \leq n < M$, there are precisely M couples $(\lfloor n\alpha \rfloor, \lfloor n\beta \rfloor)$ with $\lfloor n\alpha \rfloor = 0$ and $\lfloor n\beta \rfloor \geq 0$, containing, in total, $M + N = p + 1$ 0s.

(iii) The proof is a straightforward generalization to $p \geq 1$ of a proof included in an editorial comment to [Fraenkel 2010] stating: "...The result is so astonishing and yet easily proved that we include a short proof for the reader's pleasure." This is then followed by the above proof for the special case $p = 1$, which is itself a slight simplification of the proof given in [Fraenkel 1982].

Acknowledgments

I would like to thank Aviezri Fraenkel for providing two references that motivated generalizations of the games (in the previous version of this paper) to their current form and of course for the nice Appendix. I would also like to thank Peter Hegarty for giving valuable feedback during the earlier part of this work and Niklas Eriksen for composing parts of the caption for the figures. Lastly, I would like to thank the anonymous referee for several suggestions that helped to improve this paper.

References

[Beatty 1926] S. Beatty, "Problem 3173", *Amer. Math. Monthly* **33**:3 (1926), 159.

[Beatty et al. 1927] S. Beatty, A. Ostrowski, J. Hyslop, and A. C. Aitken, "Problems and Solutions: Solutions: 3177", *Amer. Math. Monthly* **34**:3 (1927), 159–160.

[Borwein and Borwein 1993] J. M. Borwein and P. B. Borwein, "On the generating function of the integer part: $[n\alpha + \gamma]$", *J. Number Theory* **43**:3 (1993), 293–318.

[Boshernitzan and Fraenkel 1984] M. Boshernitzan and A. S. Fraenkel, "A linear algorithm for nonhomogeneous spectra of numbers", *J. Algorithms* **5**:2 (1984), 187–198.

[Bouton 1901/02] C. L. Bouton, "Nim, a game with a complete mathematical theory", *Ann. of Math.* (2) **3**:1-4 (1901/02), 35–39.

[Connell 1959a] I. G. Connell, "A generalization of Wythoff's game", *Canad. Math. Bull.* **2** (1959), 181–190.

[Connell 1959b] I. G. Connell, "Some properties of Beatty sequences, I", *Canad. Math. Bull.* **2** (1959), 190–197.

[Conway 2001] J. H. Conway, *On numbers and games*, 2nd ed., A K Peters, Natick, MA, 2001.

[Duchêne and Gravier 2009] E. Duchêne and S. Gravier, "Geometrical extensions of Wythoff's game", *Discrete Math.* **309**:11 (2009), 3595–3608.

[Fraenkel 1969] A. S. Fraenkel, "The bracket function and complementary sets of integers", *Canad. J. Math.* **21** (1969), 6–27.

[Fraenkel 1973] A. S. Fraenkel, "Complementing and exactly covering sequences", *J. Combinatorial Theory Ser. A* **14** (1973), 8–20.

[Fraenkel 1982] A. S. Fraenkel, "How to beat your Wythoff games' opponent on three fronts", *Amer. Math. Monthly* **89**:6 (1982), 353–361.

[Fraenkel 1984] A. S. Fraenkel, "Wythoff games, continued fractions, cedar trees and Fibonacci searches", *Theoret. Comput. Sci.* **29**:1-2 (1984), 49–73.

[Fraenkel 1998] A. S. Fraenkel, "Heap games, numeration systems and sequences", *Ann. Comb.* **2**:3 (1998), 197–210.

[Fraenkel 2004] A. S. Fraenkel, "Complexity, appeal and challenges of combinatorial games: Algorithmic combinatorial game theory", *Theoret. Comput. Sci.* **313**:3 (2004), 393–415.

[Fraenkel 2010] A. S. Fraenkel, "Solution II, Problem 11365", *Amer. Math. Monthly* **117**:4 (2010), 376.

[Fraenkel and Borosh 1973] A. S. Fraenkel and I. Borosh, "A generalization of Wythoff's game", *J. Combinatorial Theory Ser. A* **15** (1973), 175–191.

[Fraenkel and Kimberling 1994] A. S. Fraenkel and C. Kimberling, "Generalized Wythoff arrays, shuffles and interspersions", *Discrete Math.* **126**:1-3 (1994), 137–149.

[Fraenkel and Peled 2009] A. S. Fraenkel and U. Peled, "Harnessing the unwieldy MEX function", preprint, 2009, http://goo.gl/3hpiLR.

[Hadad 2008] U. Hadad, *Polynomializing seemingly hard sequences using surrogate sequences*, Msc Thesis, Weizmann Institute of Science, 2008.

[Hegarty and Larsson 2006] P. Hegarty and U. Larsson, "Permutations of the natural numbers with prescribed difference multisets", *Integers* **6** (2006), A3.

[Holshouser and Reiter 2010] A. Holshouser and H. Reiter, "Three pile Nim with move blocking", preprint, 2010, http://math2.uncc.edu/~hbreiter/3-pile.pdf.

[Kimberling 1995] C. Kimberling, "Stolarsky interspersions", *Ars Combin.* **39** (1995), 129–138.

[Kimberling 2007] C. Kimberling, "Complementary equations", *J. Integer Seq.* **10**:1 (2007), Article 07.1.4.

[Kimberling 2008] C. Kimberling, "Complementary equations and Wythoff sequences", *J. Integer Seq.* **11**:3 (2008), Article 08.3.3.

[Larsson 2009] U. Larsson, "2-pile Nim with a restricted number of move-size imitations", *Integers* **9** (2009), G04, 671–690.

[O'Bryant 2002] K. O'Bryant, "A generating function technique for Beatty sequences and other step sequences", *J. Number Theory* **94**:2 (2002), 299–319.

[O'Bryant 2003] K. O'Bryant, "Fraenkel's partition and Brown's decomposition", *Integers* **3** (2003), A11.

[Rayleigh 1894] J. W. Rayleigh, *The theory of sound, I*, 2nd ed., Macmillan, London, 1894.

[Skolem 1957] T. Skolem, "Über einige Eigenschaften der Zahlenmengen $[\alpha n+\beta]$ bei irrationalem α mit einleitenden Bemerkungen über einige kombinatorische Probleme", *Norske Vid. Selsk. Forh., Trondheim* **30** (1957), 42–49.

[Smith and Stănică 2002] F. Smith and P. Stănică, "Comply/constrain games or games with a Muller twist", *Integers* **2** (2002), G3.

[Wythoff 1907] W. A. Wythoff, "A modification of the game of Nim", *Nieuw Arch. Wisk.* (2) **7** (1907), 199–202.

urban.larsson@chalmers.se *Mathematical Sciences, Chalmers University of Technology and University of Gothenburg, SE-412 96 Göteborg, Sweden*

Games of No Chance 4
MSRI Publications
Volume **63**, 2015

Computer analysis of sprouts with nimbers

JULIEN LEMOINE AND SIMON VIENNOT

Sprouts is a two-player topological game, invented in 1967 at the University of Cambridge by John Conway and Michael Paterson. The game starts with p spots, and ends in at most $3p-1$ moves. The first player who cannot play loses.

The complexity of the p-spot game is very high, so that the best hand-checked proof only shows who the winner is for the 7-spot game, and the best previous computer analysis reached $p = 11$.

We have written a computer program, using mainly two new ideas. The nimber (also known as Sprague–Grundy number) allows us to compute separately independent subgames; and when the exploration of a part of the game tree seems to be too difficult, we can manually force the program to search elsewhere. Thanks to these improvements, we have settled every case up to $p = 32$. The outcome of the 33-spot game is still unknown, but the biggest computed value is the 47-spot game! All the computed values support the *Sprouts conjecture*: the first player has a winning strategy if and only if p is 3, 4 or 5 modulo 6.

We have also used a check algorithm to reduce the number of positions needed to prove which player is the winner. It is now possible to hand-check all the games until $p = 11$ in a reasonable amount of time.

1. Introduction

Sprouts is a two-player pencil-and-paper game invented in 1967 in the University of Cambridge by John Conway and Michael Paterson [Gardner 1967]. The game starts with p spots and players alternately connect the spots by drawing curves between them, adding a new spot on each curve drawn. A new curve cannot cross or touch any existing one, leading necessarily to a *planar* graph. The first player who cannot play loses.[1]

The last rule states that a spot cannot be connected to more than 3 curves, and it induces the end of the game in a finite number of moves. If we consider

[1] The *misère* version of the game corresponds to the opposite rule, i.e., a player unable to move wins. The analysis of the misère version is more difficult, and is the object of another article [Lemoine and Viennot 2009].

Figure 1. A sample game of 2-spot Sprouts (the second player wins).

that a spot has 3 *lives* at the beginning of a game, there is a total number of $3p$ lives. Each move then consumes 2 lives and creates a spot with 1 life, globally decreasing by one the total number of lives. It follows that the game ends in at most $3p-1$ moves.

Despite the small number of moves of a given game, it is difficult to determine whether the winning strategy is for the first or the second player. The best published and complete hand-checked proof is due to Focardi and Luccio [2004], and shows who the winner is for the 7-spot game.

The first published computer analysis of Sprouts is due to Applegate, Jacobson and Sleator [Applegate et al. 1991], who computed which player wins up to the 11-spot game. They noted a pattern in their results and proposed the *Sprouts conjecture*: the first player has a winning strategy in the p-spot game if and only if p is 3, 4 or 5 modulo 6. We have written a new program, using an improved version of their original representation of Sprouts positions, and obtained significant new results with the help of two main ideas: the theoretical concept of *nimber* and the manual exploration of the game tree.

Our program enabled us to compute which player wins the p-spot game up to 32 spots, and also some values up to 47 spots. All the computed values support the Sprouts conjecture. We then performed a second computation to validate the results, in which our program tries to minimize the number of positions needed to prove the value of a given p-spot game. It is then possible to generate graphs corresponding to these minimized databases, which provides a hand-checkable winning strategy for the p-spot game.

2. Game tree

We will call *position* a graph embedded in the plane, obtained with the rules of Sprouts.

From any given Sprouts position, there is necessarily a winning strategy, either for the player to move, or for the next player. If the winning strategy is for the player to move, we will say that the *outcome* of the position is a *win* and in the other case, that it is a *loss*. The main goal of our program is then to compute the outcome of Sprouts positions, and particularly the starting positions with p spots.

We call *game tree of a position* the tree where the *root* is the given position, the nodes are the positions that can be reached by playing moves from the root,

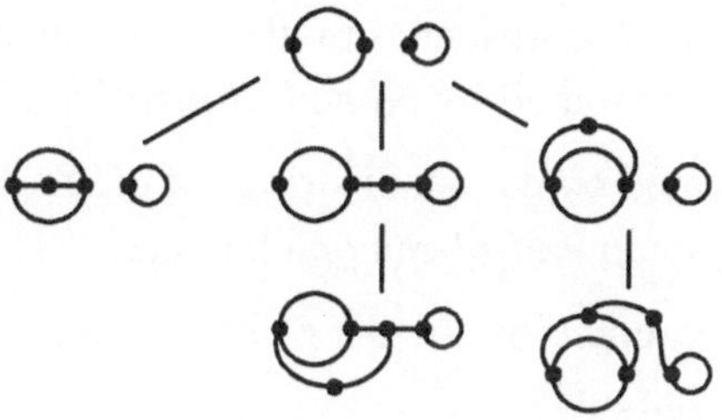

Figure 2. Game tree of a given Sprouts position.

and where an edge links two positions if the *child* is obtained from the *parent* in only one move. Figure 2 is an example of such a tree.

As we will see in Section 6, the outcome of a position is computed recursively by developing its game tree.

3. Positions representation

As a pencil-and-paper game, Sprouts is not suitable for programming. We describe here a representation of the game with strings that could be used with a computer to develop game trees, and therefore deduce winning strategies.

3.1. ***Regions and boundaries.*** In a given position, a *region* is a connected component of the plane and inside a given region, a *boundary*, is a connected component of the curves drawn by the players. An isolated spot is considered as a boundary in itself.

For example, Figure 3 contains 5 regions and 9 boundaries: there are 3 boundaries in region D, 2 boundaries in regions A and C and 1 boundary in regions B and E.

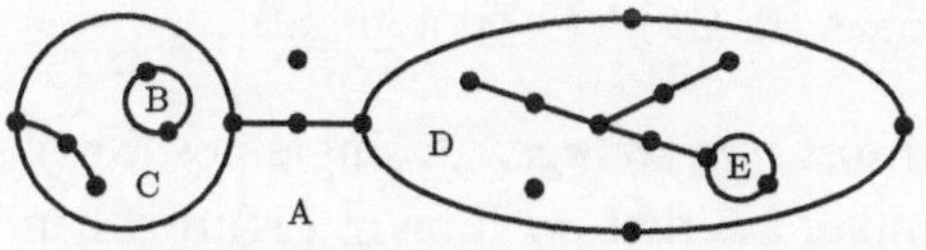

Figure 3. A Sprouts position obtained after 10 moves in a 11-spot game.

3.2. ***String representation.*** In order to perform computations with a computer, we need a way to represent the positions of Sprouts by strings of characters. We follow the representation described in [Applegate et al. 1991].[2] Basically, each spot (*vertex* in the graph theory) will be denoted by a letter, and the graph is described as follows:

[2]The equivalence between graphic and string representations is easy to see, but hard to demonstrate. This is the goal of a full 400-page report [Draeger et al. 1990].

- The complete graph is represented by the list of strings of the regions that it contains and is terminated by an *end-of-position* character: "!"
- A region is represented by the list of strings of the boundaries that it contains and is terminated by an *end-of-region* character: "}"
- A boundary is represented by the list of its vertices and is terminated by an *end-of-boundary* character: "."

For a given boundary, we write the vertices in the order they appear while following one side of the boundary. Inside a given region, the same orientation must be respected.

In our example, we turn around all boundaries clockwise, except for the (unique) boundary that surrounds a region, around which we turn counterclockwise. A string representing the above position is then:

`AL.}AL.BNMCMN.}D.COFPGQFOCM.}E.HRISJSIUKTKUIR.FQGP.}KT.}!`

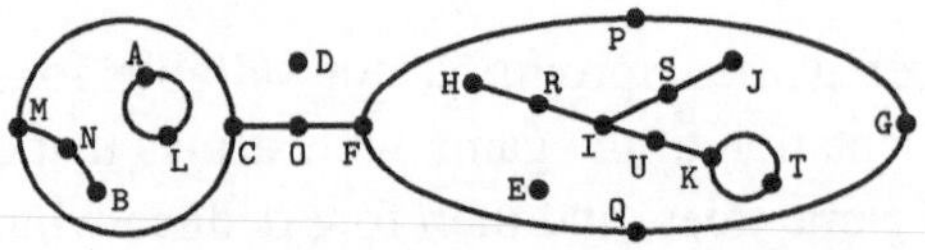

The boundary AL. appears twice, because it is inside the same region as the 4 vertices B;N;M;C, and it also surrounds a region itself. Notice how the string of the boundary containing B;N;M;C is obtained: starting at B, we follow the bottom of the boundary, meeting first N, then M, and C. Then, we continue to follow the top of the boundary, meeting M and N once again (but on the opposite side), stopping when we are back at the starting point.

Remark. The strings of the starting positions are A.}! (for the 1-spot game), A.B.}! (2-spot game), A.B.C.}! (3-spot game)....

3.3. ***Equivalent positions and strings.*** A single position can correspond to several strings. For example, the first position in Figure 4 can lead to

BDCDBE.}BE.A.}!   or   A.EB.}DCDBEB.}!.

Similarly, a single string can represent several positions: BDCDBE.}BE.A.}! can represent the two positions of Figure 4.

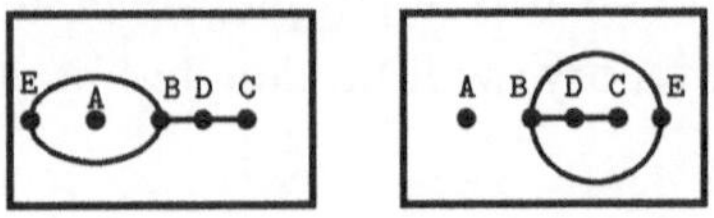

Figure 4. Two equivalent positions.

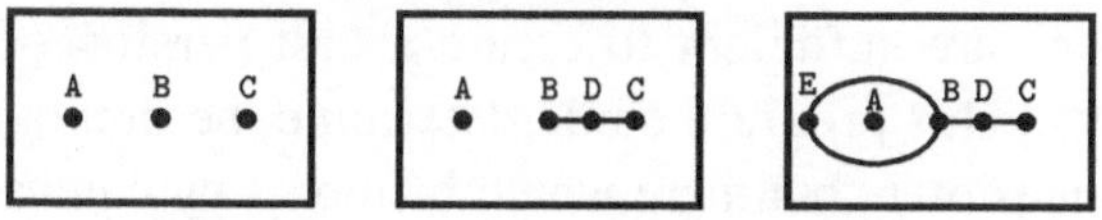

Figure 5. Two moves in a 3-spot game.

More precisely, a single string represents an infinite number of positions, as we can draw lines of multiple shapes as long as the topology is respected. Conversely, a single position corresponds to only a finite number of strings if we only allow the strings to use the first letters of the alphabet.

3.4. ***Moves.*** The main interest of the string representation described above is that moves can be easily defined and computed from it. Moves always take place in a single region but they are of two different types, depending on whether we link one boundary to itself or two different boundaries. The program we describe below will thus produce finite game trees.

Move 1. A *two-boundaries* move consists in connecting two spots of different boundaries.

Let $\mathtt{x}_1 \dots \mathtt{x}_m$ and $\mathtt{y}_1 \dots \mathtt{y}_n$ be two different boundaries in the same region, with $m \geq 2$ and $n \geq 2$. We suppose that $\mathtt{x}_i$ and $\mathtt{y}_j$ are vertices that occur two times or less in the whole string, with $1 \leq i \leq m$ and $1 \leq j \leq n$.

Then the two-boundaries move consists in merging these two boundaries in $\mathtt{x}_1 \dots \mathtt{x}_i \mathtt{z} \mathtt{y}_j \dots \mathtt{y}_n \mathtt{y}_1 \dots \mathtt{y}_j \mathtt{z} \mathtt{x}_i \dots \mathtt{x}_m$ where $\mathtt{z}$ is the new created vertex.

The same definition holds if $m = 1$ or if $n = 1$, but in these cases, $\mathtt{x}_i \dots \mathtt{x}_m$ and $\mathtt{y}_j \dots \mathtt{y}_n$ are empty boundaries.

Move 2. A *one-boundary* move consists in connecting two spots of the same boundary.

Let $\mathtt{x}_1 \dots \mathtt{x}_n$ be a boundary, with $n \geq 2$. We suppose that $\mathtt{x}_i$ and $\mathtt{x}_j$ are vertices that occur two times or less in the whole string, with $1 \leq i \leq m$, $1 \leq j \leq n$ and $i \neq j$, or that $i = j$ and $\mathtt{x}_i$ occurs only once in the whole string.

Then the one-boundary move consists in separating the other boundaries of the same region in 2 sets $\mathtt{B}_1$ and $\mathtt{B}_2$, and the original region is divided into two new regions: $\mathtt{x}_1 \dots \mathtt{x}_i \mathtt{z} \mathtt{x}_j \dots \mathtt{x}_n.\mathtt{B}_1\}$ and $\mathtt{x}_i \dots \mathtt{x}_j \mathtt{z}.\mathtt{B}_2\}$.

The same definition holds if $n = 1$, but in this case, $\mathtt{x}_j \dots \mathtt{x}_n$ is an empty boundary.

Let us give an example of the two types of moves with a 3-spot game. The initial string is `A.B.C.}!`. First, we make a two-boundaries move, which leads to `A.BDCD.}!`, and then we make a one-boundary move, which separates the position into two regions: `BDCDBE.}BE.A.}!`. Figure 5 shows three corresponding positions.

These definitions are sufficient to create a first version of a program for computing Sprouts. This program could determine the outcome of the p-spot game for a few values of p, but there would be too many equivalent strings, and the memory would saturate quite quickly.

4. Strings simplification

We will now explain several methods to simplify the strings, with two main goals. Firstly, if we go back to Figure 2, the middle and the right children of the root can be represented by the same string (we will know at the end of this section that it is `22.}]!`). These kind of simplifications will merge some equivalent branches of the game tree and thus decrease the complexity of the computation.

Secondly, the simplified strings will be more suited to perform the *canonization* step described in Section 5.

4.1. ***Deletion of the dead parts.*** First of all we delete the dead vertices (those which occur 3 times in the string). Then, we delete the empty boundaries (boundaries whose vertices are all dead) and finally, we delete the dead regions (with 0 or 1 life).

Using our previous example, we would therefore delete 5 dead vertices. Then, no boundary would be empty, but the region `KT.}!` would be dead. The new string and position would be

`AL.}AL.BNN.}D.OPGQO.}E.HRSJSUTUR.QGP.}!`

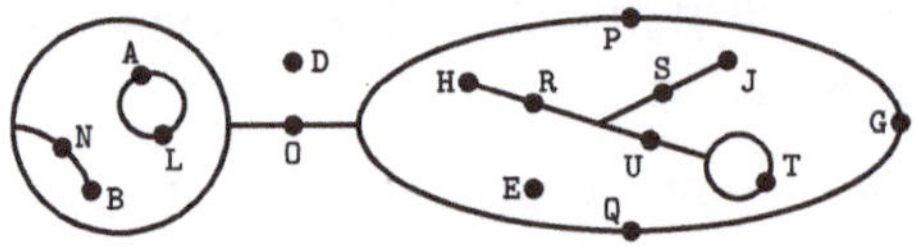

4.2. ***Generic vertices.*** We replace the vertices that only occur once and in a boundary with a single vertex, by the generic vertex “0”. We also replace the vertices that only occur once but in a boundary with several vertices, by the generic vertex “1”.

The generic vertex “2” can designate the vertices that occur twice in a row along a single boundary, just as `N` or `O` in our example. However, it can also designate the vertices that used to occur twice, and that now only occur once, because of a dead region deletion (such as `T` in our example).

`AL.}AL.12.}0.2PGQ.}0.1RS1SU2UR.QGP.}!`

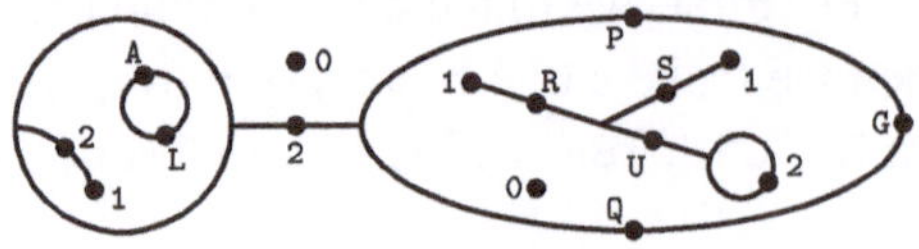

4.3. *Lands.* We cut the position into independent parts, named *lands*. A position can be cut into 2 lands when those lands have no letter in common. The end-of-land character is "`]`". Our example has 2 lands, and its string becomes:

`AL.}AL.12.}]0.2PGQ.}0.1RS1SU2UR.QGP.}]!`

4.4. *Renaming letters.* At this point, letters designate vertices that occur twice in the string, but we can distinguish between two different types: vertices that occur twice in the same boundary (which we will designate with lower-case letters) and vertices that appear in two different regions (which we will designate with upper-case letters).

We rename these vertices from "`a`" and "`A`", in the order of their appearance. We start again from "`a`" when we meet a new boundary and we start again from "`A`" when we meet a new land.

`AB.}AB.12.}]0.2ABC.}0.1ab1bc2ca.CBA.}]!`

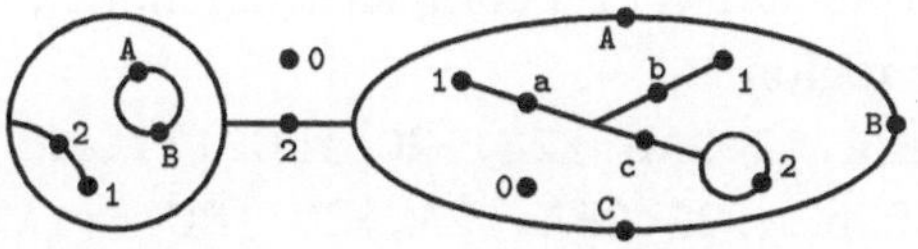

4.5. *Regions equivalences.* When a region has 3 lives or less, we have an equivalent game if we merge the boundaries. For instance, the region `A.BC.}` becomes `ABC.}`, or `2.2.}` becomes `22.}`. These equivalences are a useful trick in our program, reducing considerably the number of stored strings.

5. Strings canonization

We already know that several strings can represent the same position, e.g., the previous position could also be represented by:

`BA2C.0.}0.2ca1ab1bc.CBA.}]AB.21.}AB.}]!`

As in [Applegate et al. 1991], we call *canonization* the choice of a single string amongst all the equivalent strings. By merging equivalent branches in a game tree, canonization decreases efficiently both memory consumption and running time. On the other hand, canonization itself takes a lot of running time, and we need to take this into consideration when writing the program.

5.1. *Canonization.* First, we define an equivalence on the set of strings. Two strings are equivalent if they are equal modulo:

- the first vertex chosen in a boundary,
- the order of boundaries in a region,
- the order of regions in a land,

- the order of lands in the position,
- the orientation chosen for each region,
- a renaming of the vertices.

We can now define the term *canonization*: this is the choice of a single string in each equivalence class. This choice should be as simple as possible, therefore, we will choose the minimal string for the following lexicographical order:

`0 < 1 < 2 < a < b < ··· < A < B < ··· < . < } < ] < !`

A little reflection shows then that the canonized string of our previous position is `0.1ab1bc2ca.ABC.}0.2ABC.}]12.AB.}AB.}]!`.

Remark. We are not allowed to change the orientation of a single boundary. We can only change the orientation of a complete region, i.e., the orientation of all the boundaries in the region.

For instance, `122a2a.22.2AB.}2A.}2C.}BC.}]1122.}]!` represents a losing position, while `122a2a.22.2BA.}2A.}2C.}BC.}]1122.}]!` represents a winning position.[3]

5.2. *Pseudocanonization.* We cannot actually perform the canonization, mainly because choosing the name of the upper-case letters requires too much running time. In particular, if a land has k upper-case letters, performing the true canonization consists in choosing the minimal string for lexicographical order amongst the $k!$ possible strings.

Therefore, we only perform a pseudocanonization: the same position can be represented by several strings, for instance the strings `0.AB.CD.}0.AB.}CD.}]!` and `0.AB.CD.}0.CD.}AB.}]!` represent the same position. Our pseudocanonization algorithm renames the upper-case letters from "A" in the order of their appearance and then sorts the string. Experience shows that performing this operation twice is the most efficient for both running time and memory usage. Ultimately, we only lose a few percentages of memory compared to a true canonization, whose time complexity prevents from computing p-spot games for $p \geq 5$ or 6.

We have developed the complete game tree of the p-spot games, for $p \leq 6$, in order to evaluate the performance of our pseudocanonization. Here is an extract of the complete game tree of the 4-spot game:

[3] More obvious examples may exist, but this is the simplest that we know. However, to check this example, a long computation is necessary.

```
ABCDEF.}ABCDEG.}FG.}]!
ABCDEF.}ABCDGF.}EG.}]!
ABCDEF.}ABCGEF.}DG.}]!
ABCDEF.}ABGDEF.}CG.}]!
ABCDEF.}BCDEFG.}AG.}]!
```

Since these strings represent the same position, a true canonization would have displayed only one string instead of these five. However, this position is not needed to compute the outcome of the 4-spot game. In fact, the performance of the pseudocanonization (comparatively to the true canonization) is better in a real computation that in a complete game tree development, because in a real computation we meet strings easier to compute, with less upper-case letters.

The following table gives the number of pseudocanonized strings stored after a complete game tree development for the p-spot game.

$n =$	2	3	4	5	7
number of strings =	18	157	1796	24784	393103

This table could be useful for evaluating the performance of our canonization, in comparison with the canonization of other programs.

6. Main algorithm

6.1. *Outcome computation algorithm.* The standard recursive algorithm to compute the outcome of a position[4] can be described as follows:

Algorithm 1. function compute-win-loss(position P)

- compute the children of P
- for each child, do: if compute-win-loss(child)=Loss, return Win
- return Loss[5]

The core of our main algorithm uses this classical depth-first procedure. We also store the outcomes of the computed positions in a database (a *transposition table*), in order not to compute several times the same position. Figure 6 shows how this algorithm works with the 2-spot game.

Surrounded positions are the losing ones. As Figure 6 shows, a position is losing if all its children are winning. But on the contrary, only one losing child is sufficient to prove that a position is winning, which implies that only a part of the complete game tree is sufficient to determine the outcome of the game (only

[4] In the following, we use the term "position" even if we speak of the string that represents it.

[5] This step is reached only if no child is losing.

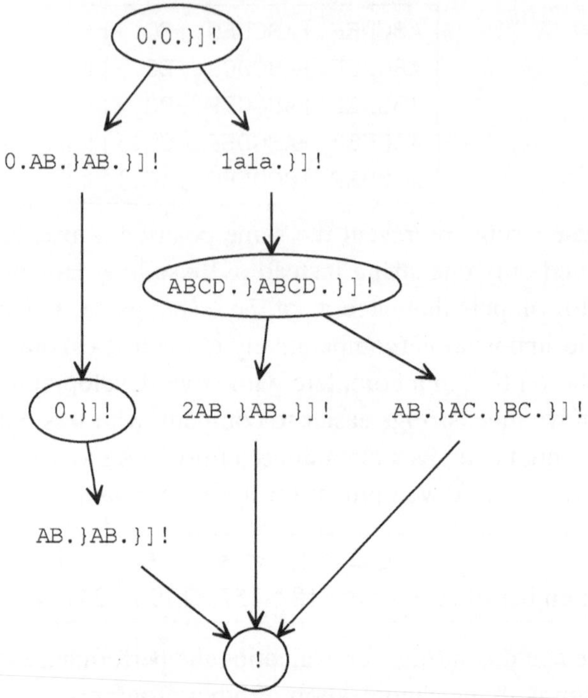

Figure 6. 2-spot game.

9 of the 18 positions of the complete game tree are needed here). We will detail this point again in Section 7.2.

6.2. ***Sum of independent games.*** Sprouts positions can frequently be separated into independent games, the *lands*. It would then be convenient to compute these lands independently, and deduce the outcome of the complete position from the outcome of the lands.

As explained in [Applegate et al. 1991], this can partially be done, because the sum of two losing games is losing, and the sum of a losing and a winning game is winning. However, the weakness of this method is that the sum of two winning games can either be winning or losing. We had implemented it in our program at first, and rather quickly, a large number of positions with several lands was saturating our databases, preventing us from improving the results of [Applegate et al. 1991] of more than 2 or 3 spots.

As described below, the concept of *nimber* can solve this problem, allowing us to compute the lands separately.

6.3. ***Nimber theory.*** A more detailed view of this theory can be found, amongst others, in [WW].

Definition. The nimber of a position P is denoted by $|P|$, and is defined as the smallest nonnegative integer that is not a nimber of a child of P.

This definition implies that $|P| = 0$ if P is losing, and $|P| \geq 1$ if P is winning. We can also see that, if we know the complete game tree of P, we can recursively compute $|P|$, but in fact it is not necessary to expand the complete game tree to compute the nimber of a position, as the Algorithm 2 shows below.

The main result of the nimber theory can be stated as follows:

Theorem. *If a position P is made of two independent positions P_1 and P_2, then $|P|$ is the "bitwise exclusive or" of $|P_1|$ and $|P_2|$, denoted by $|P_1| \wedge |P_2|$.*

For example, $|$`1AB.}AB.}]!`$| = 3$ and $|$`22.}]!`$| = 1$, so

$$|\texttt{1AB.\}AB.\}]22.\}]!}| = 3 \wedge 1 = 2.$$

6.4. ***Couples.*** In our program, instead of computing the outcome of a position, we compute the outcome of a *couple*: (position + nimber), which represents the sum of two independent games. The position part consists of the game of Sprouts for the given position. The nimber part is the game of Nim for the given nimber value. The outcome of $(\varnothing + n)$ is a loss if $n = 0$, a win if $n \geq 1$. The original position P_0 is replaced by the couple $(P_0 + 0)$.

We see that "$(P + n)$ is losing" means that $|P| = n$. As we only store losing positions in our program, it means that we will only store positions whose nimber is known (and the winning positions that we do not store correspond to positions whose nimber is only known to be different of certain values).

To determine the outcome of a couple, we can still use the Algorithm 1, by extending to couples the definition of children: the children of a couple $(P + n)$ are the children of the position part, whose form is (child(P) + n), and the children of the nimber part, whose form is $(P + m)$, with $m < n$.

6.5. ***Computation of the nimber of a position.*** With Algorithm 1 applied to a couple $(P + n)$, we already are able to determine if n is the nimber of the position P. If we need to compute the nimber $|P|$ of this position, we use the following (simple but efficient) method:

Algorithm 2. function nimber-of(position P)

- $n := 0$, *found* := *false*
- while *found* = *false* do:
  if compute-win-loss$(P + n)$ = Loss, then *found* := *true*, else $n := n + 1$
- return n

It merely consists in trying 0, 1, 2... until we find the right nimber. This algorithm will only be used on single lands, as explained below.

6.6. ***Positions with several lands.*** If a position is made of two lands, as in $(P_1]P_2]!+n)$, and if we know the nimber $|P_2|$, the above theorem shows that the outcome of $(P_1]P_2]!+n)$ will be the same as the outcome of $(P_1]!+n \wedge |P_2|)$.

So when our main algorithm meets a position with several lands, it computes with the Algorithm 2 the nimbers of all lands except one and merges the results with the nimber part of the couple. Therefore, we always compute the lands separately, and an indirect consequence is that we store only single lands in our database.

6.7. ***Main algorithm.*** With the previous ideas, Algorithm 1 is now:

Algorithm 3. function compute-win-loss(couple($P+n$))

- for the lands of P whose nimber is already stored in the database, merge their nimber with the nimber part (with bitwise xor).
- compute the nimber of all the unknown lands of P with Algorithm 2 (except for one land), and merge those nimbers with the nimber part (P is then a single land).
- for each child[6] of $(P+n)$, do:
  if compute-win-loss(child)=Loss, return Win
- store $(P+n)$ in the database and return Loss

Figure 7 shows how this algorithm works with the 3-spot game.

Couples surrounded by a rectangle are those made of several lands. Their nimber is computed from the nimber of their lands, which can be found lower in the graph.

Couples surrounded by an ellipse are losing. Only those couples are stored by the program. All their children are computed to find their position's nimber.

The remaining couples are winning. The nimber of their position part is not known. We only know the nimber to be different of one (or sometimes several) value, because this value is already the nimber of some child of the position.

We see on this example the efficiency of this algorithm. There are 157 positions in the complete game tree of the 3-spot game, but the program only needs 14 nodes to compute the outcome.

7. Computation improvement

7.1. ***Positions storage.*** As in [Applegate et al. 1991], we store only losing couples in order to reduce the memory consumption: as the complexity of the computation increases, losing couples become less frequent, so this choice allows us to reduce the size of the transposition table. The profit is variable,

[6]Children of position part, and children of nimber part.

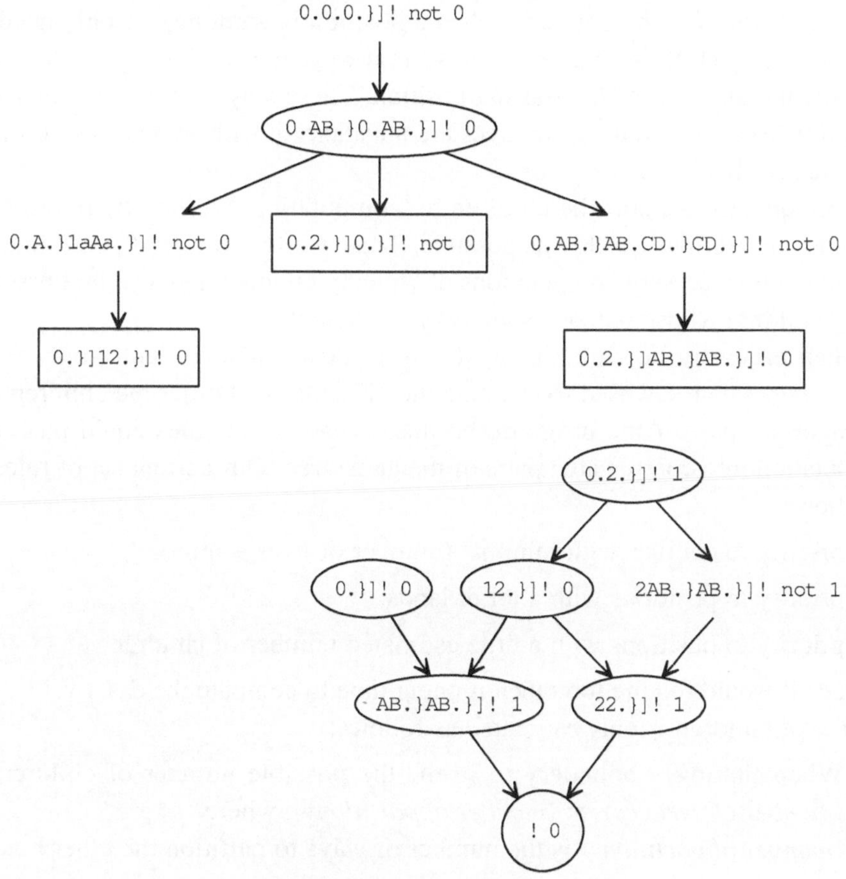

Figure 7. 3-spot game.

but we can take into consideration that the size of the database is divided by a factor between 5 and 20, which is not negligible. In exchange, we need a little more running time when computing again a winning couple that has already been computed before. Since the couple is not stored in the database, its winning outcome will be known only after a computation of its children, one of which will be found to be losing in the database.

The main difference of our database compared to [Applegate et al. 1991] is that we store losing couples (positions whose nimber is known), and not losing positions. Moreover, as explained before, we store only positions with a single land.

It is also possible in our program to export databases as text files, allowing us to analyze them manually, or resume later an incomplete computation.

7.2. ***Children ordering.*** When a position (or a couple) is losing, there is no short way to prove it: we need to compute all the children and prove that they are all winning. On the contrary, when a position is winning, we only need to find one losing child, so that there are several ways to search the game tree and compute the outcome of the starting position. These ways are not of equivalent difficulty, because Sprouts game trees are unbalanced, with some areas far more complicated than others.

Consequently we sort the children by computational difficulty, in order to find a losing child as quickly as possible. Thus, we lose the least possible time possible in unnecessary computations of winning children and also find first the losing children whose outcomes are easy to compute.

When ordering the children, we try to evaluate their difficulty only from their string. The rules used to evaluate the difficulty and order the children are an important part of the program, because a bad set of rules could push the computation into complicated parts of the game tree. Our current set of rules is as follows:

- priority to couples with minimal (number of lives + nimber);
- priority to positions with a lot of lands;
- priority to positions with a little estimated number of children.

Since it would require too much running time to compute the exact value, the number of children is only estimated as follows:

- When linking a boundary to itself, the possible number of children is $(\textit{number of vertices})^2 \times (\textit{number of partitions})$, where $(\textit{number of partitions})$ is the number of ways to partition the other boundaries into two sets.
- When linking two different boundaries, the possible number of children is the sum, for any couple (B_i, B_j) of boundaries, of $(\textit{number of vertices of } B_i) \times (\textit{number of vertices of } B_j)$.

We must also make a choice when considering a position composed of several lands. We decided to compute first the nimbers of the lands with the smallest number of lives.

These rules significantly improve the computation compared to a random exploration of the game tree, but they are still far from providing an optimal exploration of the game tree. In fact, we believe that whatever the rules used to order the children, their efficiency is eventually limited and more global search algorithms are needed.

7.3. ***Manual exploration of the game tree.*** We have implemented an interface to follow and interact in real-time with the computation process. When the computation seems to be stuck in some part of the game tree, we can manually decide to explore elsewhere.

During the computation, the program displays the list of currently studied couples. At a given instant, the program is computing the outcome of a list of couples, each one being the child of the previous one. For example, here is what the program could display during a 12-spot game computation:

level	position part	nimber part
1	`0.0.0.0.0.0.AB.}0.0.0.0.0.AB.}]!`	0
2	`0.0.0.0.0.}]0.0.0.A.}0.0.0.A.}]!`	0
3	`0.0.0.0.0.0.}]!`	1
4	`0.0.0.0.AB.}AB.}]!`	1
5	`0.0.0.0.}]!`	1
6	`0.0.1a1a.}]!`	1

With a click in the interface, it is then possible to choose another child on a given level, and so to compute another part of the game tree. For example, we could decide to compute the couple (`0.0.A.}0.0.A.}]!`+1) in the fifth level.

When the position is composed of several lands, the program computes the nimber of one of these lands with Algorithm 2 (here, it has already computed that the nimber of `0.0.0.0.0.0.}]!` is not 0, and is now trying with 1). In this case too, we can click in the interface to compute first the nimber of another land. In our example, we could decide to compute the nimber of `0.0.0.A.}0.0.0.A.}]!` in the third level.

We empirically decide where and when we should click: if the program spends too much time on a branch of the game tree, we decide to change it. If a couple on a level is almost computed to be losing (which means that almost all its children have been computed to be winning), it is often efficient to click two levels below, to try to quickly find losing children of the remaining unknown children. We can also have a look at the position to try to choose easier couples (positions are easier when they tend to be quickly cut into lands).

Being able to manually choose which part of the game tree to explore was far more efficient that any automatic selection that we imagined. For example, whereas a computation for the 12-spot game ends after having stored more than 100, 000 couples without human intervention, by clicking, somebody with some experience can end this game in less than 2, 000 couples.

7.4. ***Check computation.*** User interactions proved to be powerful, but they have a drawback: it is impossible to reproduce exactly the same computation twice. For this reason, when a computation succeeds, we perform a stand-alone check computation, which uses the previous results to guide itself in the game tree. This check computation does not authorize any interaction from the user and is of course reproducible for a given database of previous results.

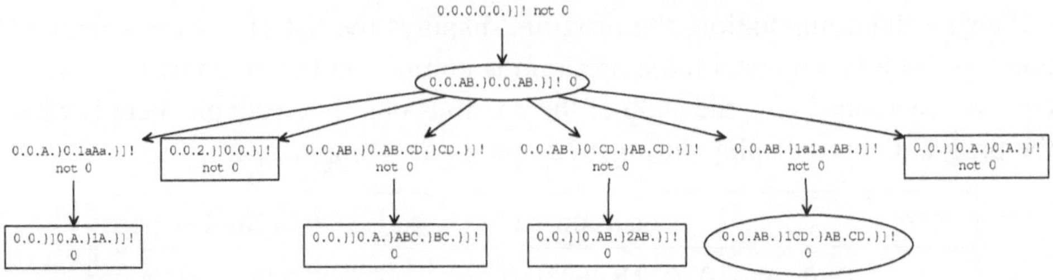

Figure 8. A solution for the 5-spot game, positions with 12 lives or more.

The check computation also reduces the number of couples needed to prove the result. Indeed, during the first computation, we compute many useless couples (the winning children of winning couples are usually useless). When we meet a new couple in the check computation, we look for its value in the database of previous results, and if it is losing (i.e., a position whose nimber is known), we compute again all its children to check it. But if it is winning, we look in the previous database which child is losing, and we compute only this one.

The result of a check computation is then a *solution tree*, providing a winning strategy without any unnecessary information. These solution trees are small enough to provide hand-checkable proofs for p-spot games with little values of p. For bigger values of p, it is mainly a way of reducing the size of the files storing the winning strategies.

During the check computation, we can generate a file compatible with the graph visualization tool Graphviz (www.graphviz.org) which enables us to draw graphs describing how the computation proceeds (e.g., Figures 6 and 7). For bigger values of p, the complete graph would be rather unreadable, but we can choose to plot only the positions with a minimal number of lives. For example, Figure 8 shows the upper part of the graph for the 5-spot game, with only the positions with 12 lives or more.

Here, we can see how cutting positions into lands simplifies the computation: to complete it, we only need to compute the nimber of 6 lands of less than 7 lives, and the outcome of a position of 12 lives. The most spectacular example is the case of $p = 17$: in our computation, every position is cut into several lands of less than 27 lives after only 3 moves from the starting position (which has 51 lives).

It is also possible to plot only reference numbers instead of the whole positions, which is useful for bigger graphs. See, for instance, Figure 9, which shows the complete 4-spot game. The numbers have no special meaning, the positions which they refer to are listed in Table 1, with their computed values.

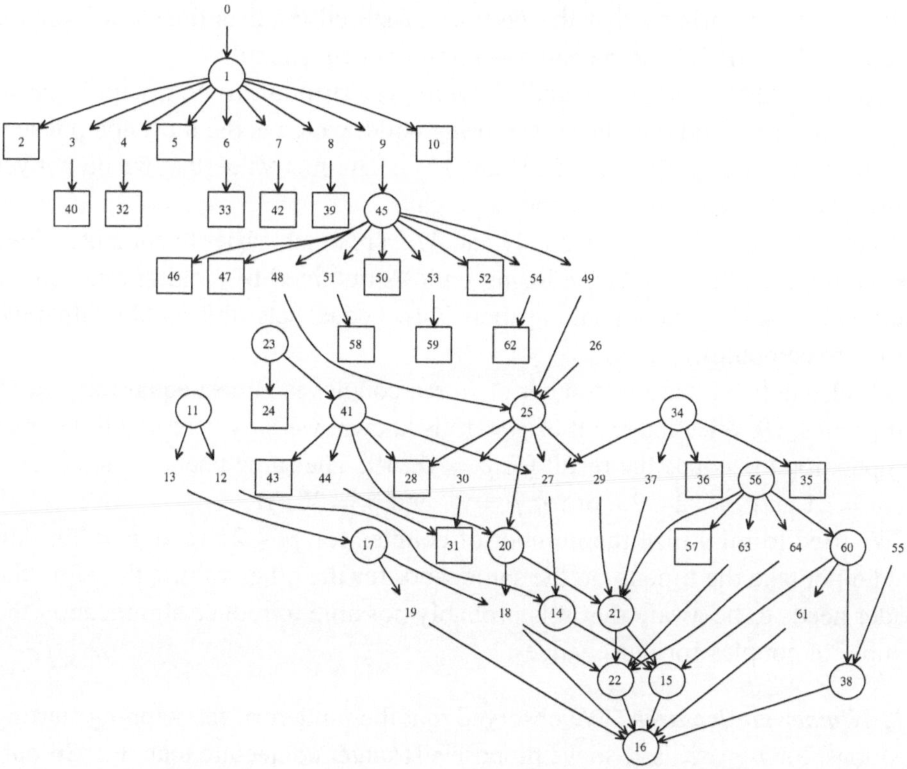

Figure 9. A complete solution for the 4-spot game.

8. Results

8.1. p-spot games already computed. The following table shows the number of couples stored in the minimal databases for all the p-spot games computed up to now. We have computed all p-spot games up to 32 spots. The 33-spot game is the first unknown one, but the biggest computed one is the 47-spot game. The results so far all support the conjecture emitted in [Applegate et al. 1991]: the first player loses if and only if p is 0, 1 or 2 modulo 6.

p	size	p	size	p	size	p	size	p	size	p	size
0	0	7	103	14	1580	21	9270	28	14813	35	18812
1	1	8	205	15	3252	22	5706	29	3414	…	?
2	3	9	63	16	1068	23	2837	30	58363	40	45782
3	6	10	140	17	471	24	9316	31	58365	41	48890
4	16	11	140	18	3233	25	9229	32	58204	…	?
5	38	12	475	19	3630	26	18567	33	?	47	54542
6	64	13	577	20	4051	27	59117	34	21107	…	?

It is rather surprising that the number of stored couples does not increase strictly with p. In fact, some patterns seem to occur modulo 6.

We found that $p = 15$, 21 and 27 were very difficult to compute, because they are winning, and only have one losing child, which is the most complicated one (its string is `0.0.0.(...)0.1a1a.}]!`). The first value that we do not yet know, $p = 33$, seems to follow the same path.

Conversely, $p = 17, 23, 29, 35, 41$ and 47 were much easier to compute. They are winning positions, and the losing child is obtained by linking one spot to itself and separating the remaining spots in two equal sets (this child is often the easiest to compute).

We also noticed that the number of stored couples is almost equal for $p = 15$ and $p = 18, 19$: once the result of $p = 15$ is known, we only needed a little more computation to deduce the result for $p = 18, 19$. The same phenomenon occurs for $p = 21$ and $p = 24, 25$, or for $p = 27$ and $p = 30, 31$.

We tried to minimize the number of couples for $p \leq 25$ (and $p = 29$), but we did not take the time to do the same work for the other values, therefore the reader needs to be aware that it is probably possible to reduce significantly the number of couples for those values.

8.2. ***Nimber conjectures.*** We observed that the nimber of the winning starting positions, for $p \leq 32$, is 1, so we propose a stronger conjecture than the "Sprouts conjecture" emitted in [Applegate et al. 1991]:

Conjecture. *The nimber for the starting position with p spots is* 0 *if p is* 0, 1 *or* 2 *modulo* 6, *and* 1 *if p is* 3, 4 *or* 5 *modulo* 6.

It is rather easy to imagine other conjectures around the nimber. For example, if we compute the nimber of the position: `222(...)222.}]!`, with p generic vertices "2", we obtain (starting from `22.}]!`):

1; 0; 2; 1; 0; 3; 1; 0; 5; 1; 0; 3; 1; 0; 3; 1; 0; 3; 1; 0; 3; 1; 0.

We also imagined another extension of the Sprouts conjecture: the nimber of the position is left unchanged by an addition of 6 boundaries "`0.`" into a given region. This conjecture is false, since `0.22.}]!` has nimber 0, and `0.0.0.0.0.0.0.22.}]!` has nimber 2. But it does work for more than 90% of the positions that we tried. The existence of such nearly true patterns tends to infirm the Sprouts conjecture.

8.3. ***Hand-checkable proofs.*** Using a computer to determine mathematical results is not completely convincing, especially because there could be a programming error, considering the size of the program. So, the most skeptic readers

#	position	nimber	#	position	nim.
0	0.0.0.0.}]!	≠0	33	0.A.}1A.}]0.}]!	0
1	0.0.AB.}0.AB.}]!	0	34	0.A.}1A.}]!	0
2	0.0.2.}]0.}]!	≠0	35	0.}]AB.}AB.}]!	≠0
3	0.0.AB.}AB.CD.}CD.}]!	≠0	36	12.}]1.}]!	≠0
4	0.0.A.}1aAa.}]!	≠0	37	1A.}ABC.}BC.}]!	≠0
5	0.0.}]0.2.}]!	≠0	38	1.}]!	1
6	0.1aAa.}0.A.}]!	≠0	39	0.AB.}2AB.}]0.}]!	0
7	0.AB.CD.}0.AB.}CD.}]!	≠0	40	0.A.}0.A.}]AB.}AB.}]!	0
8	0.AB.}0.CD.}AB.CD.}]!	≠0	41	0.A.}ABC.}BC.}]!	0
9	0.AB.}1a1a.AB.}]!	≠0	42	0.A.}ABC.}BC.}]0.}]!	0
10	0.A.}0.A.}]0.}]!	≠0	43	12.}]AB.}AB.}]!	≠0
11	0.0.}]!	0	44	ABC.}ADE.}BC.}DE.}]!	≠0
12	0.AB.}AB.}]!	≠0	45	0.AB.}1CD.}AB.CD.}]!	0
13	1a1a.}]!	≠0	46	0.2.}]1AB.}AB.}]!	≠0
14	0.}]!	0	47	0.AB.}2AB.}]1.}]!	≠0
15	AB.}AB.}]!	1	48	0.AB.}2CD.}AB.CD.}]!	≠0
16	!	0	49	0.AB.}ABC.}CDE.}DE.}]!	≠0
17	ABCD.}ABCD.}]!	0	50	0.AB.}AB.}]12.}]!	≠0
18	2AB.}AB.}]!	≠0 1	51	0.A.}1B.}aAaB.}]!	≠0
19	AB.}AC.}BC.}]!	≠0	52	0.}]1AB.}2AB.}]!	≠0
20	0.2.}]!	1	53	1aAa.}1BC.}ABC.}]!	≠0
21	12.}]!	0	54	1AB.}AB.CD.} CD.EF.}EF.}]!	≠0
22	22.}]!	1			
23	0.A.}0.A.}]!	1	55	1AB.}AB.}]!	≠1
24	0.}]12.}]!	0	56	1AB.}2AB.}]!	1
25	0.AB.}2AB.}]!	0	57	1.}]22.}]!	0
26	0.0.2.}]!	≠0	58	0.}]1.}]AB.}AB.}]!	0
27	0.A.}2A.}]!	≠0	59	12.}]1A.}2A.}]!	0
28	0.}]22.}]!	≠0	60	1A.}2A.}]!	0
29	1aAa.}2A.}]!	≠0	61	2A.}2A.}]!	≠0
30	2AB.}AB.CD.}CD.}]!	≠0	62	1AB.}2AB.}]AB.}AB.}]!	0
31	22.}]AB.}AB.}]!	0	63	2AB.}2AB.}]!	≠1
32	0.0.}]12.}]!	0	64	2A.}ABC.}BC.}]!	≠1

Table 1. Correspondence for Figure 9.

could use the program to generate files that would provide them with hand-checkable proofs of the simplest results. For example, Figure 9 provides a proof that the first player wins in the 4-spot game.

We printed the result of a check computation for the 9-spot game and manually checked 66 losing couples (positions of known nimber), and 258 winning couples (positions whose nimber is known to be different of certain values). For the losing couples, we checked that we obtained the same sets of children as our program, and for the winning ones, we had to check only one child, very seldom two or more. This took us a few hours.

The table in Section 8.1 implies that we could use this method to check in a reasonable amount of time the value of the p-spot game with $p \leq 11$. The program can also point in the right direction someone who would like to create a totally manual proof, like the one in [Focardi and Luccio 2004].

Conclusion

The main obstacle for computing higher p-spot games is neither memory nor computation time, but rather human time for the manual exploration of the game tree. This human intervention is an embryo of a best-first search, which is probably more suited for the game of Sprouts than the depth-first search currently performed by our program. Therefore, implementing classical best-first search algorithms, such as the PN-search, would probably lead to better results.

Distributed computing is also another solution to compute higher p-spot games, and the check computation, by reducing the size of the databases, would be really efficient for this. It is likely that in the coming years, we will know who wins in the game of Sprouts when starting with more than fifty spots.

The program that we used for the computations is available with its source code on our website, sprouts.tuxfamily.org, under a GNU license, together with several resulting databases.

Acknowledgements

We wish to thank Jean-Paul Delahaye for his decisive contribution to the continuation of our work.

References

[Applegate et al. 1991] D. Applegate, G. Jacobson, and D. Sleator, "Computer analysis of sprouts", Computer Science Technical Report CMU-CS-91-144, Carnegie Mellon University, 1991, http://www.cs.cmu.edu/~sleator/papers/sprouts.pdf.

[Draeger et al. 1990] J. Draeger, S. Hahndel, G. Köstler, and P. Rossmanith, "Sprouts – Formalisierung eines topologischen Spiels", Tech. Report TUM-I9015, Technische Universität München, Institut für Informatik, 1990.

[Focardi and Luccio 2004] R. Focardi and F. L. Luccio, "A modular approach to Sprouts", *Discrete Appl. Math.* **144**:3 (2004), 303–319.

[Gardner 1967] M. Gardner, "Mathematical games: Of sprouts and Brussels sprouts, games with a topological flavor", *Scientific American* **217** (1967), 112–115.

[Lemoine and Viennot 2009] J. Lemoine and S. Viennot, "Analysis of misère Sprouts game with reduced canonical trees", preprint, 2009. arXiv 0908.4407

[WW] E. R. Berlekamp, J. H. Conway, and R. K. Guy, *Winning ways for your mathematical plays*, vol. 1, 2nd ed., A K Peters Ltd., 2001, Natick, MA.

julien.lemoine@gmail.com *France*

sviennot@jaist.ac.jp *Department of Information Science/Artificial Intelligence, Japan Advanced Institute of Science and Technology, 1-1 Asahidai, Nomi 923-1211, Japan*

Games of No Chance 4
MSRI Publications
Volume **63**, 2015

Navigating the MAZE

NEIL MCKAY, RICHARD NOWAKOWSKI AND ANGELA SIEGEL

The game of MAZE was introduced in 2006 by Albert, Nowakowski and Wolfe, and is an instance of an option-closed game and as such each position has reduced canonical form equal to a number or a switch. It was conjectured that because of the 2-dimensional structure of the board there was a bound on the denominator of the numbers which appeared as numbers or in the switches. We disprove this by constructing, for each number and each switch, a MAZE position whose reduced canonical form is that value. Surprisingly, we can also restrict the interior walls to be in one direction only, seemingly giving an advantage to one player. This also gives a linear time algorithm that determines the best move up to an infinitesimal.

1. Introduction

MAZE was introduced in [Albert et al. 2007], but apart from a few scattered observations, nothing was known about the values of the game. In the original article, MAZE is played on a rectangular grid oriented 45° to the horizontal.

The token starts at the top of the board and highlighted edges are *walls* that may not be crossed. Left is allowed to move a token any number of cells in a southwesterly direction and Right is allowed to move similarly in a southeasterly direction. However, for ease of referring to specific places in the position, we re-orient the sides parallel to the page so that Left moves downward and Right moves to the right; see Figure 1. One interesting feature is that any number of consecutive Left (Right) moves also can be accomplished in one move. This feature had been noted in several games, including HACKENBUSH strings [Berlekamp et al. 2001], and given the name of *option-closed* in [Nowakowski and Ottaway 2011], a reference we henceforth abbreviate as [NO]. Siegel [2011] notes that the partial order of option-closed games born on day n forms a planar lattice.

For a game G, G^L (G^R) is a left (right) option of G; $G^{\mathcal{L}}$ ($G^{\mathcal{R}}$) is the set of all left (right) options of G; and for a set of games S, $S^{\mathcal{L}}$ ($S^{\mathcal{R}}$) is the set of all left (right) options that can be reached from any game in S (i.e. $G^{\mathcal{L}\mathcal{L}}$ is the set

Keywords: option-closed, reduced canonical form, HACKENBUSH string, CRICKET PITCH, MAZE.

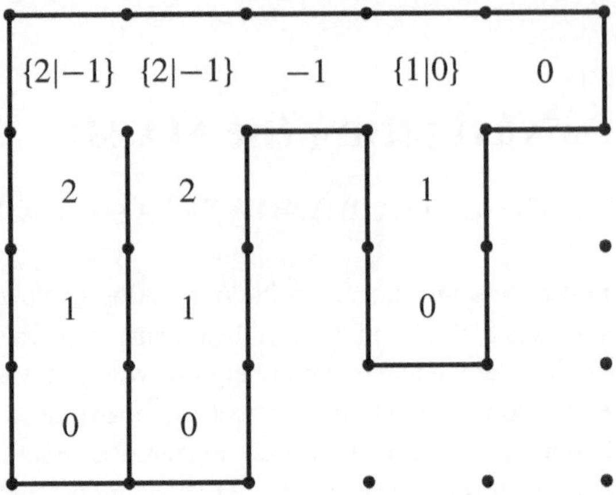

Figure 1. A MAZE board rotated so that Left and Right move down and right, respectively. Reduced canonical form of values included on board.

of all options that can be reached in two consecutive left moves from G). Most of combinatorial game theory considers the canonical forms of games, those where $G^{\mathcal{L}}$ and $G^{\mathcal{R}}$ have had dominated options removed and reversible options bypassed. We need the form of the game $G = \{G^{\mathcal{L}} | G^{\mathcal{R}}\}$ with all the options, even the bad ones. We call this the *literal* form of the game.

Definition 1 [NO]. A game G is called *option-closed* if, in the literal form, $G^{\mathcal{LL}} \subset G^{\mathcal{L}}$, $G^{\mathcal{RR}} \subset G^{\mathcal{R}}$ and, recursively, all the followers of G are option-closed.

For example, even though the canonical form of $G = \{0 \mid \{1 \mid 0\}, 0\}$ and $H = \{0 \mid \{1 \mid -1\}, 0\}$ are the same, G is option-closed but $H = \{0 \mid \{1 \mid -1\}, 0\}$ is not since Right can move to -1 in two moves but not in one. The fact that a game is option-closed is intrinsic to the game and is not necessarily identifiable from the canonical form [NO].

The canonical form of an option-closed game can be quite complicated. However, often the differences between many of the options in the canonical form are infinitesimals. The *reduced canonical form* (simplest game infinitesimally close, [Grossman and Siegel 2009]) should be much less complicated and indeed, in [NO], it was shown that the reduced canonical form of an option-closed game is either a number or a switch $\{a|b\}$ of numbers $a > b$.

In [NO], the authors noted that MAZE is an option-closed game and asked for an analysis of the game. It was thought that the two-dimensionality of the board would restrict the powers of 2 that could occur as a denominator of numbers that occur in reduced canonical forms. We show that this is false and give an algorithm to construct all numbers and all switches. Surprisingly, the algorithm constructs a position which is rectangular and all interior walls are vertical, that is, they prevent only Right moves.

In the next section, we present an overview of the required theory of the reduced canonical form and option-closed games before moving on to prove our main result in Section 4. Also, in Section 5 (Theorem 14), we obtain a linear time algorithm to find the reduced canonical form of any MAZE position.

We follow the combinatorial game theory notation and definitions of [Albert et al. 2007; Berlekamp et al. 2001].

2. Reduced canonical form and option-closed background.

Definition 2 [Grossman and Siegel 2009]. Two games G and H are *equalish*, written $G =_I H$, if $G - H$ is an infinitesimal. A game G is *numberish* if there is a dyadic rational x such that $G - x$ is an infinitesimal. A game G is *infinitesimally-dominated* by H if there is some integer n, such that $H + n.\uparrow - G \geq 0$; written as $H \geq_I G$. A left option, G^L, of G is *infinitesimally-reversible* if there is some $G^{LR} \leq_I G$.

Definition 3 [Grossman and Siegel 2009]. A game G is in *reduced canonical form*, denoted rcf(G), if for any follower H of G, either (i) H is a number in simplest form, or (ii) H is neither a number nor numberish and H contains no infinitesimally-dominated or infinitesimally-reversible options.

Lemma 4 [Grossman and Siegel 2009]. *[Thm 4.8] If G is not numberish, then rcf(G) is obtained by (i) replacing options with simpler options infinitesimally close to the original option, (ii) eliminating infinitesimally-dominated options, and (iii) bypassing infinitesimally-reversible options.*

Lemma 5 [NO]. *If a and b are numbers with $a \geq b$, then $a \geq_I \{a \mid b\} \geq_I b$.*

To illustrate the above concepts, we sketch the proof of Lemma 5. Let a and b be as in Lemma 5. Note that a and $\{a \mid b\}$ are incomparable. Consider $a - \{a \mid b\} + 3\cdot\uparrow = a + \{-b \mid -a\} + 3\cdot\uparrow$. Left wins by moving to $a - b + 3\cdot\uparrow > 3\cdot\uparrow > 0$. By the Number Avoidance Theorem, if Right can win then she must have a good move in either $\{a \mid b\}$ or in $3\cdot\uparrow$. But $a - a + 3\cdot\uparrow > 0$ while moving in $3\cdot\uparrow$ leaves $a + \{-b \mid -a\} + \Uparrow *$, in which case Left wins by responding to $a - b + \Uparrow * \geq \Uparrow * > 0$.

We need two other relatively trivial but necessary results about the transitivity of $\geq_I$ that we will use without mention.

Lemma 6. *If $x \geq_I y$ and $y \geq_I z$ then $x \geq_I z$.*

Lemma 7. *If $x \geq_I y$ and $y > z$ with y and z numbers, then $x > z$.*

3. The structure of MAZE

The main result of [NO] is that the reduced canonical form of an option-closed game is either a number or a switch.

Lemma 8 [NO]. *If G is an option-closed game, then $rcf(G) = rcf(\{a \mid b\})$ for some numbers a and b.*

In the special case of MAZE, this result is a corollary of the results needed to prove that our construction of Section 4 works.

Let G be an option-closed game where it and all of its followers are given in their literal forms. A right option G' is a *first right option* of G if $G' \in G^{\mathcal{R}} \setminus G^{\mathcal{RR}}$, i.e., can be reached in one right move but not in two right moves. A *right-option-closed sequence* of G is a sequence of right options of G, $\alpha = \langle y_1, y_2, \ldots, y_n \rangle$ where y_1 is a first right option of G, y_{i+1} is a first right option of y_i, for $0 < i < n$, and the value of y_n is a number. A *first left option* and a *left-option-closed sequence* are defined analogously. Let α be a right (left)-option-closed sequence. The *norm* of α, written $\overline{\alpha}$ is the minimum (maximum) of all numbers in the sequence. The norm of $\langle \varnothing \rangle$ is not defined.

Lemma 9. *Let H be an option-closed game $\alpha = \langle y_0, y_1, \ldots, y_n \rangle$ be a right- or left-option-closed sequence. Let i be the least index such that y_i is a number. Then $y_i = \overline{\alpha}$.*

Proof. Suppose that α is a right-option-closed sequence and let i be the least index such that y_i is a number. Suppose y_j, $j > i$ is a number then y_j is a right option of y_i and hence $y_i < y_j$. The proof for a left-option-closed sequence is similar. □

Let M be a MAZE position. Then listing the options in order, $\langle G^{\mathcal{L}} \rangle$ and $\langle G^{\mathcal{R}} \rangle$ are, respectively, a left- and a right-option-closed sequence. This is an important fact but it doesn't define MAZE. Extra conditions must hold, if, from a cell, Left can move p cells and then Right q and reversing the order gives a legal move then the two resulting positions are the same.

If M is not a number then $\overline{\langle G^{\mathcal{L}} \rangle}$ is the left stop of M and $\overline{\langle G^{\mathcal{R}} \rangle}$ is the right stop and in those terms we could invoke results from [NO]. However, we need to know more details about the order relations between the options in the special case of MAZE. The following results can be and are generalized to games where $G^{\mathcal{L}}$ and $G^{\mathcal{R}}$ contain more than one option-closed sequence.

Theorem 10. *Let G be a MAZE position and set $\langle G^{\mathcal{R}} \rangle = \langle x_1, x_2, \ldots, x_m \rangle$ and $\langle G^{\mathcal{L}} \rangle = \langle y_1, y_2, \ldots, y_n \rangle$. Then*

(i) $rcf(G) = rcf\big(\{\overline{\langle G^{\mathcal{L}} \rangle} \mid \overline{\langle G^{\mathcal{R}} \rangle}\}\big)$;

(ii) *If G is a number, then $G^L < G < G^R$ for all $G^L \in G^{\mathcal{L}}$ and $G^R \in G^{\mathcal{R}}$. Moreover, $G = \{\langle\overline{G^{\mathcal{L}}}\rangle \,|\, \langle\overline{G^{\mathcal{R}}}\rangle\}$.*

(iii) *If G is not a number, then $\langle\overline{G^{\mathcal{L}}}\rangle \geq_I G \geq_I \langle\overline{G^{\mathcal{R}}}\rangle$.*

Proof. We induct on the size, i.e. number of cells, of the MAZE position. If the position has one cell (i.e. $G^{\mathcal{L}} = G^{\mathcal{R}} = \varnothing$), then the claims are trivially true. Suppose that the result is true for all positions with at most s cells.

Let G be a position with s cells. We may assume that G^R is not empty and that $\langle\overline{G^{\mathcal{R}}}\rangle = x_k$ for some $1 \leq k \leq m$.

(i) If $k = 1$, then x_1 is a number and by (ii), $x_1 < x_i$ for all i. If $k \neq 1$, then for $1 \leq i < k$, x_i is not a number and by induction from (iii), $x_i \geq_I x_k$. For $i > k$, $x_k < x_i$ by (ii). Combining these results gives $x_k \leq_I x_i$ for all i. Thus, $\mathrm{rcf}(G) = \mathrm{rcf}(\{\langle\overline{G^{\mathcal{L}}}\rangle \,|\, x_k\})$.

(ii) If G is a number, then $G < x_k$ since $x_k \in G^{\mathcal{R}}$. For all i we have $G < x_k \leq_I x_i$ from above, and so by Lemma 7, $G < x_i$. If $G^{\mathcal{L}} = \varnothing$, then $G = \{.\,|\, x_k\}$. Otherwise, $\langle\overline{G^{\mathcal{L}}}\rangle < G < x_k$, so $G = \{\langle\overline{G^{\mathcal{L}}}\rangle \,|\, x_k\}$.

(iii) If G is not a number, then $G^{\mathcal{L}} \neq \varnothing$ and $\langle\overline{G^{\mathcal{L}}}\rangle = y_j$ for some $1 \leq j \leq n$, and $\mathrm{rcf}(G) = \mathrm{rcf}(\{y_j | x_k\})$. If $y_j < x_k$, then $y_i \leq_I y_j < x_k < x_l$ for all i, l, so $G = \{y_j \,|\, x_k\}$ is a number, which is a contradiction. Hence, $y_j \geq x_k$. Combining the above results, $G \equiv_I \{y_j \,|\, x_k\} \geq_I x_k = \langle\overline{G^{\mathcal{R}}}\rangle$ by Lemma 5. Thus, $G \equiv_I \{y_j \,|\, x_k\}$.

Thus, $y_j \geq x_k$ and $G \equiv_I \{y_j \,|\, x_k\}$ and so $\geq_I x_k$ by Lemma 5. □

In the case that G is not a number, then $\langle\overline{G^R}\rangle$ and $\langle\overline{G^L}\rangle$ are the right and left stops of G. In this language, Theorem 10 is an extension of [NO], applied to the specific case of Maze.

4. The construction

All positions will be rectangular mazes with vertical walls plus the horizontal walls on the lower edge of the rectangle. For brevity, we refer to such a position as a *vertical* position. An *interior* wall has its endpoints non-adjacent along the outside boundary. In any MAZE layout, each cell has a value corresponding to that position where the token is on that cell. The value of the cell at the top left of the maze the *value of the rectangle*. With a number a, we associate $m(a)$, any MAZE position with a reduced canonical form of value a. First, we must show how to adjust the height of any MAZE position without changing its value.

Lemma 11. *Let M be a* MAZE *position. Let M' be the* MAZE *position obtained by: deleting the bottom and right-hand walls of the position; adding another row at the bottom and a column on the right hand side with walls on the bottom*

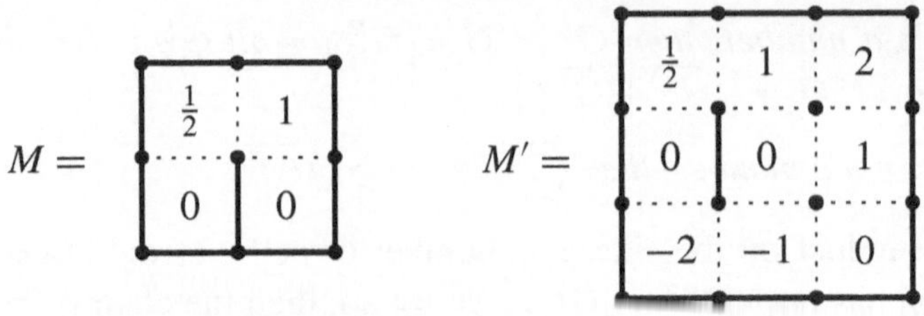

Figure 2. MAZE boards of equivalent value.

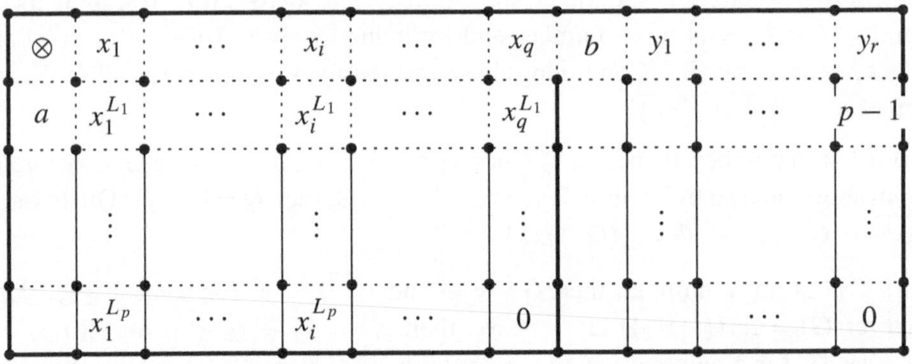

Figure 3. Outline of the Construction. Values without subscripts are actual values. Thick black lines represent existing walls, dashed lines walls that do not exist and thin lines walls that may or may not exist.

of the new row and on the right-hand side of the new column. (Each dimension has increased by 1.) *The values of M and M' are equal.*

Proof. Let the rows of M be indexed $1, 2 \dots, p$ and the columns $1, 2, \dots, q$ and let the rows and columns of M' be indexed $0, 1, 2 \dots, p$ and $0, 1, 2 \dots, q$ respectively, where (0,0) and (1,1) are the bottom-right corners of M' and M, respectively.

The cells on the bottom row of M' have the values $0, -1, -2, \dots, -q$, from right to left, and the right-hand cells have values $0, 1, 2, \dots, q$, from bottom to top.

The cell (1,1) in M' has value $\{-1 \mid 1\} = 0$, which equals the value of the (1,1) cell in M. We now proceed by induction on $i + j$. Note that the value of (i, j) is $\leq i$ since Left has at most i moves and similarly the value is $\geq -j$. Therefore, any moves from (i, j) to the zero row or to the zero column are dominated and the values of the cells then the values in the rest of M' are the same as those of M. □

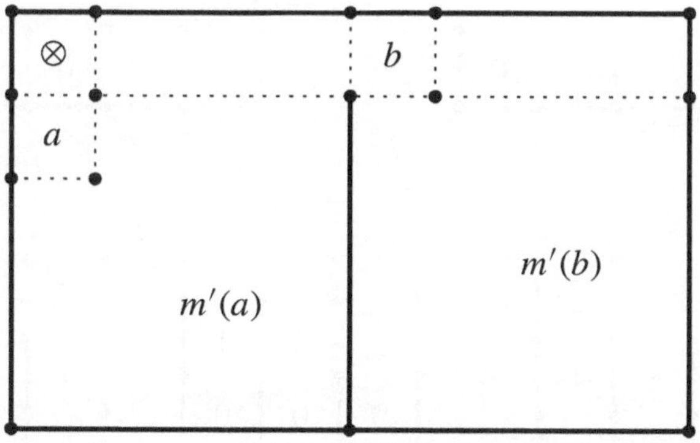

Figure 4. Outline of the Construction. Conventions are as in the previous figure.

The Construction. The idea is the following: if we wish to construct G with $\text{rcf}(G) = \text{rcf}(\{a \mid b\})$, a and b numbers, then take the MAZE positions $m(a)$ and $m(b)$ obtained via the Construction, adjust the heights and adjoin the two positions as in Figure 4. We need one more piece of notation. Let $x = y/2^m$ where y is odd and $m > 0$, put $\|x\| = m$. If $x = \{a \mid b\}$), the naming convention of numbers only allows the power of 2 in the denominator of x to to increase slightly over that of a and b.

Lemma 12. *If p and q, $p < q$, are dyadic rationals and $x = \{p \mid q\}$ then $\|x\| \leq \max\{\|p\|, \|q\|\} + 1$. If x is in canonical form then $\|x\| = \max\{\|p\|, \|q\|\} + 1$ and $\|x\| < \|s\|$ for any dyadic rational $p < s < q$.*

Theorem 13. *Let g be a MAZE position in canonical form.*

(1) *Suppose g is an integer. If $g = 0$. then let M be a single cell; if $g > 0$ then let M be a row of $g+1$ cells; if $g < 0$ then let M be a column of $g+1$ cells. In all cases, M has no interior walls.*

(2) *Suppose g is not integer. Then $g = \{a \mid b\}$ where a and b are dyadic rationals. Take the MAZE positions obtained via the Construction for a and b. Adjust the height, if necessary, to obtain $m'(a)$, a $p \times (q+1)$ rectangle and $m'(b)$, a $(p+1) \times (r+1)$ rectangle with the same values as $m(a)$ and $m(b)$ respectively. Let $m''(a)$ be formed from $m'(a)$ by removing the top outside walls and adding a new top row with no interior walls. Form the MAZE position M by concatenating the rows of $m''(a)$ and $m'(b)$.* (See Figure 4.)

The value of rcf(M) is g.

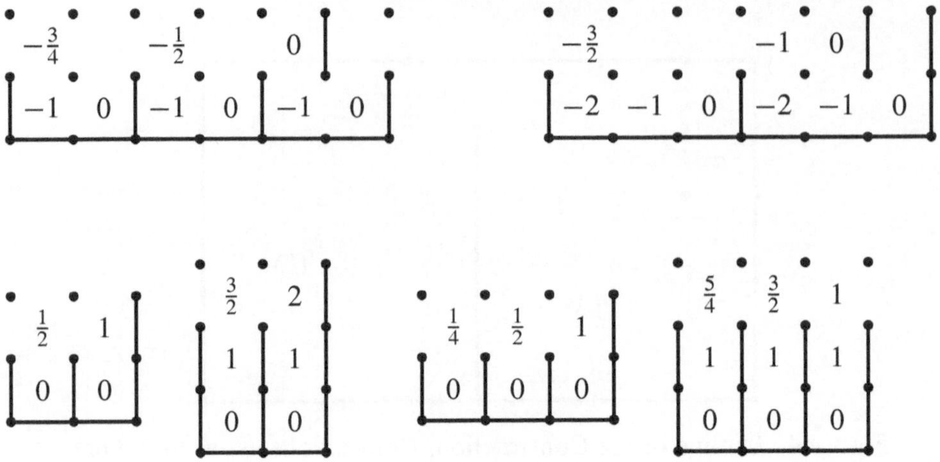

Figure 5. MAZE boards for $g = n + 1/2^k$.

Proof. (1) Suppose g is an integer then $M = \text{rcf}(M) = g$ is clear.
(2) Let $g = n + 1/2^k$, n an integer and $k \geq 1$. Then g can be constructed as follows. We only put in representative cases as the general case is clear but tedious.

Note that for any entry x of M other than the top left cell (which has value g) then $\|x\| < \|g\| = k$.

Suppose now that g is not an integer nor is $g = n + 1/2^k$ (i.e. a not an integer). Let $g = \{a \mid b\}$ in canonical form, with a and b be dyadic rationals. Let M be the MAZE position whose reduced canonical form is claimed to be g. Note that both $m''(a)$ and $m'(b)$ are obtained by the Construction and so contain no interior walls in their top rows. Moreover, if c is a cell in $m'(a)$, other than the top-left cell, then $\|c\| < \|a\|$. See Figure 3 for the naming of the values of the cells. Since there are no walls in the top row of $m(b)$, $\langle x_1, x_2, \ldots, x_q, b, y_1, y_2, \ldots, y_r = p\rangle$ is a right-option-closed sequence, and the values in the first column, a and below, form a left-option-closed sequence. Thus, $\overline{\langle \otimes^{\mathcal{L}} \rangle} = a$ by Lemma 9 and since $\langle b, y_1, y_2, \ldots, y_r\rangle$ is also a right-option-closed sequence, then $y_i > b$ by Theorem 10. Thus,

$$\text{rcf}(\otimes) = \text{rcf}\left(\{a \mid \overline{\langle x_1, x_2, \ldots, x_q, b, y_1, y_2, \ldots, y_r\rangle}\}\right) = \left(\{a \mid \overline{\langle x_1, x_2, \ldots, x_q, b\rangle}\}\right)$$

and we need to show that

$$\overline{\langle x_1, x_2, \ldots, x_q, b\rangle} = b.$$

Now, if no x_i is a number, then by Lemma 9, $\overline{\langle x_1, x_2, \ldots, x_q, b\rangle} = b$.

Therefore, we may let i be the greatest index such that x_i is a number. Necessarily then, $x_i < b$ since b is a right option of x_i. Since there are no walls in the top row of $m'(a)$, i.e. between a and any $x_i^{L_1}$, $\langle x_1^{L_1}, \ldots, x_{q-1}^{L_1}, x_q^{L_1} = p-1\rangle$ is a right-option-closed sequence. Also, $x_i^{L_1}$ is a Right option of a and so by Theorem 10, $x_i^{L_1} > a$.

Now suppose $a > b$. We have the inequalities $x_i^{L_1} > a > b > x_i$ which is a contradiction since for any game G, $G^L \not> G$. Therefore, there is no i such that x_i is a number and thus $\mathrm{rcf}(\otimes) = \{a \mid b\}$.

Now suppose $a < b$ and let $g = \{a \mid b\} = \frac{y}{2^k}$ for some integers y odd, and $k \geq 1$. The case a is an integer is already covered, thus we may assume that $\|a\| \geq 1$. Repeated use of Theorem 10 gives $a < x_i^{L_1} < x_1 < b$. Since $\{a \mid b\}$ is in canonical form, then $\|x_i\| > k$ by Lemma 12. Since x_i comes from a cell in a MAZE position then, by Theorem 10,

$$x_i = \left\{\langle\overline{x_i^{\mathcal{L}}}\rangle \,\middle|\, \langle\overline{x_i^{\mathcal{R}}}\rangle\right\} = \left\{\langle\overline{x_i^{\mathcal{L}}}\rangle \,\middle|\, b\right\} = \{c \mid b\}$$

for some c in $m'(a)$ and so $\|c\| < \|a\|$ by induction. By Lemma 12, $\|a\| < \|g\|$ and so $\|c\| < \|g\| = k$. If $a < c < b$, then by Lemma 12 $\|c\| \geq \|g\|$, which is a contradiction. Thus, $c < a$.

Hence no x_i is a number and thus $\mathrm{rcf}(\otimes) = g = \{a \mid b\}$.

Note that the entries in $m'(a)$ and $m'(b)$ are the same as in $m(a)$ and $m(b)$ except for the addition of integers around the bottom and right-hand sides. Hence, the values other than a in $m'(a)$ have norms less than $\|a\|$. The values in $m''(a)$ not in $m'(a)$ are the x_i, $i = 1, \ldots, p$ and none of them are numbers, so $\|\otimes\| = \max\{\|a\|, \|b\|\} + 1$ which occurs only at the top-left cell. □

5. Evaluating MAZEs

Grossman & Siegel [Grossman and Siegel 2009] note that for most situations the reduced canonical form is sufficient to evaluate a position since the infinitesimals, at most, change the parity of who gets the last move. Calculating the reduced canonical form can be done in linear time with regard to the number of cells in the position.

Theorem 14. *The reduced canonical form of an $p \times q$ MAZE position can be calculated in $O(pq)$ time.*

Note that for ease of describing the proof, the MAZE position is implicitly embedded in a rectangle and the reduced canonical form of every cell is found regardless of whether it can be reached from the top left hand cell.

Proof. The evaluation goes row by row starting at the bottom and always starting at the righthand cell. By Theorem 10, we only need look for the nearest number down a column and to the right along a row. To that end, let S_{ij} be the value of the cell (i, j) where $(1, 1)$ refers to the bottom-right corner.

(1) Define variables R_i, $i = 1, 2, \ldots, p$ and and C_j, $j = 1, 2, \ldots, q$. Initially, set $R_i = -pq$ and all $C_j = pq$ for all i and j.

(2) For j from 1 to q do
 For i from 1 to p do
 For cell S_{ij}, if there is a right wall from (i, j) then set $C_j = \infty$; if there is a left wall from (i, j) then set $R_i = -\infty$; set $S_{ij} = \{R_i | C_j\}$ and if $R_i < C_j$ then set $R_i = C_j = \{R_i | C_j\}$.

Note that for each pair (i, j), R_i is the norm of the left-option-closed sequence starting at $S_{i-1,j}$ the first left option of S_{ij} and C_j is the norm of the right-option-closed sequence. That $\text{rcf}(C_{ij}) = \{R_i | C_j\}$ follows from Theorem 10. Each cell is looked at and there is a constant number of operations associated with each cell. (Note that evaluating $\{a \mid b\}$ for numbers a and b, $a < b$ is linear.) □

See [McKay et al. 2010] for a fanciful interpretation of this method.

6. Open questions

For any numbers a and b we have constructed, G, a MAZE position with $\text{rcf}(G) = \{a \mid b\}$.

Question 15. *For numbers a and b, is there a* MAZE *position G with $G = \{a \mid b\}$?*

From [NO] we know that for any option closed game G, $\Downarrow + * < G - \text{rcf}(G) < \Uparrow + *$. It is easy to see that the Construction also gives games of the form $*n$.

Question 16. *What infinitesimals occur in* MAZE?

References

[Albert et al. 2007] M. Albert, R. J. Nowakowski, and D. Wolfe, *Lessons in play: An introduction to combinatorial game theory*, A K Peters Ltd., Wellesley, MA, 2007.

[Berlekamp et al. 2001] E. R. Berlekamp, J. H. Conway, and R. K. Guy, *Winning ways for your mathematical plays, I*, 2nd ed., A K Peters Ltd., Natick, MA, 2001.

[Grossman and Siegel 2009] J.P. Grossman and A. N. Siegel, "Reductions of partizan games", pp. 427–445 in *Games of no chance 3*, edited by M. H. Albert and R. J. Nowakowski, Math. Sci. Res. Inst. Publ. **56**, Cambridge: Cambridge University Press, 2009.

[McKay et al. 2010] N. A. McKay, R. J. Nowakowski, and A. A. Siegel, "The gods play MAZE", pp. 153–159 in *Recreational Mathematics Colloquium, I* (Evora, 2009), edited by J. N. Silva, Assoc. Ludus, Lisbon, 2010.

[Nowakowski and Ottaway 2011] R. J. Nowakowski and P. Ottaway, "Option-closed games", *Contrib. Discrete Math.* **6**:1 (2011), 142–153.

[Siegel 2011] A. A. Siegel, *On the structure of games and their posets*, Ph.D. thesis, Dalhousie University, 2011, http://hdl.handle.net/10222/13499.

nmckay@mathstat.dal.ca *Mathematics and Statistics, Dalhousie University, Halifax, NS B3H 3J5, Canada*

rjn@mathstat.dal.ca *Mathematics and Statistics, Dalhousie University, Halifax, NS B3H 3J5, Canada*

siegel@mathstat.dal.ca *Mathematics and Statistics, Dalhousie University, Halifax, NS B3H 3J5, Canada*

Games of No Chance 4
MSRI Publications
Volume **63**, 2015

Evaluating territories of Go positions with capturing races

TEIGO NAKAMURA

In analysing capturing races, or *semeai*s, we have been focusing on the method to find which player wins the race so far, because whether to win or to lose the capturing race largely affects the territory score and it somtimes can decide the outcome of the game. But in order to evaluate the state of the game properly, we usually have to count the territory score precisely regardless of which player wins the race. Sometimes the loser of a capturing race has good moves although the moves don't affect the status of winning or losing the race. In this paper, we propose a method for evaluating territory score in each decomposed subgame of a capturing race considering the status of the winner of the race.

1. Introduction

Combinatorial game theory has been applied to many kinds of existing games and has produced many excellent results. In the case of the game of Go, applications of CGT have been focused on endgames [Berlekamp and Wolfe 1994; Berlekamp 1996; Müller et al. 1996; Nakamura and Berlekamp 2003; Spight 2003] and eyespace values [Landman 1996] so far. But it can be applied to any situations that involve counting. Recently, we developed a new genre of application of CGT to Go, that is, to count liberties in capturing races [Nakamura 2003; Nakamura 2009; Nakamura 2006].

Capturing races, or *semeai* is a particular kind of life and death problem in which two adjacent opposing groups are each fighting to capture the opponent's group. A player's strength in Go depends on their skills in winning capturing races as well as opening and endgame skills. In order to win a complicated capturing race, various techniques in counting liberties, taking away the opponent's liberties, and extending self-liberties, are required in addition to broad and deep reading. Human expert players usually count liberties for each part of the blocks involved in semeai, sum them, and decide the outcome. A position of capturing races can also be decomposed into independent subpositions, as in the cases of endgames and eyespaces, and we can apply CGT to analyse the capturing races. We propose

Keywords: none.

a method of analysing capturing races that have no shared liberty or have only simple shared liberties, and then, using combinatorial game values of external liberties, give an evaluation formula to find the outcome of the capturing races.

Past methods of analysing capturing races have focused on ways to find which player wins the race, because winning or losing the capturing race largely affects the territory score and sometimes it can decide the outcome of the game. But in order to evaluate the state of the game properly, we usually have to count the territory score precisely regardless of which player wins the race. Sometimes the loser of a capturing race has good moves although the moves don't affect the status of winning or losing the race. In this paper, we propose a method for evaluating territory score in each decomposed subgame of a capturing race that takes into account the status of the winner of the race.

2. Analysing capturing races using CGT

2.1. ***How to decide the winner.*** In order to model capturing races, we define the Liberty Counting Game (LCG), which has the same rules as Go except for scoring. I briefly explain LCG below. More details can be found in [Nakamura 2003; Nakamura 2009; Nakamura 2006].

In LCG, the terminal score is basically the number of liberties of essential blocks,[1] but it is exactly the number of opponent's moves that are required to take away all the liberties of essential blocks. By convention, Black is Left and White is Right, Black scores are positive and White scores are negative.

Figure 1 shows some examples of CGT values of LCGs. In part (a), White's essential block[2] has three liberties, but Black cannot directly attack White's external liberty b, because if he simply fills the liberty b, Black's attacking block gets to be

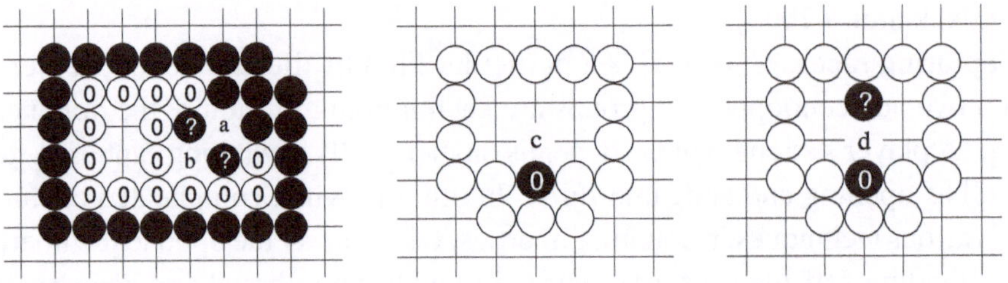

(a) -4 (b) $\{3 \mid 0\}$ (c) $\{4 \mid 0\}$

Figure 1. CGT values of LCGs.

[1] A block of connected stones involved in a capturing race is called an essential block, if capturing the block immediately decides the race.

[2] The block of circled stone denotes an essential block. In this example, the circled block is not involved in semeai, but in LCG we just count liberties of essential blocks.

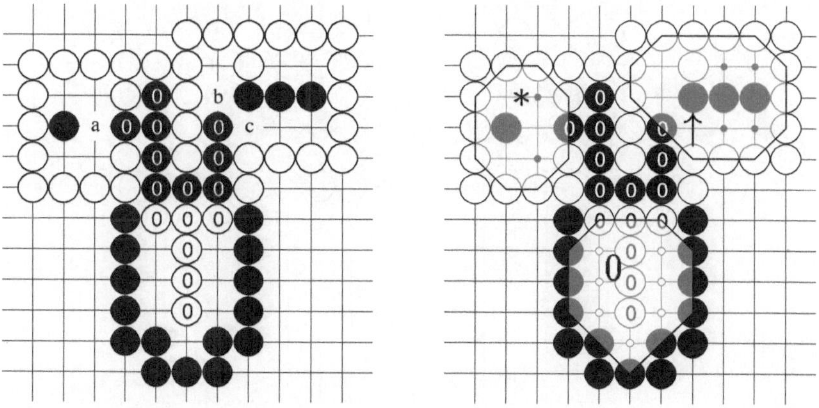

Figure 2. Left: an example problem. Right: its analysis using CGT.

in *atari*, and White can capture Black's three stones by playing to *a* and White's essential block becomes alive. So, Black needs to spend one move to protect *a* prior to attack. Generally, liberty scores in external liberty regions are greater than or equal to the number of liberties of essential blocks. In part (b) of the figure, if White plays first, the score is zero, but if Black plays to *c* first, the number of liberties becomes 3. In (c), Black can connect his two stones of an essential block and a neutral block playing to *d* and the score becomes 4, if he plays first.

Figure 2, left, is an example semeai problem. The right side of the figure shows an analysis of each subgame. The upper left subgame is $\{4 \mid 0\}$, the upper right subgame is $\{6 \mid \{4 \mid 0\}\}$ and the lower subgame is -7. We cool these subgames by two degrees and obtain $2*$, $4\uparrow$ and -7, respectively. The total value is $-1\uparrow*$ and it is incomparable to -1. If Black plays first, he can round the value up to 0 and wins the race by one move. If White plays first, he can round the value down to -2 and wins the race by three moves. So the first player wins the race of Figure 2. The Black's only winning move is move *a*. After a sequence of Black *a*, White *b* and Black *c*, the number of liberties of Black's essential block is 8 and the number of liberties of White's essential block is 7 and Black definitely wins the race by one move. On the other hand, White has two winning moves of move *a* and move *b*. After White *a*, even if Black moves to *b*, White can win the race by two moves.[3] Alternatively, after White *b* for his first move, move *a* and move *c* are *miai* and White can win the race by three moves. In consequence, White *a* loses one liberty count in semeai compared to White *b*. But this loss is not only in liberty count but also in territory score. We will discuss the territory score involved in semeais in the next section.

---

[3] We assume the winner should play the last move in showing how many moves ahead, even if he doesn't need to play any moves to win. In this case, we take a White's extra move into account.

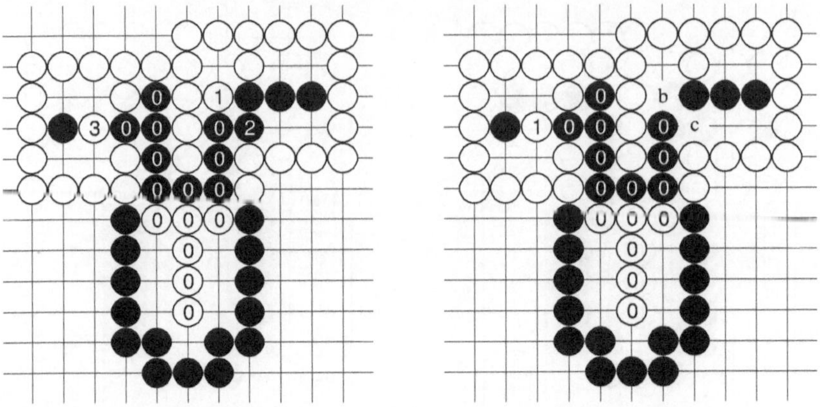

Figure 3. Two winning sequences for white.

2.2. ***Territory scores.*** Figure 3, left, shows a winning sequence for White after White *b* in 2.

Black's 13 stones of the essential block are bound to be captured eventually. White has six territory points in the upper left subregion, one point in the upper right subregion and 26 points as capturing Black's 13 stones. Black has no territory points because all the empty points in the lower subregion are *dame*. So the total is 33 points for White. In Figure 2, right, on the other hand, if White *b* and Black *c* are played after White 1, the resulting position is identical to the position of the left part of the figure and the total is also 33 points. But if Black plays the move *b*, Black can reduce White's territory points although the move doesn't change the status of the capturing race. After Black *b*, White's territory score is 32 $(= 6+26)$ points. So Black's move *b* takes away one White's territory point compared with the case of White *b*.[4] If White plays both the move *b* and the move *c* in Figure 3, right, the total score increases to 35 points.[5]

Generally, this kind of phenomena in capturing races is called *semedori*. As in the above analysis in terms of territory score, if we search all the possible winning sequences of an entire capturing race after we analyse the status of the capturing race, we can figure out the final territory score. But the procedure is not efficient because of combinatorial explosion. In the next section, we will show a method to evaluate territory score on each subgame and to combine it taking into account the status of capturing races and semedori.

[4]This reduction is also against Figure 3, left.

[5]White's moves of *b* and *c* cut off Black's three stones in the upper right region and the territory score of the upper right subregion becomes $1 + 10 = 11$. Since the number of stones of Black's essential block is 9, the total score is $6 + 11 + 18 = 35$.

3. Evaluating territory scores of subgames involved in capturing races

For each subgame we introduce three auxiliary games in order to evaluate, without a full search, the territory score of a position involved in capturing races:

G_{lc}: The Liberty Counting Game (LCG).

G_{win}: A game whose score is partial territory score, assuming that the defender (player who owns the essential block in the region) is the winner of the entire capturing race.

G_{lose}: A game whose score is partial territory score, assuming that the defender (player who owns the essential block in the region) is the loser of the entire capturing race.

We use G_{lc} to decide the status of the capturing race. Either G_{win} or G_{lose} are used to evaluate the territory score according to the status of the capturing race.

Although each subgame should be independent in Combinatorial Game Theory, some subgames may depend on each other in the game of Go. Müller proposed a framework of Conditional Combinatorial Game (CCG) to describe games with some dependency [Müller 2003].

3.1. *Conditional Combinatorial Game.* The Conditional Combinatorial Game (CCG) is described as follows.

$$G \stackrel{\text{def}}{=} \{L^1_{C_1},\ L^2_{C_2},\ L^3_{C_3},\ \dots \mid R^1_{D_1},\ R^2_{D_2},\ R^3_{D_3},\ \dots\}.$$

Here L^i is a game to which Left moves from G and R^i is a game to which Right moves from G. Each C_i and D_i is some kind of predicate to check each move is legal or not using global information.

For our purposes, we can describe G_{win} and G_{lose} as a game of CCG using G_{lc} as the conditional predicates.

3.2. *Territory scores of subgames.* We show some subgames of capturing races and their corresponding games of G_{lc}, G_{win} and G_{lose} in Table 1.

The Territory score is the points of territory other than the essential block in each region. For example, G_{lose} of the subgame A is $\{-4 \mid -6\}$. We calculate the value as follows.

> If White cuts off Black's one stone, White's territory score becomes six points in the region. On the other hand, if Black connects the stone and the essential block, White has to play four more moves in this region in order to to capture Black's block. So, White's territory score becomes four points, that is, two captured stone, other than the essential block, and two territory points.

	liberty score		territory score			
	G_{lc}	cooled value	G_{win}	G_{lose}	cooled value	baseline
A	$4 \mid 0$	$2*$	$0 \mid -6$	$-4 \mid -6$	$-5*$	-7
B	$6 \| 4 \mid 0$	$4\uparrow$	$0 \| -1 \mid -11$	$-8 \| -9 \mid -11$	$-9\uparrow$	-13
C	$0 \mid -4$	$-2*$	$8 \mid 0$	$8 \mid 6$	$7*$	9
D	$0 \| -2 \mid -6$	$-2\uparrow$	$13 \| 6 \mid 0$	$13 \| 12 \mid 10$	$12\uparrow$	14
E	-7	-7	0	0	0	7
F	$-5 \mid -9$	$-7*$	$8 \mid 0$	$8 \mid 6$	$7*$	14
G	$-5 \mid -8$	$-6\frac{1}{2}$	$6 \mid 0$	$6 \mid 5$	$5\frac{1}{2}$	12
H	$8 \mid 4 \| 2 \||| 0$	$2\Downarrow*$	$0 \mid -6 \| -9 \||| -12$	$-8 \mid -10 \| -11 \||| -12$	$-11\Downarrow*$	-13
I	$6 \mid 3 \| 1$	$2\frac{3}{4}$	$0 \mid -1 \| -2 \mid -8 \||| -15$	$-13 \mid -14 \| -15$	$-14\frac{1}{4}$	-17
J	$3 \mid 0$	$1\frac{1}{2}$	$0 \mid -2 \| -5$	$-4 \mid -5$	$-4\frac{1}{2}$	-6

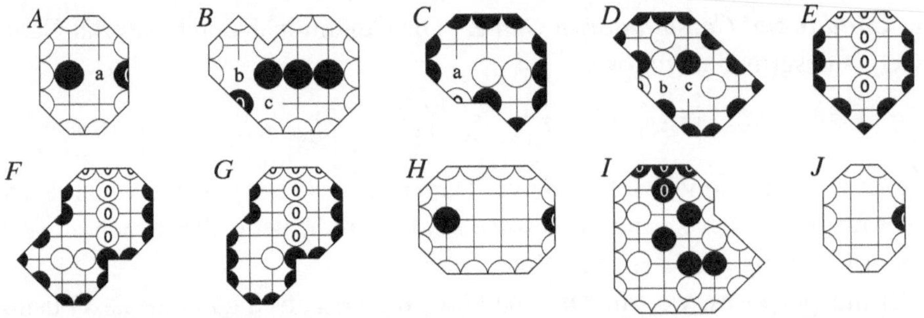

Table 1. Evaluation of liberty and territory for subgames A–J.

The "baseline" column means the case where we assume that the subregion is not involved in *semeai* and the opponent's essential block is already dead. The territory score of the dead essential block is excluded from the baseline value.

A formula to calculate the cooled value of G_{lose} is

$$\mathrm{Cool}(G_{\mathrm{lose}}, 1) = \text{baseline value} + \mathrm{Cool}(G_{\mathrm{lc}}, 2), \tag{1}$$

$\mathrm{Cool}(G, t)$ being the game G cooled by t degrees. Formula (1) works as follows:

> The winner of capturing races has to fill all the liberties of the opponent's essential block and capture it eventually. In case of *semedori*,

the liberty filling moves are played in his territory region. The number of liberty filling moves equals to the number of liberty count of the opponent's essential block in the subregion. The moves reduce his own territory points and we need to subtract the number from the baseline of the territory points. The winner's own territory region is the subregion of G_{lose} and the number of liberty count G_{lc} is the number of liberties of opponent's essential block, so positive and negative are reversed. Consequently we should add the baseline and the cooled value of G_{lc} in order to take into account the reduction.

We can evaluate the territory score taking into account the winner of the capturing race as follows.

Figure 2, left, is an example position of a capturing race combined A, B and E in Table 1. If White plays first, White can win the race and the total territory score is the sum of G_{lose} of A, G_{lose} of B and G_{win} of E. The cooled value of the total is

$$-5* + -9\uparrow + 0 = -14\uparrow*.$$

Considering the number of stones of each essential block, that is, Black's nine stones and White's six stones, we can figure out the final score is

$$-14\uparrow* - 18 = -32\uparrow*.$$

On the other hand, if Black plays first, the situation is more complicated. The total territory score is the sum of G_{win} of A, G_{win} of B and G_{lose} of E. White's moves in the region of A and B can be effective attacking moves to Black's essential block and the status of the capturing race can be changed. In that case, whether White's attacking moves and Black's responses are good or bad should be evaluated in terms of not only the gain in territory points in G_{win} but also the result of G_{lc} as the condition of CCG at the highest priority. As a result, White b is sente in terms of the status of the capturing race after Black's move a in Figure 2, so Black has to respond the move c immediately. In consequence, G_{win} of B becomes -1 from $\{0 \mid \{-1 \mid -11\}\}$ and the final score is -1.

If two subgames have the same property of G_{lc}, we can compare them in terms of territory. G_{lc} of the subgame A and G_{lc} of the subgame C have the same infinitesimal part of $*$. So, in case that Black is the winner, we compare G_{win} of A with G_{lose} of C and it will turn out that A is hotter than C. On the contrary, in case that White is the winner, we compare G_{lose} of A with G_{win} of C and it will turn out that C is hotter than A. We can also compare B with D because G_{lc} of B and G_{lc} of D have the same infinitesimal part of $\uparrow$. In case that Black is the winner, we compare G_{win} of B with G_{lose} of D and it will turn out that B is preferred to D. On the contrary, in case that White is the winner, we compare G_{lose} of B with G_{win} of D and it will turn out that D is hotter than B.

3.3. ***More examples.*** We can combine subgames A, B, C and D in Table 1, add some extra liberties and construct semeai problems that have same liberty counts. For example, all the semeai games of $A+B-7$, $A+D-1$, $C+B-3$ and $C+D+3$ have the same liberty count of $-1{\uparrow}{*}$, because

$$\begin{aligned}
(1)\ A+B-7 &= 2{*} + 4{\uparrow} - 7 = -1{\uparrow}{*},\\
(2)\ A+D-1 &= 2{*} - 2{\uparrow} - 1 = -1{\uparrow}{*},\\
(3)\ C+B-3 &= -2{*} + 4{\uparrow} - 3 = -1{\uparrow}{*},\\
(4)\ C+D+3 &= -2{*} - 2{\uparrow} + 3 = -1{\uparrow}{*}.
\end{aligned}$$

Although we can conclude that first player wins for all the semeai games, it is not easy to figure out the optimal path in terms of territory score for each of the semeai games from (1) to (4). Figure 4 shows the canonical game tree of ${\uparrow}{*}$. But it's not true in semeai games of the sum of G_{lc}s. It's enough for the attacker to win the race, because whether to win or to lose usually makes a big difference in territory scores and he doesn't need to gain extra liberties at the risk of losing territory scores. Canonicalization process for semeai games may prune some effective branch in terms of territory score.

Figure 5 shows the game tree of the semeai games from (1) to (4). Heavy lines denote winning paths in semeais for each player and numbers at the leaf nodes are the liberty score. Boxed numbers under the leaf nodes denote the territory score for each of semeai games and italic numbers in heavy boxes mean the nodes should be selected in terms of territory score by priority. For example, in semeai game (1), if Black plays first, Black should play move a. At this point, White b is sente and Black has to respond move c immediately. The territory score becomes -1 as described in Section 3.2. But if White plays first, White b

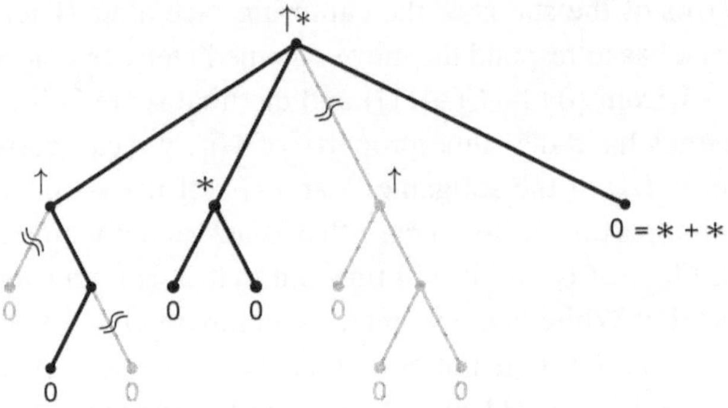

Figure 4. Canonical game tree of ${\uparrow}{*}$.

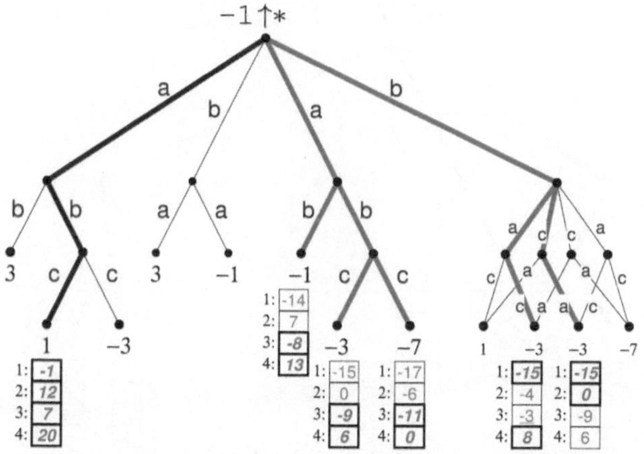

Figure 5. Game tree of $-1{\uparrow}*$ semeai.

will be preferred to a. Then a and c are miai and the score will be -15. In case of the semeai game (3), White's optimal move is different from the above case. White should play move a and the chilled territory score becomes $-9{\uparrow}$.

If we give some more liberties for Black, the situation will be changed as shown in Figures 6 and 7.

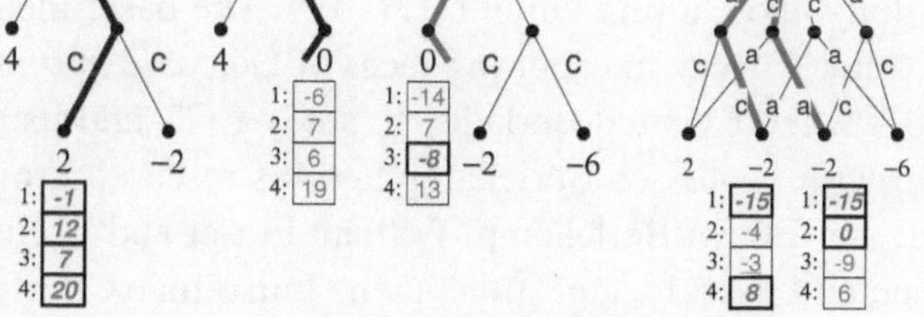

Figure 6. Game tree of ${\uparrow}*$ semeai.

4. Summary

We proposed a method for evaluating territory score in each decomposed subgame of a capturing race considering the status of the winner of the race. We introduced three different kinds of games, G_{lc}, G_{lose} and G_{win}, for each subgame and showed

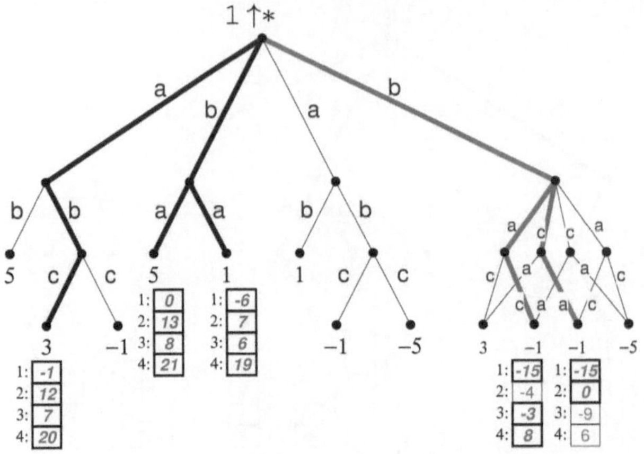

Figure 7. Game tree of $1{\uparrow}*$ semeai.

a method to evaluate territory score of a position involved in capturing races without entire search.

G_{lose} can be calculated from G_{lc} using the formula (1), but it is difficult to evaluate G_{win} in combination with other subgames because the canonicalization process for G_{lc} may prune some effective branch in terms of territory score. So future work includes how to select good moves in G_{win} for each players efficiently keeping the status of the entire semeais.

Acknowledgements

This research is partially supported by Hayao Nakayama Foundation for Science & Technology and Culture Grant (H17-A7). The basic idea for applying Combinatorial Game Theory to capturing races of Go came into my mind in the summer of 2001, when I visited Berkeley to study CGT and its application to the game of Go with Professor Berlekamp and his research group. I am very grateful to Professor Elwyn Berlekamp, William Fraser and William Spight for their kind advice and useful suggestions then. I also thank Professor Richard Nowakowski for his kind advice and encouragement.

References

[Berlekamp 1996] E. Berlekamp, "The economist's view of combinatorial games", pp. 365–405 in *Games of no chance* (Berkeley, CA, 1994), edited by R. J. Nowakowski, Math. Sci. Res. Inst. Publ. **29**, Cambridge Univ. Press, 1996.

[Berlekamp and Wolfe 1994] E. Berlekamp and D. Wolfe, *Mathematical Go: Chilling gets the last point*, A K Peters, Wellesley, MA, 1994.

[Landman 1996] H. A. Landman, "Eyespace values in Go", pp. 227–257 in *Games of no chance* (Berkeley, CA, 1994), edited by R. J. Nowakowski, Math. Sci. Res. Inst. Publ. **29**, Cambridge Univ. Press, Cambridge, 1996.

[Müller 2003] M. Müller, "Conditional combinatorial games and their application to analyzing capturing races in Go", *Inform. Sci.* **154**:3-4 (2003), 189–202.

[Müller et al. 1996] M. Müller, E. Berlekamp, and W. Spight, "Generalized thermography: Algorithms, implementation, and application to Go endgames", TR–96–030, International Computer Science Institute, 1996, http://citeseerx.ist.psu.edu/viewdoc/summary?doi=10.1.1.34.6699.

[Nakamura 2003] T. Nakamura, "Counting liberties in capturing races using combinatorial game theory", *IPSJ SIG-GI 2003–GI–9–5* **2003**:35 (2003), 27–34. In Japanese.

[Nakamura 2006] T. Nakamura, "On analysing capturing races using combinatorial game theory", pp. 1–16 in *Proceedings of the 4th International Conference on Baduk*, Korea Baduk Association, Baduk, 2006.

[Nakamura 2009] T. Nakamura, "Counting liberties in Go capturing races", pp. 177–196 in *Games of no chance 3*, edited by M. H. Albert and R. J. Nowakowski, Math. Sci. Res. Inst. Publ. **56**, Cambridge: Cambridge University Press, 2009.

[Nakamura and Berlekamp 2003] T. Nakamura and E. Berlekamp, "Analysis of composite corridors", pp. 213–229 in *Computers and games* (Edmonton, 2002), edited by J. Schaeffer et al., Lecture Notes in Computer Science **2883**, Springer, Berlin, 2003.

[Spight 2003] W. L. Spight, "Evaluating Kos in a neutral threat environment: Preliminary results", pp. 413–428 in *Computers and games* (Edmonton, 2002), edited by J. Schaeffer et al., Lecture Notes in Computer Science **2883**, Springer, Berlin, 2003.

teigo@ai.kyutech.ac.jp

Department of Artificial Intelligence, Kyushu Institute of Technology, 680-4 Kawazu Iizuka, Fukuoka 820-8502, Japan

Games of No Chance 4
MSRI Publications
Volume **63**, 2015

Artificial intelligence for Bidding Hex

SAM PAYNE AND ELINA ROBEVA

We present a Monte Carlo algorithm for efficiently finding near optimal moves and bids in the game of Bidding Hex. The algorithm is based on the recent solution of Random-Turn Hex by Peres, Schramm, Sheffield, and Wilson together with Richman's work connecting random-turn games to bidding games.

1. Introduction

Hex is well-known for the simplicity of its rules and the complexity of its play. Nash's strategy-stealing argument shows that a winning strategy for the first player exists, but finding such a strategy is intractable by current methods on large boards. It is not known, for instance, whether the center hex is a winning first move on an odd size board. The development of artificial intelligence for Hex is a notoriously rich and challenging problem, and has been an active area of research for over thirty years [Davis 1975/76; Nishizawa 1976; Anshelevich 2000; 2002a; 2002b; Cazenave 2001; Rasmussen and Maire 2004], yet the best programs play only at the level of an intermediate human [Melis and Hayward 2003]. Complete analysis of Hex is essentially intractable; the problem of determining which player has a winning strategy from a given board position is PSPACE-complete [Reisch 1981], and the problem of determining whether a given empty hex is *dead*, or irrelevant to the outcome of the game, is NP-complete [Björnsson et al. 2007]. Some recent research has focused on explicit solutions for small boards [Hayward et al. 2004; Hayward et al. 2005], but it is unclear whether such techniques will eventually extend to the standard 11×11 board.

Bidding Hex is a variation on Hex in which, instead of alternating moves, the players bid for the right to move. Suppose, for example, that Alice plays against Bob, and both start with 100 chips. If Alice bids seventeen for the first move and Bob bids nineteen, then Bob gives nineteen chips to Alice and makes a move. Now Alice has 119 chips and Bob has 81, and they bid for the second move. The total number of chips in the game remains fixed, and the chips have no value

Supported by the Clay Mathematics Institute, NSF DMS-1068689, NSF CAREER DMS-1149054, and a Stanford University VPUE grant.

at the end of the game — the goal is simply to win the game of Hex. The main goal of this paper is to present an efficient algorithm for near-optimal play of Bidding Hex. We have implemented this algorithm and it is overwhelmingly effective — undefeated against human opponents. Code is available from the authors on request.

2. Bidding games

Bidding games are variations on traditional two-player games in which, instead of alternating moves, the players bid for the right to move (or for the right to determine who moves next, in situations where moving may not be desirable). Richman developed an elegant theory of such games with real-valued bidding in the late 1980s; his theory is presented in [Lazarus et al. 1996; Lazarus et al. 1999]. For recreational play, integer valued bidding is preferred. Bidding Chess has also achieved some popularity among fans of Chess variations [Beasley 2008a; 2008b].

For mathematical purposes, the coarseness of integer bidding with small numbers of chips creates additional subtleties. For instance, the set of optimal first moves in Bidding Tic-Tac-Toe depends on the total number of chips — the center is the unique optimal first move when the total number of chips is greater than 26, but the center is not an optimal first move when the total number of chips is five [Develin and Payne 2010, Theorem 6.17].

One of Richman's insights was the fundamental connection between bidding games with real-valued bidding and random-turn games in which players flip a coin to determine who moves next. Say G is a finite, loop-free game played by Alice and Bob. Let $R(G)$ be the critical threshold, between zero and one, such that Alice has a winning strategy for G with real-valued bidding if her proportion of the total bidding resources is greater than $R(G)$, and she does not have a winning strategy if her proportion is less than $R(G)$. Let $P(G)$ be the probability that Alice wins G as a random-turn game, assuming optimal play. Intuitively, one expects that if the game favors Alice then $R(G)$ will be closer to zero and $P(G)$ will be closer to one, but the precise relation discovered by Richman remains surprising.

Richman's Theorem. *Let G be a finite, loop-free game. Then*

$$R(G) = 1 - P(G).$$

Furthermore, the set of optimal moves for G with real-valued bidding is the same as the set of optimal moves for random-turn play, and optimal bids may be determined as follows. For any position v in G, let G_v denote the game played starting from v. Define

$$R^+(v) = \max\{R(G_w) \mid \text{ Bob can move from } v \text{ to } w\},$$
$$R^-(v) = \min\{R(G_{w'}) \mid \text{ Alice can move from } v \text{ to } w'\}.$$

The function R is then determined by the relation

$$R(G_v) = \big(R^+(v) + R^-(v)\big)/2,$$

together with the conditions that $R(G_v)$ is equal to zero if v is a winning position for Alice, and is equal to one if v is a winning position for Bob. From position v, an optimal bid for both players is $\delta(v) = \big|R^+(v) - R^-(v)\big|/2$, which may be thought of, roughly speaking, as the value of moving from position v.

Since $R^+(v)$ and $R^-(v)$ can be interpreted, through Richman's Theorem, as probabilities of winning certain random-turn games, results and insights from random-turn games can be applied directly to bidding games with real-valued bidding. For discrete bidding, such probabilistic methods are also useful, though less directly. For any fixed positive ϵ, if Alice's proportion of the total number of chips is at least $R(G) + \epsilon$ and the total number of chips is sufficiently large with respect to G and ϵ, then Alice has a winning strategy in which all of her moves are optimal for real-valued bidding and hence also for random-turn play [Develin and Payne 2010, Theorem 3.10]. The bounds on the number of chips required are astronomical for most interesting games and grow exponentially with $1/\epsilon$ and the number of possible turns in the game but, in practice, following real-valued bidding strategy is highly effective even when the number of chips is small.

3. Hex background

The game of Hex never ends in a draw. In other words, the only way a player can block the other is by completing a winning chain. We sketch a proof of this fact following [Gale 1979] in the form of an algorithm for determining the winner of a completed game. This algorithm is a key ingredient in the implementation of step (1) in our artificial intelligence program for Bidding Hex, presented in Section 5.

Suppose Alice and Bob place amber and blue hexes, respectively, until the board is full. Color the regions northwest and southeast of the board amber, and the regions northeast and southwest of the board blue, so Alice wins if she makes a chain of amber hexes connecting the two amber regions, and Bob wins if he creates a chain of blue hexes connecting the two blue regions (Figure 1).

Now the boundary between the amber and blue regions of the board is a finite union of loops, plus two paths whose four endpoints are at the north, south, east,

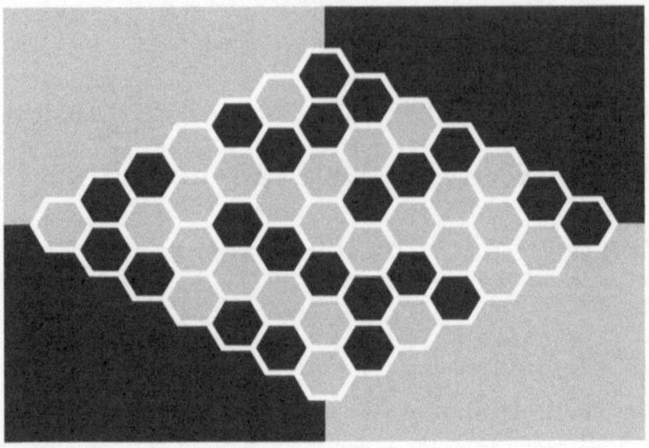

Figure 1. Alice, playing amber, has successfully blocked Bob, playing blue.

and west corners of the board. The directed path beginning at the west corner of the board always has amber on its left and blue on its right, so its endpoint must be at either the north or the south corner of the board. If its endpoint is at the north corner, the blue hexes adjacent to this path on the right form a winning chain for Bob; if its endpoint is south, as in Figure 2, the amber hexes adjacent to this path on the left form a winning chain for Alice.

In summary, to determine the winner of a completed game, start at the west corner and follow the boundary between amber and blue regions. If the path leads to the north corner of the board, then Bob wins. Otherwise, the path leads to the south corner of the board, and Alice wins.

If the board is only partially completed, we can determine the state of the game as follows. First, color all of the empty hexes amber and determine the winner. If Bob is the winner, then he already has a winning chain, so the game is over. Otherwise, color all empty hexes blue to check whether Alice already has a winning chain. If not, then neither player has won yet, and the game continues.

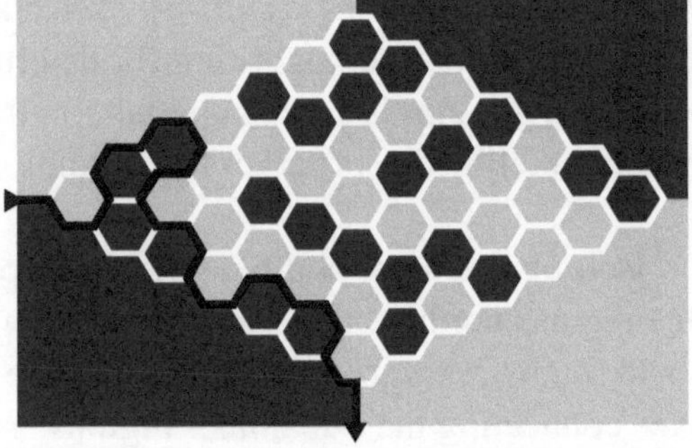

Figure 2. It follows that Alice has a winning chain joining her two external regions, given by following the component of the boundary between amber and blue that starts at the left corner.

4. Random-Turn Hex

Random-Turn Hex has recently been solved by the probabilists Peres, Schramm, Sheffield, and Wilson; their study was apparently motivated by suggestions of Werner pointing toward a conformal invariance property for Hex played on unusually shaped boards with vanishingly small hexes, based on relations with percolation theory. One of their main results is the Random-Turn Hex Theorem below.

In a completely filled hex board, a hex is *critical* if changing the color of that hex would change the outcome of the game. Equivalently, a hex is critical if every winning chain contains that hex. If the board is only partially filled in, then the *probability* of an open hex being critical is the number of fillings of the remainder of the board for which the given hex is critical, divided by the total number of possible fillings, which is 2^N, where N is the number of empty hexes.

Random-Turn Hex Theorem [Peres et al. 2007]. *The set of optimal moves from any position in Random-Turn Hex is the same for both players, and is exactly the set of open hexes whose probability of being critical is maximal.*

The Random-Turn Hex Theorem gives an exact solution to Random-Turn Hex. From any position, fill in the empty hexes in every possible way and count the number of times that each hex is critical. The hexes that are critical most often are the optimal moves. When the number of empty hexes is large, the number of possible fillings is astronomical, so finding the optimal moves exactly is practically impossible. Nevertheless, the Random-Turn Theorem suggests a Monte Carlo algorithm — if one fills in the board at random a large number of times, then the hexes that are critical most often in this sampling are likely to be optimal or near-optimal. This algorithm has been implemented by David Wilson in the program Hexamania [Wilson 2007]. Wilson's program beats a skilled human player more than half the time, but a person can still beat the computer fairly often, by randomly winning most of the coin flips. We have implemented a similar algorithm for Bidding Hex, finding not only near-optimal moves but also near-optimal bids, and the resulting program plays exceedingly well.

Cohen [2004a; 2004b] independently experimented with an analogous Monte Carlo method for playing traditional alternating move Hex, in which random hexes are filled in with alternating colors, so the end filling has equal numbers of hexes of both colors if the board size is even, and first player has exactly one extra hex in the final filling if the board size is odd.

5. Bidding Hex

We now describe our artificial intelligence program for playing Bidding Hex. Given a partially filled board, we must efficiently find a near optimal move and a near optimal bid. Start with a partially filled board. For an empty hex H, let L_H be the probability that H is filled with the losing color, when the remainder of the board is filled in at random.

Proposition 5.1. *The probability that H is not critical is $2L_H$.*

Proof. In any filling, if H is of the losing color then it is not critical. Furthermore, there is a natural bijection between fillings in which H is of the losing color and those in which H is of the winning color but not critical, given by switching the color of H. Therefore, the number of fillings in which H is not critical is exactly twice the number of fillings in which H is of the losing color. □

The proposition shows that the optimal moves are those hexes H such that L_H is as small as possible. An optimal bid can also be determined in terms of L_H, as follows.

Proposition 5.2. *Let H be an open hex such that L_H is minimal, from position v. Then an optimal bid for real-valued bidding is the proportion*

$$\delta(v) = \tfrac{1}{2} - L_H$$

of the total bidding resources.

Proof. The probability that H is critical is $1-2L_H$, and is equal to $R^+(v)-R^-(v)$, the difference between the conditional probability that Alice wins if she gets H and the probability that she wins if Bob gets H. Therefore, $\delta(v) = (1-2L_H)/2$, as required. □

Here is our algorithm for finding near optimal bids and moves from a partially filled Hex board.

(1) Fill in the remainder of the board at random a large number of times, and determine the outcome of the game each time.

(2) Keep track of the number of times that each open hex is of the losing color.

(3) If H is the open hex that is least often of the losing color, bid the integer part of $\frac{1}{2} - L_H$ times the total number of chips in the game.

(4) If this bid wins, move in hex H. Otherwise, the opponent has probably overbid. Take his chips, and let him move wherever he wants.

A moderately fast desktop computer, by 2008 standards, can check roughly fifty thousand random fillings of an 11×11 board per second. A few hundred thousand fillings per move suffice to consistently defeat the best human players today.

References

[Anshelevich 2000] V. V. Anshelevich, "The game of Hex: An automatic theorem proving approach to game programming", pp. 189–194 in *Proceedings of the Seventeenth National Conference on Artificial Intelligence* (Austin, TX, 2000), AAAI Press, Menlo Park, CA, 2000.

[Anshelevich 2002a] V. V. Anshelevich, "The game of Hex: The hierarchical approach", pp. 151–165 in *More games of no chance* (Berkeley, CA, 2000), edited by R. Nowakowski, Math. Sci. Res. Inst. Publ. **42**, Cambridge Univ. Press, Cambridge, 2002.

[Anshelevich 2002b] V. V. Anshelevich, "A hierarchical approach to computer Hex", *Artificial Intelligence* **134**:1–2 (2002), 101–120.

[Beasley 2008a] J. Beasley, "Bidding Chess", *Variant Chess* **7**:57 (2008), 42–44.

[Beasley 2008b] J. Beasley, "Isolated pawns", *Variant Chess* **7**:58 (2008), 71.

[Björnsson et al. 2007] Y. Björnsson, R. Hayward, M. Johanson, and J. van Rijswijck, "Dead cell analysis in Hex and the Shannon game", pp. 45–59 in *Graph theory in Paris*, edited by A. Bondy et al., Birkhäuser, Basel, 2007.

[Cazenave 2001] T. Cazenave, "Retrograde analysis of patterns versus metaprogramming", pp. 57–75 in *Computational intelligence in games*, edited by N. Baba and L. C. Jain, Studies in Fuzziness and Soft Computing **62**, Physica-Verlag, Heidelberg, 2001.

[Cohen 2004a] B. Cohen, "Board evaluation in hex", 2004, http://www.advogato.org/person/Bram/diary/112.html.

[Cohen 2004b] B. Cohen, "Hex", 2004, http://www.advogato.org/person/Bram/diary/116.html.

[Davis 1975/76] M. Davis, "On artificial machine learning: Some ideas in search of a theory", *Internat. J. Comput. Math.* **5**:4 (1975/76), 315–329.

[Develin and Payne 2010] M. Develin and S. Payne, "Discrete bidding games", *Electron. J. Combin.* **17**:1 (2010), R85, 40.

[Gale 1979] D. Gale, "The game of Hex and the Brouwer fixed-point theorem", *Amer. Math. Monthly* **86**:10 (1979), 818–827.

[Hayward et al. 2004] R. Hayward, Y. Björnsson, M. Johanson, M. Kan, N. Po, and J. van Rijswijck, "Solving 7 × 7 Hex: Virtual connections and game-state reduction", pp. 261–278 in *Advances in computer games* (Graz, 2003), edited by H. J. van den Herik et al., IFIP Int. Fed. Inf. Process. **135**, Kluwer, Boston, 2004.

[Hayward et al. 2005] R. Hayward, Y. Björnsson, M. Johanson, M. Kan, N. Po, and J. van Rijswijck, "Solving 7 × 7 Hex with domination, fill-in, and virtual connections", *Theoret. Comput. Sci.* **349**:2 (2005), 123–139.

[Lazarus et al. 1996] A. J. Lazarus, D. E. Loeb, J. G. Propp, and D. Ullman, "Richman games", pp. 439–449 in *Games of no chance* (Berkeley, CA, 1994), edited by R. J. Nowakowski, Math. Sci. Res. Inst. Publ. **29**, Cambridge Univ. Press, Cambridge, 1996.

[Lazarus et al. 1999] A. J. Lazarus, D. E. Loeb, J. G. Propp, W. R. Stromquist, and D. H. Ullman, "Combinatorial games under auction play", *Games Econom. Behav.* **27**:2 (1999), 229–264.

[Melis and Hayward 2003] G. Melis and R. Hayward, "SIX wins Hex tournament", *ICGA* **26**:4 (2003), 277–280.

[Nishizawa 1976] T. Nishizawa, "An improved version of the program for Hex", pp. 40–61 in *Studies on puzzles by games and computers*, Research Institute for Mathematical Sciences, Kyoto, 1976.

[Peres et al. 2007] Y. Peres, O. Schramm, S. Sheffield, and D. B. Wilson, "Random-turn hex and other selection games", *Amer. Math. Monthly* **114**:5 (2007), 373–387.

[Rasmussen and Maire 2004] R. Rasmussen and F. Maire, "An extension of the H-search algorithm for artificial Hex players", pp. 646–657 in *AI 2004: Advances in artificial intelligence*, edited by G. I. Webb and X. Yu, Lecture Notes in Comput. Sci. **3339**, Springer, Berlin, 2004.

[Reisch 1981] S. Reisch, "Hex ist PSPACE-vollständig", *Acta Inform.* **15**:2 (1981), 167–191.

[Wilson 2007] D. Wilson, "Hexamania", 2007, http://research.microsoft.com/~dbwilson/hex/.

sam.payne@yale.edu
Mathematics Department, Yale University, 10 Hillhouse Ave., New Haven, CT 06511, United States

robeva@stanford.edu
Department of Mathematics, Stanford University, Stanford, CA 94305, United States

Games of No Chance 4
MSRI Publications
Volume **63**, 2015

Nimbers in partizan games

CARLOS PEREIRA DOS SANTOS AND JORGE NUNO SILVA

"The chess board is too small for two queens."
–Victor Korchnoi, challenger for World Chess Championship

We propose some classes of games particularly useful for constructing nimbers in partizan games. We exhibit a $*4$ in Amazons with the help of a particular algebraic table.

Before reading this article one can consider the following Amazons position:

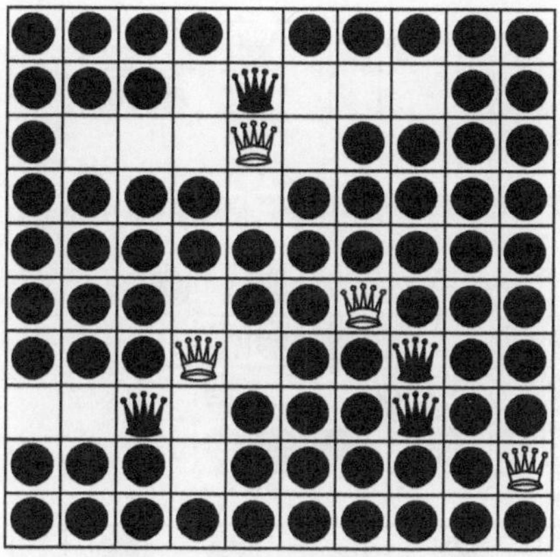

In Amazons there are a lot of hot positions: the players want to play to gain some territory. However, the position shown is not of this kind. If we use [Siegel 2011] to analyze it, this position proves to have value $*3 + *2 = *$. We will attempt to analyze the options of such a position and construct nimbers in partizan games. For instance, $*3 = \{0, *, *2 \mid 0, *, *2\}$, however, when we study the players options with [Siegel 2011] we see that the options are not $\{0, *, *2\}$. In a partizan game we have a bigger number of possible options to construct a nimber than in an impartial game. For instance, we know that a game like $\{\uparrow \mid 0\}$ has value $*$ too. When we think about higher stars the number of possibilities is just gigantic. So it's important to make some mathematical considerations to classify the games that "can act as nimbers". In this article, we prove some useful results about the construction of nimbers and show some interesting examples in Amazons.

Some values in Amazons. In [Berlekamp 2000] we can see the first interesting game values in Amazons:

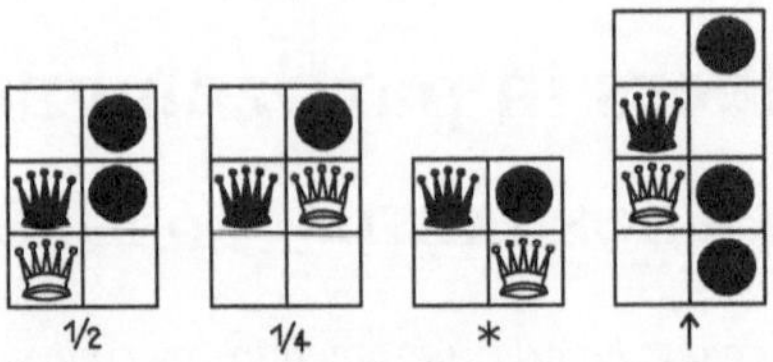

In [Snatzke 2002b], we can see more values and the first $*2$:

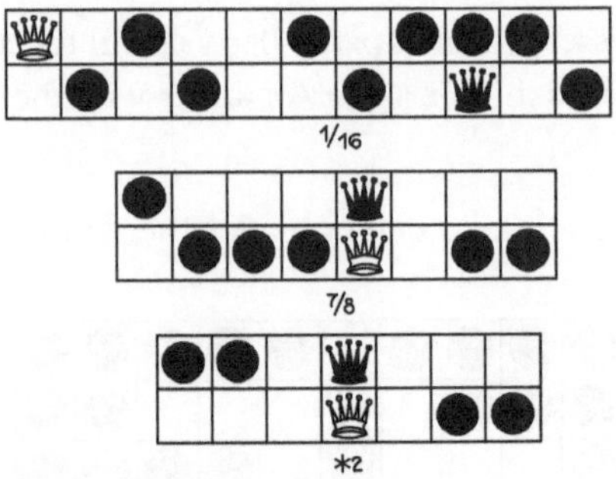

In [Snatzke 2002a] and [Tegos 2002], we can see a vast list of values:

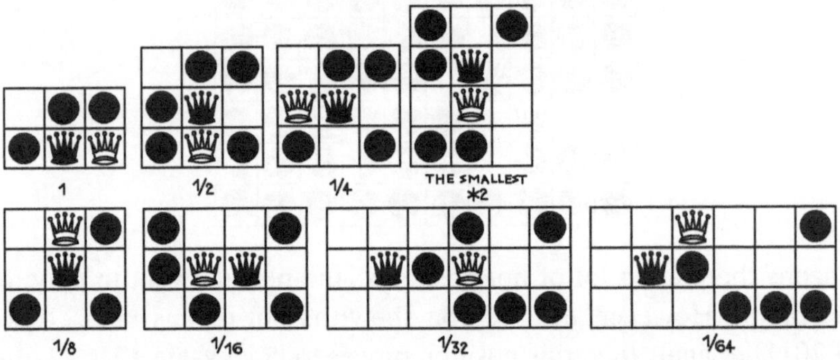

to which we can add some infinitesimals...

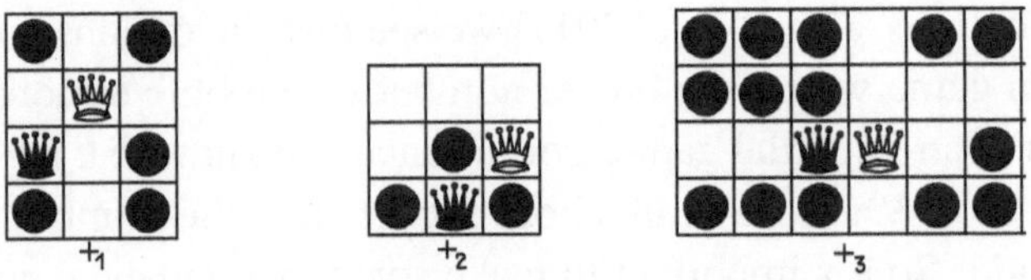

So, we can say that Amazons is a rich combinatorial game with a vast number of interesting examples. However, there is a little drawback: Amazons is very counterintuitive. For instance, in the last picture, the game of value $+_1$ is winning for Left (black). It's curious that a position with such centralized Queen is lost for Right (white). So it's very difficult to analyze an Amazons game in chess-like fashion, using general strategic principles.

R-classes. In partizan games there is the reversibility phenomena. A reversible move for Left is one which Right can promise to respond to in such a way that prospects are at least as good as they were before. That is, a left option A of G is reversible if A has a right option, A^R, with $A^R \leq G$. In any context Right promises "if you ever choose option A of G then I will immediately move to A^R". So, Left just chooses the option A if he intends to follow up Right's move to A^R with an immediate response to one of A^R's left options. If he plans some other move elsewhere, he might just as well start with that. Reversibility is a simplification principle of combinatorial games.

For instance, with the reversibility simplification principle we can see why $\{\uparrow \mid 0\}$ has value $*$. We know that $\uparrow = \{0 \mid *\}$. It's easy to observe that $\{\uparrow \mid 0\} + * \geq 0$ (the game $\{\uparrow \mid 0\} + *$ is a previous player win so its value is 0). When Left chooses the option $\uparrow$, then Right will immediately move to $*$, which guarantees prospects at least as good as before. But Left has the option 0 immediately available so $\{\uparrow \mid 0\} = \{0 | 0\} = *$. Left "can go to 0 in *two tempos*".

This theoretical background is very important, but we need a more explicit way to recognize the options that can "act" like nimbers by reversibility. We will propose some classes of games. First we need some notation:

Definition. Consider a game $G = \{G^L \mid G^R\}$. A set of games $\Delta \subseteq G^L$ has the type $\text{Mate}(G_1, \ldots, G_n)$ if, for all $i \in \{1, \ldots, n\}$, there exists $L \in \Delta$ with $L + G_i \geq 0$. A set of games $\Delta \subseteq G^R$ has the type $\text{Mate}(J_1, \ldots, J_n)$ if, for all $i \in \{1, \ldots, n\}$, exists $R \in \Delta$ with $R + J_i \leq 0$. If the set G^L has the type $\text{Mate}(G_1, \ldots, G_n)$, we write $G = \{\text{Mate}(G_1, \ldots, G_n) \mid G^R\}$. Respectively, if the set G^R has the type $\text{Mate}(J_1, \ldots, J_n)$, we write $G = \{G^L \mid \text{Mate}(J_1, \ldots, J_n)\}$.

We can now define some classes:

Definition. For $n, K \in \mathbb{N}_0$ and $n < K$ (we are preparing the construction of a $*K$), we define by recurrence the set of games as follows: $+_K R_n^{(N)}$

$$+_K R_n^{(0)} = *n,$$

and for $N > 0$, the class $+_K R_n^{(N)}$ has the form

$$\Big\{\text{Mate}(0, *, \ldots, *(n-1)) \,\|\, \{G' \in +_K R_n^{(N-1)}, \underbrace{\Delta_1}_{\trianglelefteq *K} \mid \text{Mate}(0, \ldots, *(K-1))\}, \underbrace{\Delta_2}_{\trianglelefteq *n}\Big\}.$$

The games in the set Δ_1 are smaller or confused with $*K$. The games in the set Δ_2 are smaller or confused with $*n$.

Remarks and examples. (1) When we say $G \in +_K R_n$ we mean that $\exists N_0 : G \in +_K R_n^{(N_0)}$. When we say $G \in -_K R_n^{(N)}$ we mean that $-G \in +_K R_n^{(N)}$. When we say $G \in -_K R_n$ it means that $\exists N_0 : G \in -_K R_n^{(N_0)}$.

(2) A game can be an element of a lot of classes. For instance,

$$+_1 = \{0 \mid \{0 \mid -1\}\} \in +_K R_0^{(1)}.$$

If we consider the game $G = \{2 \mid \{0 \mid -3\}, \{+_1 \mid -20\}\}$ then

$$G \in +_K R_0^{(1)} \cap +_K R_0^{(2)}.$$

(3) If $G \in +_K R_n$, we call

$$\text{depth}_{(n,K)}(G) = \text{Max}\{N \in \mathbb{N}_0 : G \in +_K R_n^{(N)}\}.$$

Lemma 1. *If $G \in +_K R_n$ then $G + *n \geq 0$.*

Proof (complete induction in $\text{depth}_{(n,K)}(G)$). If $\text{depth}_{(n,K)}(G) = 0$ then $G = *n$. In this case the result is trivial.

Suppose $G + *n \geq 0$ for all $G \in +_K R_n$ and $\text{depth}_{(n,K)}(G) \leq N-1$. If $J \in +_K R_n$ and $\text{depth}_{(n,K)}(J) = N$ then J has the form

$$\Big\{\text{Mate}(0, *, \dots, *(n-1)) \,\|\, \{G' \in +_K R_n^{(N-1)}, \underbrace{\dots}_{\trianglelefteq *K} \mid \text{Mate}(0, \dots, *K-1)\}, \underbrace{\dots}_{\trianglelefteq *n}\Big\}.$$

Let's analyze the right options of the game $J + *n$. Right can move to

$$\Big\{G' \in +_K R_n^{(N-1)}, \underbrace{\dots}_{\trianglelefteq *K} \mid \text{Mate}(0, \dots, *(K-1))\Big\} + *n$$

or Right can move to $J + *i$ $(i < n)$. In the first case, Left chooses

$$G' \in +_K R_n^{(N-1)}) + *n,$$

and wins (induction hypothesis). In the second case, Left wins because J^L is $\text{Mate}(0, \dots, *(n-1))$. □

Lemma 2. *If $G \in +_K R_n$ then $G + *K$ is fuzzy.*

Proof (complete induction $\text{depth}_{(n,K)}(G)$). If $\text{depth}_{(n,K)}(G) = 0$ then $G = *n$. In this case the game $*n + *K$ is an easy win for next player.

Suppose $G + *K$ is fuzzy for all $G \in +_K R_n$ with $\text{depth}_{(n,K)}(G) \leq N-1$. If $J \in +_K R_n$ and $\text{depth}_{(n,K)}(J) = N$ then we must analyze

$$\Big\{\text{Mate}(0, *, \dots, *n-1) \,\|\, \{G' \in +_K R_n^{(N-1)}, \underbrace{\dots}_{\trianglelefteq *K} \mid \text{Mate}(0, \dots, *K-1)\}, \underbrace{\dots}_{\trianglelefteq *n}\Big\} + *K.$$

If Left moves first, then he moves $*K$ to $*n$ and wins (Lemma 1). If Right moves first then he plays to

$$\Big\{G' \in +_K R_n^{(N-1)}, \underbrace{\ldots}_{\trianglelefteq *K} \mid \mathrm{Mate}(0, \ldots, *K-1)\Big\} + *K.$$

Now, Left must choose $G' \in +_K R_n^{(N-1)} + *K$ and Right wins (I. H.). □

Construction theorem. *If* $G_i \in +_K R_i$ *and* $J_i \in -_K R_i$, $i \in \{0, \ldots, K-1\}$ *then* $\{G_0, \ldots, G_{K-1} \mid J_0, \ldots, J_{K-1}\} = *K$.

Proof. Let's play $\{G_0, \ldots, G_{K-1} \mid J_0, \ldots, J_{K-1}\} + *K$.

If Left moves to $\{G_0, \ldots, G_{K-1} \mid J_0, \ldots, J_{K-1}\} + *i$, $i < K$ then Right moves to $J_i + *i$ and wins (Lemma 1). If Left plays to $G_i + *K$ then the game is fuzzy by Lemma 2, so Right wins. If Right plays first, the argument is the same and Left wins, so $\{G_0, \ldots, G_{K-1} \mid J_0, \ldots, J_{K-1}\} + *K = 0$. □

Examples of stars in partizan games:

We have $\uparrow \in_1 R_0$ therefore, $\{\uparrow \mid 0\} = *$.

Note that $\uparrow \notin_2 R_0$ and $\{\uparrow, * \mid 0, *\} \neq *2$.

Example in Amazons. We can use the theorem just proved to analyze canonical forms of nimbers in partizan games. Consider one important move on the top of the first position of this text (G):

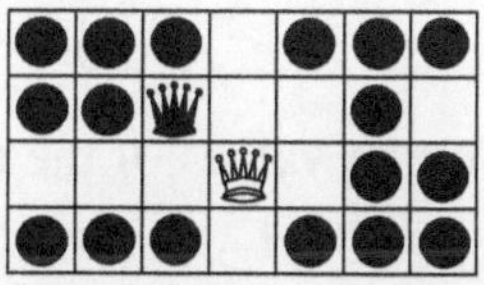

If one canonicalizes this in CGSuite one gets the horrible canonical form

$$\Big\{\tfrac{1}{2} \,\Big|\, 0, \big\{*, \{1 \mid *, \pm\{1, \{2 \mid 0\}\}\}\big\} \,\Big|\, \{0 \mid -1\}, \{*, \pm 1 \mid -1\}, \big\{0, * \,\|\, -\tfrac{1}{2}* \mid -3\big\},$$
$$\Big\{\uparrow, \uparrow * \,\Big|\, \{0 \mid -1\}, \{* \mid -\tfrac{1}{4}\}\Big\},$$
$$\Big\{0, *, *2 \,\Big|\, *, \{*, \{1* \mid -\tfrac{1}{2}\} \mid -\tfrac{1}{2}\},$$
$$\{*, \{1*, \{2 \mid 0\} \mid -\tfrac{1}{2}, \{\pm 1, \tfrac{1}{2}, \{\tfrac{1}{2} \mid 0\} \mid -1\}\} \mid -\tfrac{1}{2}\}\Big\}, \quad (=: J)$$
$$\Big\{0, \uparrow *, \{\tfrac{1}{2}, \{\tfrac{1}{2} \mid 0\} \mid -1\} \,\Big|\, -1\Big\}\Big\}.$$

Now we can make some considerations:

(A) $\frac{1}{2}$ is a left option so the set of left options is $\mathrm{Mate}(0, *)$.

(B) Consider the right option occupying two lines; call it J. We can observe that $*$ and $\{*, \{1* \mid -\frac{1}{2}\} \mid -\frac{1}{2}\}$ are right options, so J^R is Mate(0, $*$, $*2$).

(C) All the other right options of G are smaller or confused with $*2$.

(D) In the right option of item (B), all the left options are smaller or confused with $*3$. $*2$ is a left option.

From (A), (B), (C) and (D) we can conclude that G has the form

$$\Big\{\underbrace{\tfrac{1}{2}}_{\text{Mate}(0,\,*)}, \dots \,\Big\|\, \{\underbrace{*2}_{\in +_3 R_2^{(0)}},\ \underbrace{0, *}_{\trianglelefteq *3} \mid \underbrace{*, \{*, \{1* \mid -\tfrac{1}{2}\} \mid -\tfrac{1}{2}\}, \dots\}}_{\text{Mate}(0,\,*,\,*2)}, \underbrace{\dots}_{\trianglelefteq *2}\Big\},$$

thus, $G \in +_3 R_1^{(1)}$. In a similar way we can conclude that $G \in +_3 R_2^{(1)}$. It's very interesting because the same option can act like a $*$ and like a $*2$.

4 *in Amazons. Theodore Tegos [2002] organized and studied a big database of Amazons game positions. About nimbers, he wrote "The current challenge is to find an Amazons position whose combinatorial value is $*4$, if such position exists."

To construct a $*4$ we will use an idea involving an algebraic table:

+	$*2$	$*3$
$*2$	0	$*$
0	$*2$	$*3$

If we observe carefully the resulting values of the table we see that those values are the "needed stuff" to construct a $*4$.

In a partizan game we can construct nimbers with other options than nimbers. So, there is a gigantic number of "good algebraic tables". For instance, we know that

$$+_1 + *2 = \{\underbrace{0, *, *2}_{\text{Mate}(0,\,*)} \,\|\, \underbrace{*2}_{\in +_4 R_2^{(0)}} \mid \underbrace{-1*2}_{\text{Mate}(0,\,*)}\} \in +_4 R_2^{(1)}$$

and

$$+_2 + *3 = \{\underbrace{0, *, *2, *3}_{\text{Mate}(0,\,*,\,*2)} \,\|\, \underbrace{*3}_{\in +_4 R_3^{(0)}} \mid \underbrace{-2*3}_{\text{Mate}(0,\,*,\,*2)}\} \in +_4 R_3^{(1)}.$$

When we want to construct a $*4$, the games $+_1 + *2$ and $+_2 + *3$ acts like $*2$ and $*3$. So, the following table is also very useful to construct a $*4$:

+	$*2$	$*3$
$*2$	0	$*$
$+_1$ or $+_2$	$\{0, *, *2 \parallel *2 \mid -1*2\} \in +_4R_2^{(1)}$	$\{0, *, *2, *3 \parallel *3 \mid -2*3\} \in +_4R_3^{(1)}$

We will produce this table on the Amazons board. First, we introduce two fundamental positions:

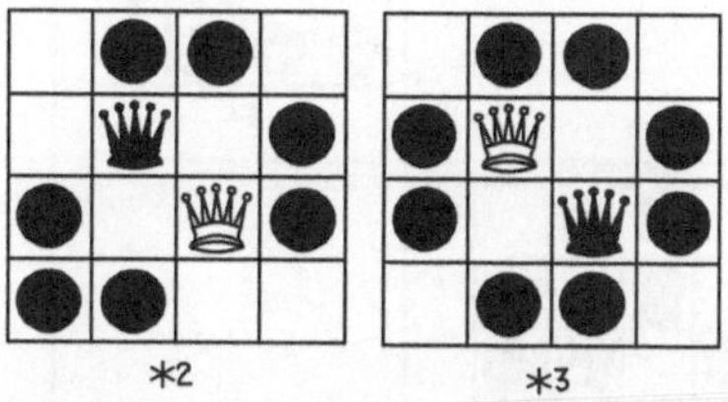

Second, we introduce some similar positions:

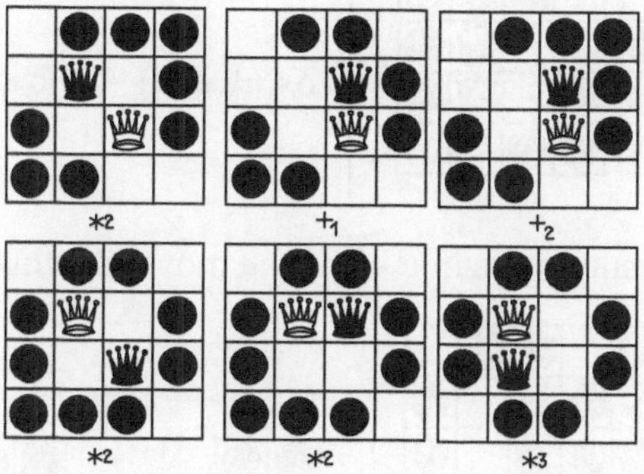

Using a *collage* idea we can join the two fundamental positions obtaining the following position (G):

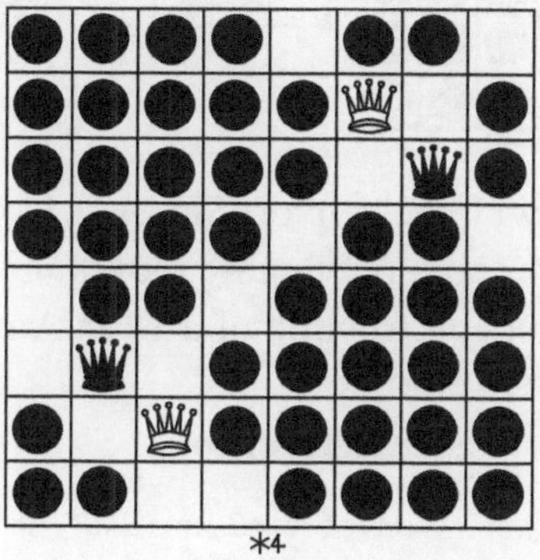

*4

With this construction it's possible to have all the results of the algebraic table under consideration:

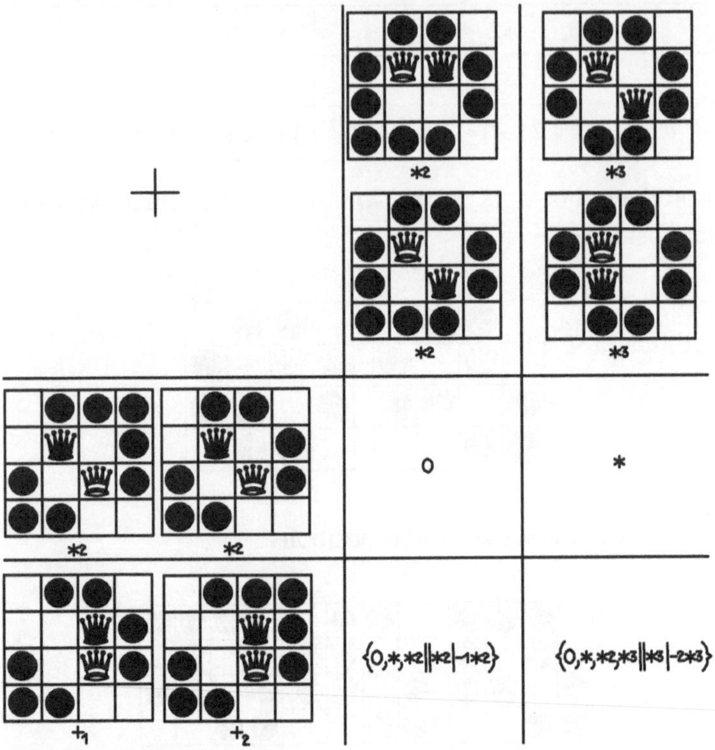

For instance, if we want to obtain a $*$ we can move like this:

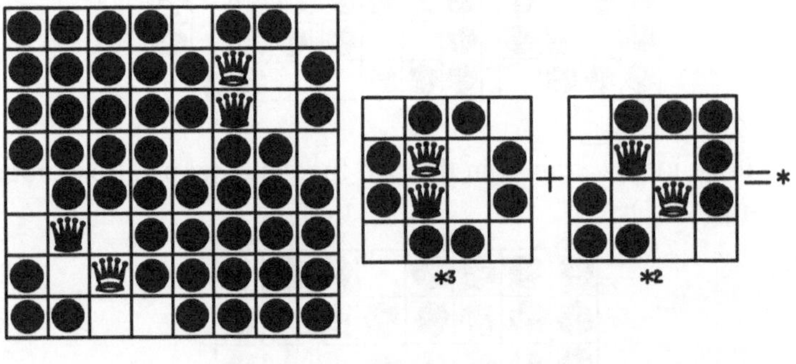

With exhaustive analysis (with the help of [Siegel 2011]) it is possible to see that all the other possible moves are not winning moves in the game $G + *4$. So the exposed position is the first construction of a $*4$ in Amazons.

References

[Berlekamp 2000] E. R. Berlekamp, "Sums of $N \times 2$ Amazons", pp. 1–34 in *Game theory, optimal stopping, probability and statistics*, edited by F. T. Bruss and L. Le Cam, IMS Lecture Notes Monogr. Ser. **35**, Inst. Math. Statist., Beachwood, OH, 2000.

[Siegel 2011] A. Siegel, "Combinatorial game suite", 2011, http://cgsuite.sourceforge.net/.

[Snatzke 2002a] G. Snatzke, "New results in exhaustive search in the game Amazons", preprint, University of Jena, 2002.

[Snatzke 2002b] R. G. Snatzke, "Exhaustive search in the game Amazons", pp. 261–278 in *More games of no chance* (Berkeley, CA, 2000), edited by R. Nowakowski, Math. Sci. Res. Inst. Publ. **42**, Cambridge Univ. Press, 2002.

[Tegos 2002] T. Tegos, *Shooting the last arrow*, Masters thesis, University of Alberta, 2002, https://webdocs.cs.ualberta.ca/~jonathan/PREVIOUS/Grad/tegos.msc.pdf.

carlos.santos@isec.universitas.pt *Centro de Estruturas Lineares e Combinatórias, Faculdade de Ciências da Universidade de Lisboa, Campo Grande, Edifício 6, Piso 2, 1749-016, Lisboa, Portugal*

jnsilva@cal.berkeley.edu *Departamento de Matemática, Universidade de Lisboa, Campo Grande, C6, Piso 2, 1749-016 Lisboa, Portugal*

Games of No Chance 4
MSRI Publications
Volume **63**, 2015

Misère canonical forms of partizan games

AARON N. SIEGEL

We show that partizan games admit canonical forms in misère play. The proof is a synthesis of Conway's simplest form theorems for normal-play partizan games and misère-play impartial games. As an immediate application, we show that there are precisely 256 games born by day 2, and obtain a bound on the number of games born by day 3.

1. Introduction

Disjunctive compounds of short combinatorial games have been studied for many years under a variety of assumptions. A structure theory for normal-play impartial games was established in the 1930s by the *Sprague–Grundy theorem* [Grundy 1939; Sprague 1935; 1937]. Every such game G is equivalent to a Nim-heap, and the size of this heap, known as the *nim value* of G, completely describes the behavior of G in disjunctive sums. The Sprague–Grundy theorem underpins virtually all subsequent work on impartial combinatorial games.

Decades later, Conway generalized the Sprague–Grundy theorem in two directions [Berlekamp et al. 2003; Conway 2001]. First, he showed that every partizan game G can be assigned a *value* that exactly captures its disjunctive behavior, and this value is represented by a unique simplest form for G. Conway's game values are partizan analogues of nim values, and his simplest form theorem directly generalizes the Sprague–Grundy theorem.

Conway also introduced a misère-play analogue of the Sprague–Grundy theorem. He showed that every impartial game G is represented by a unique *misère* simplest form [Conway 2001]. Unfortunately, in misère play such simplifications tend to be weak, and as a result the canonical theory of misère games is less useful in practice than its normal-play counterparts.

In each case — normal-play impartial, normal-play partizan, and misère-play impartial — the identification of simplest forms proved to be a key result, at once establishing a structure theory and opening the door to further investigations. In this paper, we prove an analogous simplest form theorem for the misère-play partizan case. The proof integrates techniques drawn from each of Conway's advances, together with a crucial lemma from [Mesdal and Ottaway 2007].

As an immediate application of this theorem, we also calculate the number of partizan misère games born on day 2. As it turns out, all 256 nonisomorphic games born by day 2 are already in simplest form, so there are no day-2 reductions at all! This sobering result is evidence that partizan misère games are yet more unruly than their impartial cousins.

2. Misère equivalence

We denote by $o(G)$ the misère outcome class of G:

$$o(G) = \begin{cases} \mathscr{L} & \text{if Left can win no matter who plays first,} \\ \mathscr{R} & \text{if Right can win no matter who plays first,} \\ \mathscr{P} & \text{if second player (the Previous player) can win,} \\ \mathscr{N} & \text{if first player (the Next player) can win,} \end{cases}$$

with misère play assumed in all cases. These four outcome classes are naturally partially ordered by *favorability to Left*:

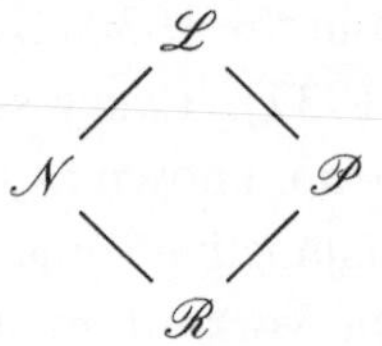

This induces the usual partial order on all misère games.

$$G \geq H \quad \text{if and only if} \quad o(G+X) \geq o(H+X) \text{ for all games } X.$$

We also define equality in the usual manner.

$$G = H \quad \text{if and only if} \quad G \geq H \text{ and } H \geq G.$$

These definitions are exactly the same as the corresponding definitions for the normal-play theory: the only difference is in the meaning of $o(G)$. The normal-play theory was studied by Berlekamp, Conway and Guy [Berlekamp et al. 2003; Conway 2001] with great success. They showed that there is a simple recursive test for normal-play equality:

Fact 2.1. *In normal play, $G \geq H$ if and only if no $G^R \leq H$ and $G \leq$ no H^L.*

From this it follows, by a straightforward induction, that $G = 0$ for every $\mathscr{P}$-position G (still assuming normal play). These facts are crucial ingredients in the following key result.

Fact 2.2 (simplest form theorem [Berlekamp et al. 2003; Conway 2001]). *In normal play, suppose $G = H$, and assume that neither G nor H has any dominated or reversible options. Then for every H^L there is a G^L such that $G^L = H^L$, and vice versa; and likewise for Right options.*

Remarkably, the statement of the simplest form theorem is *exactly* the same in misère play, with *exactly* the same definitions of dominated and reversible options! However the misère analogue of Fact 2.1 is badly false. A recent result due to Mesdal and Ottaway shows just how decisively it fails:

Fact 2.3 [Mesdal and Ottaway 2007]. *Let G be any game. Then $G \neq 0$ in misère play, unless G is identically zero.*

The following proposition is the nearest misère analogue to Fact 2.1.

Proposition 2.4. *In misère play, $G \geq H$ if and only if the following two conditions hold:*

(i) *for all X with $o(H+X) \geq \mathscr{P}$, we have $o(G+X) \geq \mathscr{P}$; and*
(ii) *for all X with $o(H+X) \geq \mathscr{N}$, we have $o(G+X) \geq \mathscr{N}$.*

Note that $o(G) \geq \mathscr{P}$ is equivalent to "Left can win playing second on G," and $o(G) \geq \mathscr{N}$ is equivalent to "Left can win playing first on G."

Proof of Proposition 2.4. $\Rightarrow$ is immediate. For the converse, we must show that $o(G+X) \geq o(H+X)$, for all X. If $o(H+X) = \mathscr{R}$, then there is nothing to prove; if $o(H+X) = \mathscr{P}$ or $\mathscr{N}$, it is immediate from (i) or (ii), respectively. Finally, if $o(H+X) = \mathscr{L}$, then by (i) and (ii) we have $o(G+X) \geq$ both $\mathscr{P}$ and $\mathscr{N}$, whence $o(G+X) = \mathscr{L}$. □

3. Ends and adjoints

In normal play, $G + (-G)$ is a $\mathscr{P}$-position, and hence equal to 0, for every game G. This is emphatically false in misère play: by Fact 2.3, $G + (-G) \neq 0$ for *every* game G except 0 itself. Nonetheless, we can give an explicit example of a game G°, the *adjoint* of G, such that $G + G^\circ$ is a misère $\mathscr{P}$-position. Readers familiar with the impartial theory will recognize it as the partizan analogue of Conway's *mate* [Conway 2001].

Games of the form $\{G^L \mid\}$ and $\{\mid G^R\}$, with options for just one player, will feature prominently in the succeeding analysis. Recall that in normal play, every such game is equal to an integer [Berlekamp et al. 2003]. In misère play, by contrast, such games emerge as a crucial and complicated pathology. We will sometimes use a dot to indicate "no moves", when it is useful to clarify the notation; for example, we may write $\{0 \mid \cdot\}$ in place of $\{0 \mid\}$.

We now focus exclusively on misère play, and we won't consider normal play again until Section 7. All further uses of the notation $o(G)$, and the relations $\geq$ and $=$, refer unambiguously to misère play.

Definition 3.1. G is a *Left (Right) end* if G has no Left (Right) option.

Definition 3.2. The *adjoint* of G, denoted G°, is given by

$$G^\circ = \begin{cases} * & \text{if } G = 0, \\ \{(G^R)^\circ \mid 0\} & \text{if } G \neq 0 \text{ and } G \text{ is a Left end}, \\ \{0 \mid (G^L)^\circ\} & \text{if } G \neq 0 \text{ and } G \text{ is a Right end}, \\ \{(G^R)^\circ \mid (G^L)^\circ\} & \text{otherwise.} \end{cases}$$

Note that G° always has at least one option for each player, and hence is never an end.

Proposition 3.3. *$G + G^\circ$ is a $\mathscr{P}$-position.*

Proof. By symmetry, it suffices to show that Left can win $G + G^\circ$ moving second. G° is not a Right end, so $G + G^\circ$ has at least one Right option. There are two cases:

- If Right moves to $G^R + G^\circ$ or $G + (G^L)^\circ$, Left makes the mirror image move on the other component, which wins by induction on G.
- If G is a Left end and Right moves to $G + 0$, then Left has no move, and so wins a priori. □

Theorem 3.4. *If $G \not\geq H$, then:*

(a) *there is some T such that $o(G + T) \leq \mathscr{P}$ but $o(H + T) \geq \mathscr{N}$; and*

(b) *there is some U such that $o(G + U) \leq \mathscr{N}$ but $o(H + U) \geq \mathscr{P}$.*

Proof. We know by Proposition 2.4 that one of (a) or (b) must hold, so it suffices to show that (a) $\Rightarrow$ (b) and (b) $\Rightarrow$ (a). The arguments are identical, so we will show that (a) $\Rightarrow$ (b).

Fix T so that $o(G + T) \leq \mathscr{P}$ and $o(H + T) \geq \mathscr{N}$, and put

$$U = \{(H^R)^\circ \mid T\}.$$

Now from $G+U$, Right has a winning move, to $G+T$. Therefore $o(G+U) \leq \mathscr{N}$. Likewise, consider $H + U$. It is certainly not a Right end, since Right has a move from U to T. Now if Right moves to $H^R + U$, Left has a winning response to $H^R + (H^R)^\circ$. If instead Right moves to $H + T$, then since $o(H + T) \geq \mathscr{N}$, Left wins *a priori*. Therefore $o(H + U) \geq \mathscr{P}$, as needed. □

The following lemma is a straightforward generalization of the Mesdal–Ottaway result (Fact 2.3), and it proves to be a crucial piece of the analysis.

Lemma 3.5. *If H is a Left end and G is not, then $G \not\geq H$.*

Proof. Put

$$T = \{(H^R)^\circ \parallel \cdot \mid (G^L)^\circ\}.$$

Consider $H+T$. If Right moves to H^R+T, then Left can respond to $H^R+(H^R)^\circ$, winning by Proposition 3.3; if Right moves to $H+\{\cdot \mid (G^L)^\circ\}$, then Left wins outright, since he has no further move. Therefore $o(H+T) \geq \mathscr{P}$.

Now consider $G+T$. Right has a move to $G+\{\cdot \mid (G^L)^\circ\}$. Left's only response is to $G^L+\{\cdot \mid (G^L)^\circ\}$, which must exist since G is not a Left end. But Right may then respond to $G^L+(G^L)^\circ$, winning. Therefore $o(G+T) \leq \mathscr{N}$.

This shows that $o(G+T) \not\geq o(H+T)$, so in fact $G \not\geq H$. □

4. Dominated and reversible options

Definition 4.1. Let G be a game.

(a) A Left option G^L is said to be *dominated* if $G^{L'} \geq G^L$ for some other Left option $G^{L'}$.

(b) A Right option G^R is said to be *dominated* if $G^{R'} \leq G^R$ for some other Right option $G^{R'}$.

(c) A Left option G^L is said to be *reversible* if $G^{LR} \leq G$ for some Right option G^{LR}.

(d) A Right option G^R is said to be *reversible* if $G^{RL} \geq G$ for some Left option G^{RL}.

Lemma 4.2. *Suppose G^{L_1} is dominated by G^{L_2}, and let G' be the game obtained by eliminating G^{L_1} from G. Then $G = G'$.*

Proof. Since the Left options of G' are a subset of those of G, and since G' still has at least one Left option (namely, G^{L_2}), we trivially have $G' \leq G$. Thus it suffices to show that $G' \geq G$.

By Proposition 2.4, we must show that for all X,

(i) if $o(G+X) \geq \mathscr{P}$, then $o(G'+X) \geq \mathscr{P}$;

(ii) if $o(G+X) \geq \mathscr{N}$, then $o(G'+X) \geq \mathscr{N}$.

The proof is by induction on X.

(i) Suppose $o(G+X) \geq \mathscr{P}$. Then every $o(G^R+X) \geq \mathscr{N}$ and every $o(G+X^R) \geq \mathscr{N}$. Now G and G' have exactly the same Right options, so every $o((G')^R+X) \geq \mathscr{N}$. Also, by induction on X, we have

$$o(G'+X^R) = o(G+X^R) \geq \mathscr{N}$$

for all X^R. This shows that $o(G'+X) \geq \mathscr{P}$, except possibly in the case where $G'+X$ is a Right end. But since G and G' have the same Right options, this would imply that $G+X$ is a Right end, contradicting the assumption $o(G+X) \geq \mathscr{P}$.

(ii) Suppose $o(G+X) \geq \mathscr{N}$. We know that $G+X$ is not a Left end (since G is assumed to have a Left option), so either some $o(G+X^L) \geq \mathscr{P}$ or some $o(G^L+X) \geq \mathscr{P}$. If $o(G+X^L) \geq \mathscr{P}$, then by induction $o(G'+X^L) \geq \mathscr{P}$, so we are done. If $o(G^L+X) \geq \mathscr{P}$ and G^L is a Left option of G', then the conclusion is immediate. So the only remaining case is when $G^L = G^{L_1}$. But since $G^{L_2} \geq G^{L_1}$, we have

$$o(G^{L_2}+X) \geq o(G^{L_1}+X) \geq \mathscr{P}.$$

In all cases, $o(G'+X) \geq \mathscr{N}$. □

Lemma 4.3. *Suppose G^{L_1} is reversible through $G^{L_1R_1}$, and let G' be the game obtained by bypassing G^{L_1}:*

$$G' = \{G^{L_1R_1L}, G^{L'} \mid G^R\},$$

where $G^{L'}$ is understood to range over all Left options of G except G^{L_1}. Then $G = G'$.

Proof. By Proposition 2.4, we must show that for all X,

(i) $o(G+X) \geq \mathscr{P}$ if and only if $o(G'+X) \geq \mathscr{P}$;

(ii) $o(G+X) \geq \mathscr{N}$ if and only if $o(G'+X) \geq \mathscr{N}$.

The proof is by induction on X.

(i) If either $G+X$ or $G'+X$ is a Right end, then both must be, since G and G' have the same Right options. If both are Right ends, then necessarily $o(G+X) \leq \mathscr{N}$ and $o(G'+X) \leq \mathscr{N}$, so (i) is satisfied. So assume neither is a Right end; then by induction on X we have

$$\begin{aligned} o(G+X) \geq \mathscr{P} &\iff \text{every } o(G^R+X) \geq \mathscr{N} \text{ and every } o(G+X^R) \geq \mathscr{N} \\ &\iff \text{every } o((G')^R+X) \geq \mathscr{N} \text{ and every } o(G'+X^R) \geq \mathscr{N} \\ &\iff o(G'+X) \geq \mathscr{P}. \end{aligned}$$

(ii) First suppose $o(G+X) \geq \mathscr{N}$. We know that G is not a Left end (since it has G^{L_1} as a Left option), so either some $o(G+X^L) \geq \mathscr{P}$, or else some $o(G^L+X) \geq \mathscr{P}$. There are three subcases.

- If $o(G+X^L) \geq \mathscr{P}$, then $o(G'+X^L) \geq \mathscr{P}$, by induction on X.
- If $o(G^L+X) \geq \mathscr{P}$ for $G^L \neq G^{L_1}$, then G^L is also a Left option of G' and the conclusion is immediate.

- Suppose instead that $o(G^{L_1} + X) \geq \mathscr{P}$. Then in particular,

  $$o(G^{L_1 R_1} + X) \geq \mathscr{N}.$$

  So either $o(G^{L_1 R_1 L} + X) \geq \mathscr{P}$ or $o(G^{L_1 R_1} + X^L) \geq \mathscr{P}$. In the first case we are done, since $G^{L_1 R_1 L}$ is a Left option of G'. But in the second case, we have $G \geq G^{L_1 R_1}$ (by hypothesis) so

  $$o(G + X^L) \geq o(G^{L_1 R_1} + X^L) \geq \mathscr{P},$$

  and therefore $o(G' + X^L) \geq \mathscr{P}$, by induction on X.

Finally, suppose $o(G' + X) \geq \mathscr{N}$. Then again we have three subcases: either $o(G^L + X) \geq \mathscr{P}$, or $o(G' + X^L) \geq \mathscr{P}$, or $o(G^{L_1 R_1 L} + X) \geq \mathscr{P}$. The first two subcases are just the same as before, so suppose $o(G^{L_1 R_1 L} + X) \geq \mathscr{P}$. Then in particular, $o(G^{L_1 R_1} + X) \geq \mathscr{N}$. But $G \geq G^{L_1 R_1}$ so

$$o(G + X) \geq o(G^{L_1 R_1} + X) \geq \mathscr{N}. \qquad \square$$

5. The simplest form theorem

Theorem 5.1 (simplest form theorem). *Suppose $G = H$, and assume that neither G nor H has any dominated or reversible options. Then for every H^L there is a G^L such that $G^L = H^L$, and vice versa; and likewise for Right options.*

In order to prove Theorem 5.1, we generalize some machinery from Conway's proof [2001, Theorem 76] that impartial games admit misère simplest forms.

Definition 5.2. (a) G is *downlinked to H (by T)* if

$$o(G + T) \leq \mathscr{P} \quad \text{and} \quad o(H + T) \geq \mathscr{P}.$$

(b) G is *uplinked to H (by T)* if $o(G + T) \geq \mathscr{P}$ and $o(H + T) \leq \mathscr{P}$.

Theorem 5.3. *$G \geq H$ if and only if the following four conditions hold.*

(i) *G is downlinked to no H^L;*

(ii) *no G^R is downlinked to H;*

(iii) *if H is a Left end, then so is G;*

(iv) *if G is a Right end, then so is H.*

Proof. For $\Rightarrow$ (i), fix any game T. If $o(G+T) \leq \mathscr{P}$, then certainly $o(H+T) \leq \mathscr{P}$ as well. Therefore $o(H^L + T) \leq \mathscr{N}$, so T cannot downlink G to H^L. $\Rightarrow$ (ii) is similar, and $\Rightarrow$ (iii) and (iv) are just restatements of Lemma 3.5 (and its mirror image).

We now prove $\Leftarrow$. By Proposition 2.4, we must show that for all T,

- if $o(H + T) \geq \mathscr{P}$, then $o(G + T) \geq \mathscr{P}$; and

- if $o(H+T) \geq \mathscr{N}$, then $o(G+T) \geq \mathscr{N}$.

The proof proceeds by induction on T. First suppose (for contradiction) that $o(H+T) \geq \mathscr{P}$ but $o(G+T) \leq \mathscr{N}$. Then either

$$o(G+T^R) \leq \mathscr{P} \quad \text{or} \quad o(G^R+T) \leq \mathscr{P},$$

or else $G+T$ is a Right end. If $o(G+T^R) \leq \mathscr{P}$, then by induction on T we have $o(H+T^R) \leq \mathscr{P}$, contradicting the assumption that $o(H+T) \geq \mathscr{P}$. If $o(G^R+T) \leq \mathscr{P}$, then T downlinks G^R to H, contradicting (ii). Finally, if $G+T$ is a Right end, then in particular G is a Right end, so by (iv) H is a Right end. Therefore $H+T$ is a Right end, contradicting the assumption that $o(H+T) \geq \mathscr{P}$.

This shows that $o(H+T) \geq \mathscr{P}$ implies $o(G+T) \geq \mathscr{P}$. The proof that $o(H+T) \geq \mathscr{N}$ implies $o(G+T) \geq N$ is identical, with (i) and (iii) used in place of (ii) and (iv). □

Theorem 5.4. *G is downlinked to H if and only if no $G^L \geq H$ and $G \geq$ no H^R.*

Proof. Suppose T downlinks G to H, so that $o(G+T) \leq \mathscr{P}$ and $o(H+T) \geq \mathscr{P}$. Then necessarily $o(G^L+T) \leq \mathscr{N}$ and $o(H^R+T) \geq \mathscr{N}$, so T witnesses both $G^L \not\geq H$ and $G \not\geq H^R$.

Conversely, suppose that no $G^L \geq H$ and $G \geq$ no H^R. Then for each G_i^L, Theorem 3.4 yields an X_i such that

$$o(G_i^L+X_i) \leq \mathscr{P} \quad \text{and} \quad o(H+X_i) \geq \mathscr{N}.$$

Likewise, for each H_j^R, there is some Y_j such that

$$o(G+Y_j) \leq \mathscr{N} \quad \text{and} \quad o(H_j^R+Y_j) \geq \mathscr{P}.$$

Put

$$T = \begin{cases} * & \text{if } G = H = 0, \\ \{0 \mid (H^L)^\circ\} & \text{if } G = 0 \text{ and } H \text{ is a nonzero Right end,} \\ \{(G^R)^\circ \mid 0\} & \text{if } H = 0 \text{ and } G \text{ is a nonzero Left end,} \\ \{Y_j, (G^R)^\circ \mid X_i, (H^L)^\circ\} & \text{otherwise.} \end{cases}$$

We claim that G is downlinked to H by T. We will show that $o(G+T) \leq \mathscr{P}$; the proof that $o(H+T) \geq \mathscr{P}$ is identical.

We first show that $G+T$ has a Left option. If G has a Left option, this is automatic. If G or H has a Right option, then T necessarily has a Left option. This exhausts every case except when $G = 0$ and H is a Right end; but then Left's move to 0 is built into the definition of T.

Thus $G+T$ is not a Left end, and it therefore suffices to show that every Left option is losing. If Left moves to G_i^L+T, Right can respond to $G_i^L+X_i$, which wins by choice of X_i. If Left moves to $G + (G^R)^\circ$, Right can respond to $G^R + (G^R)^\circ$, which wins by Proposition 3.3. Left's move to $G + Y_j$ loses automatically, by choice of Y_j. The only remaining possibility is Left's additional move to 0 in the first two cases of the definition of T. But that move is only available when $G = 0$, so it ends the game immediately. □

Proof of Theorem 5.1. Fix H^L. Since $G \geq H$, Theorem 5.3 implies that G is not downlinked to H^L. By Theorem 5.4, either $G^L \geq H^L$ or $G \geq H^{LR}$. The latter would imply that $H \geq H^{LR}$, contradicting the assumption that H has no reversible options. So we must have $G^L \geq H^L$.

An identical argument, using the fact that $H \geq G$, shows that $H^{L'} \geq G^L$ for some $H^{L'}$. Therefore

$$H^{L'} \geq G^L \geq H^L.$$

Since H has no dominated options, we must have $H^{L'} = H^L$, so that

$$H^{L'} = G^L = H^L.$$

The same argument suffices for the remaining cases. □

6. Games born by day 2

There are four games born by day 1; and they are familiar from the normal-play theory:

$$0 = \{\cdot \mid \cdot\}, \quad * = \{0 \mid 0\}, \quad 1 = \{0 \mid \cdot\}, \quad \bar{1} = \{\cdot \mid 0\}.$$

Remarkably, they are pairwise incomparable.

Proposition 6.1. *The four games* 0, $*$, 1, *and* $\bar{1}$ *are pairwise incomparable.*

Proof. Theorem 5.3(iii) and (iv) immediately yield $*, 1 \not\geq 0, \bar{1}$ and $0, 1 \not\geq *, \bar{1}$. Since 0 is downlinked to 0 (by $*$), (i) and (ii) furthermore imply $0 \not\geq 1$ and $\bar{1} \not\geq 0$.

Now as a trivial consequence of Theorem 5.4, we have that $\bar{1}$ is downlinked to 0 and 0 is downlinked to 1. It therefore follows from Theorem 5.3(i) that $\bar{1} \not\geq 1, *$ and from (ii) that $\bar{1}, * \not\geq 1$. This exhausts all possibilities. □

Theorem 6.2. *There are* 256 *games born by day* 2.

Proof. There are 16 subsets of $\{0, *, 1, \bar{1}\}$. This gives 256 isomorphism types for games born by day 2, so it suffices to show that every (formal) game born by day 2 is already in simplest form.

So fix such a game G. By Proposition 6.1, G has no dominated options, so it suffices to show that G has no reversible options. Consider some G^{LR}. Since G is born by day 2, G^{LR} is born by day 0, so necessarily $G^{LR} = 0$. But G^{LR} is a Left end and G is not (since it has G^L as a Left option), so by Lemma 3.5 $G \not\geq G^{LR}$. Likewise, by symmetry, $G \not\leq G^{RL}$ for any G^{RL}. □

Let $\mathbb{P}$ denote the set of games born by day 2. We next describe the partial order structure of $\mathbb{P}$. Define

$$\mathbb{P}^+ = \{G \in \mathbb{P} : G \text{ is a nonzero Left end}\},$$
$$\mathbb{P}^- = \{G \in \mathbb{P} : G \text{ is a nonzero Right end}\},$$
$$\mathbb{P}^0 = \{G \in \mathbb{P} : G \text{ is not an end.}\}$$

Then we can write $\mathbb{P}$ as a disjoint union

$$\mathbb{P} = \mathbb{P}^+ \cup \mathbb{P}^- \cup \mathbb{P}^0 \cup \{0\}.$$

Now let $\mathbb{B}_4$ denote the complete Boolean lattice of dimension 4. Let $\mathbb{B}_4^+$ be the partial order obtained by removing the largest element from $\mathbb{B}_4$, and likewise delete the smallest element to obtain $\mathbb{B}_4^-$. We will show that

$$\mathbb{P}^+ \cong \mathbb{B}_4^+, \quad \mathbb{P}^- \cong \mathbb{B}_4^-, \quad \mathbb{P}^0 \cong \mathbb{B}_4^+ \times \mathbb{B}_4^-. \tag{†}$$

In order to characterize the structure of $\mathbb{P}$, it will then suffice to describe the interaction between components.

We first show that there are certain order relations $G \geq H$ that can be inferred from the top-level structure of G and H.

Definition 6.3. We say that G *simply exceeds* H, and write $G \geq_S H$, if:

(i) every H^L is equal to some G^L;

(ii) every G^R is equal to some H^R;

(iii) if H is a Left end, then so is G;

(iv) if G is a Right end, then so is H.

Lemma 6.4. *If $G \geq_S H$, then $G \geq H$.*

Proof. Certainly G is downlinked to no G^L, since whenever Right can win $G+T$ moving second, then he can necessarily win $G^L + T$ moving first. So by (i) G is downlinked to no H^L. Likewise, no H^R can be downlinked to H, so by (ii) no G^R is downlinked to H. The conclusion now follows from Theorem 5.3. □

Theorem 6.5. *Fix G and H satisfying Definition 6.3*(iii) *and* (iv), *and assume that G and H are both born by day* 2. *If $G \geq H$, then $G \geq_S H$.*

Proof. We show that every Left option of H is a Left option of G; the argument that every Right option of G is a Right option of H is identical.

So fix an H^L; by Theorem 5.3, G is not downlinked to H^L. By Theorem 5.4, either $G^L \geq H^L$ for some G^L, or else $G \geq H^{LR}$. Now since H has the Left option H^L, it is not a Left end, whence by assumption neither is G. Since H is born by day 2, we know that every $H^{LR} = 0$, so by Lemma 3.5 we cannot have $G \geq H^{LR}$. Therefore $G^L \geq H^L$ for some G^L.

But G^L and H^L are both born by day 1. By Proposition 6.1, this implies $G^L = H^L$. □

Now if G and H are in the same component of $\mathbb{P}$, then they necessarily satisfy Definition 6.3(iii) and (iv). Therefore, *on each component*, the relations $\geq$ and $\geq_S$ coincide. But this immediately establishes †. For example, for the isomorphism $\mathbb{P}^+ \to \mathbb{B}_4^+$, we can regard $\mathbb{B}_4$ as the powerset lattice of $\{0, *, 1, \bar{1}\}$; then each G maps to its set of Right options.

To complete the picture of $\mathbb{P}$, we must characterize the interaction between the four components. We are concerned specifically with the case where H is a Right end, but G is not; or where G is a Left end, but H is not (the converses are ruled out by Lemma 3.5).

Theorem 6.6. *The ordering of* $\mathbb{P}$ *is generated by its restrictions to* $\mathbb{P}^+$, $\mathbb{P}^-$, *and* $\mathbb{P}^0$, *together with the following four relations and their mirror images*:

$$\{\cdot \mid *, 1\} \geq 0, \quad \{* \mid *, 1\} \geq \{* \mid \cdot\}, \quad \{\bar{1} \mid *, 1\} \geq \{\bar{1} \mid \cdot\}, \quad \{*, \bar{1} \mid *, 1\} \geq \{*, \bar{1} \mid \cdot\}.$$

Proof. It is easy to verify each of the four stated relations. To prove the theorem, we must show that they collectively generate every relation $G \geq H$, for games $G, H \in \mathbb{P}$.

It suffices to prove the following facts (together with their mirror images, which follow by symmetry):

(i) If $G > 0$, then $G \in \mathbb{P}^+$ and $G \geq \{\cdot \mid *, 1\}$;

(ii) If $G \geq H$, $H \in \mathbb{P}^-$, and $G \in \mathbb{P}^0$, then one of the following three inequalities holds:

$$G \geq \{* \mid *, 1\} \geq \{* \mid \cdot\} \geq H,$$
$$G \geq \{\bar{1} \mid *, 1\} \geq \{\bar{1} \mid \cdot\} \geq H, \quad \text{or}$$
$$G \geq \{*, \bar{1} \mid *, 1\} \geq \{*, \bar{1} \mid \cdot\} \geq H.$$

(i) Suppose $G \geq 0$. By Theorem 5.3, G is necessarily a Left end, and furthermore no G^R is downlinked to 0. By Theorem 5.4, the Right options of G must therefore be a subset of $\{*, 1\}$. So either $G = 0$, or $G \geq_S \{\cdot \mid *, 1\}$.

(ii) Suppose $G \geq H$, H is a Right end, and G is not. Consider any G^R. By Theorem 5.3, G^R is not downlinked to H. Since H is a Right end, it is necessarily the case that $G^{RL} \geq H$, by Theorem 5.4. In particular, G^R is not a Left end. Therefore $G^R \in \{*, 1\}$. This shows that the Right options of G form a subset of $\{*, 1\}$.

Furthermore, G is born by day 2, so G^{RL} is born on day 0. Therefore $G^{RL} = 0$ and we have $H \leq 0$. By (i), this implies that $H \leq_S \{*, \bar{1} \mid \cdot\}$. This shows that the Left options of H form a subset of $\{*, \bar{1}\}$.

Finally, by Theorem 6.5, we know that $G \geq_S H$. So the Left options of G form a superset of those of H.

There are now three possibilities for the Left options of H: either $\{*\}$, $\{\bar{1}\}$, or $\{*, \bar{1}\}$. In each case we obtain one of the three inequalities described in (ii) (since the Right options of G form a subset of $\{*, 1\}$, and the Left options of G form a superset of those of H). □

Antichains by day 2. Since $\mathbb{P}$ has such a clean structure, we can get a tight bound on the number of antichains. From a standard reference such as [OEIS], we find that $\mathbb{B}_4$ has 168 antichains. This shows that $\mathbb{P}^+$ (and hence $\mathbb{P}^-$ as well) has precisely 167, since there is a unique antichain containing the largest element of $\mathbb{B}_4$.

Consider $\mathbb{P}^0$. We have $\mathbb{B}_4 \times \mathbb{B}_4 \cong \mathbb{B}_8$, trivially. Again from a standard reference, we find that $\mathbb{B}_8$ has 56130437228687557907788 antichains. Since every antichain of $\mathbb{P}^0$ is an antichain of $\mathbb{B}_8$, this gives an upper bound for the number of antichains of $\mathbb{P}^0$.

Finally, $\{0\}$ trivially admits just two antichains, $\varnothing$ and $\{0\}$. Since every antichain on $\mathbb{P}$ restricts to an antichain on each component, this gives an upper bound of

$$M = 2 \times 167 \times 167 \times 56130437228687557907788$$

for the number of antichains of $\mathbb{P}$. Thus we obtain an upper bound of M^2 games born by day 3. This number is large, indeed (roughly 2^{183}), but it is much smaller than 2^{512}, the number of nonisomorphic game trees of height 3.

7. Relationships to other theories

In this section we consider how the partizan misère theory relates to other theories of combinatorial games. To avoid confusion, we now abandon the notation $o(G)$, and denote explicitly by $o^+(G)$ and $o^-(G)$ the normal-play and misère-play outcome classes of G. We also write

$$G \geq^+ H \quad \text{if and only if} \quad o^+(G+X) \geq o^+(H+X) \text{ for every game } X;$$

$$G \geq^- H \quad \text{if and only if} \quad o^-(G+X) \geq o^-(H+X) \text{ for every game } X.$$

$\geq^-$ is the relation that we have been calling $\geq$, and o^- is the same mapping as o; for this section only, we include the minus signs for clarity. $\geq^+$ is, of course, the usual Berlekamp–Conway–Guy inequality for normal-play partizan games. The following result shows that $\geq^+$ is a coarsening of $\geq^-$.

Theorem 7.1. *If $G \geq^- H$, then $G \geq^+ H$.*

Proof. We must show that Left can get the last move in $G - H$. Suppose Right plays to $G^R - H$ (the argument is the same if he plays to $G - H^L$). Since

$G \geq^- H$, Theorem 5.3 implies that G^R is not downlinked to H. By Theorem 5.4, either $G^{RL} \geq^- H$ or $G^R \geq^- H^R$. By induction, we may assume that either $G^{RL} \geq^+ H$ or $G^R \geq^+ H^R$. In the first case, Left wins by moving to $G^{RL} - H$; in the second, by moving to $G^R - H^R$. □

In addition to the usual equivalences $=^+$ and $=^-$, obtained by symmetrizing $\geq^+$ and $\geq^-$, we have two further equivalences when G and H are impartial.

$$G =_I^+ H \text{ if and only if } o^+(G+X) = o^+(H+X) \text{ for every impartial game } X;$$
$$G =_I^- H \text{ if and only if } o^-(G+X) = o^-(H+X) \text{ for every impartial game } X.$$

It is a well-known fact that $=_I^+$ is just the restriction of $=^+$ to impartial games. It is worth pointing out that the analogous statement does *not* hold for misère games, as the following proposition demonstrates.

Proposition 7.2. *There exist impartial games G and H such that* $G =_I^- H$ *but* $G \neq^- H$.

Proof. It is well-known that $* + * =_I^- 0$ (see [Conway 2001]). However, $* + * \neq^- 0$, by Lemma 3.5. □

Therefore $=_I^-$ is a strict coarsening of $=^-$. This highlights an interesting difference between normal and misère play: there exist impartial games G and H that are distinct in partizan misère play, but that are not distinguishable by any impartial game. This behavior arises in other theories, as well; for example, there exist impartial loopy games G and H that are distinct (in normal play), but that are not distinguishable by any impartial loopy game [Siegel 2009]. Indeed, the coincidence of $=^+$ and $=_I^+$ appears to be an artifact of the special nature of short games in normal play: it is the exception rather than the rule.

8. Partizan misère quotients

Recently Thane Plambeck [2005] observed that, if $\mathscr{A}$ is any set of impartial games, then its misère-play structure can often be simplified by localizing the misère equivalence relation to $\mathscr{A}$. Plambeck showed that many important aspects of the theory can be generalized to the local setting, and the structure theory of such quotients has been explored in detail; see [Plambeck and Siegel 2008; Siegel 2007].

It is not our intention to replicate that analysis here, but merely to remark that a partizan generalization exists. The construction is exactly the same, but instead of a bipartite monoid $(\mathcal{Q}, \mathcal{P})$, we now have a *tetrapartite monoid* $(\mathcal{Q}, \Pi)$, where $\Pi : \mathcal{Q} \to \{\mathscr{L}, \mathscr{R}, \mathscr{P}, \mathscr{N}\}$ is the *outcome partition* for $\mathcal{Q}$.

Intriguingly, such monoids have an induced partial order structure, given by

$$x \geq y \text{ if and only if } \Pi(xz) \geq \Pi(yz) \text{ for all } z \in \mathcal{Q}.$$

If $G \geq H$, then it is certainly true that $\Phi(G) \geq \Phi(H)$. However, the quotient may also gain new order-relations that are not present in the universe of games. We include one example to illustrate the rich possibilities. In the previous section we remarked that 1 and $\bar{1}$ are incomparable, and we have also seen that $1 + \bar{1} \neq 0$. In $\mathcal{Q}(1, \bar{1})$, however, the expected inequalities hold:

$$\Phi(\bar{1}) > \Phi(0) > \Phi(1) \quad \text{and} \quad \Phi(\bar{1})\Phi(1) = \Phi(0);$$

and indeed we have $\mathcal{Q}(1, \bar{1}) \cong \mathbb{Z}$, equipped with the usual partial-order structure. We leave it to the reader to verify these assertions.

In the time since this manuscript was first prepared, Meghan Allen has obtained some exciting results on partizan misère games, including identifying several interesting quotients. These results represent the current state of the art on the subject.

Acknowledgements

I wish to thank Richard Nowakowski and his students for their careful reading of the manuscript, and Dan Hoey for many helpful comments and suggestions. I also wish to thank the referee for many insightful comments.

References

[Berlekamp et al. 2003] E. R. Berlekamp, J. H. Conway, and R. K. Guy, *Winning ways for your mathematical plays, II*, 2nd ed., A K Peters, Natick, MA, 2003.

[Conway 2001] J. H. Conway, *On numbers and games*, 2nd ed., A K Peters, Natick, MA, 2001.

[Grundy 1939] P. M. Grundy, "Mathematics and games", *Eureka* **2** (1939), 6–8. Reprinted in *Eureka*, **27**, 1974: 9–11.

[Mesdal and Ottaway 2007] G. A. Mesdal and P. Ottaway, "Simplification of partizan games in misère play", *Integers* **7** (2007), G06, 12.

[OEIS] N. J. A. Sloane, A000372 – Dedekind numbers or Dedekind's problem: number of monotone Boolean functions of n variables..., *Online encyclopedia of integer sequences*, http://oeis.org/A000372.

[Plambeck 2005] T. E. Plambeck, "Taming the wild in impartial combinatorial games", *Integers* **5**:1 (2005), G5, 36.

[Plambeck and Siegel 2008] T. E. Plambeck and A. N. Siegel, "Misère quotients for impartial games", *J. Combin. Theory Ser. A* **115**:4 (2008), 593–622.

[Siegel 2007] A. N. Siegel, "The structure and classification of misère quotients", preprint, 2007. arXiv math/0703070

[Siegel 2009] A. N. Siegel, "Coping with cycles", pp. x+575 in *Games of no chance 3*, vol. 56, edited by M. H. Albert and R. J. Nowakowski, Mathematical Sciences Research Institute Publications, Cambridge University Press, Cambridge, 2009.

[Sprague 1935] R. P. Sprague, "Über mathematische Kampfspiele", *Tôhoku Math. J.* **41** (1935), 438–444.

[Sprague 1937] R. P. Sprague, "Über zwei Abarten von Nim", *Tôhoku Math. J.* **43** (1937), 351–359.

aaron.n.siegel@gmail.com *San Francisco, CA 94188, United States*

Games of No Chance 4
MSRI Publications
Volume **63**, 2015

The structure and classification of misère quotients

AARON N. SIEGEL

A *bipartite monoid* is a commutative monoid $\mathcal{Q}$ together with an identified subset $\mathcal{P} \subset \mathcal{Q}$. In this paper we study a class of bipartite monoids, known as *misère quotients*, that are naturally associated to impartial combinatorial games.

We introduce a structure theory for misère quotients with $|\mathcal{P}| = 2$, and give a complete classification of all such quotients up to isomorphism. One consequence is that if $|\mathcal{P}| = 2$ and $\mathcal{Q}$ is finite, then $|\mathcal{Q}| = 2^n + 2$ or $2^n + 4$.

We then develop computational techniques for enumerating misère quotients of small order, and apply them to count the number of nonisomorphic quotients of order at most 18. We also include a manual proof that there is exactly one quotient of order 8.

1. Introduction

An *impartial combinatorial game* Γ is a two-player game with no hidden information and no chance elements, in which both players have exactly the same moves available at all times. When Γ is played under the *misère play convention*, the player who makes the last move loses.

Thirty years ago, Conway [] showed that the misère-play combinatorics of such games are often frighteningly complicated. However, new techniques recently pioneered by Plambeck [2005] have reinvigorated the subject. At the core of these techniques is the *misère quotient*, a commutative monoid that encodes the additive structure of an impartial combinatorial game (or a set of such games). See [Siegel 2015] for a gentle introduction to misère quotients, and [Plambeck and Siegel 2008] for a more rigorous one; see [Plambeck 2009] for a survey of the theory.

The introduction of misère quotients opens up a fascinating new area of study: the investigation of their algebraic properties. Such investigations are intrinsically interesting, and also have the potential to reveal new insights into the misère-play structure of combinatorial games. In this paper, we introduce several new results that expose quite a bit of structure in misère quotients.

Bipartite monoids and misère quotients. We recall some basic facts about misère quotients from Plambeck and Siegel [2008]. A *bipartite monoid* is a commutative monoid $\mathcal{Q}$ together with an identified subset $\mathcal{P} \subset \mathcal{Q}$. Two elements $x, y \in \mathcal{Q}$ are *indistinguishable* if, for all $z \in \mathcal{Q}$, we have $xz \in \mathcal{P} \Leftrightarrow yz \in \mathcal{P}$. A bipartite monoid $(\mathcal{Q}, \mathcal{P})$ is *reduced* if the elements of $\mathcal{Q}$ are pairwise distinguishable. In [Plambeck and Siegel 2008] it is shown that every bipartite monoid has a unique reduced quotient (up to isomorphism), the *reduction* of $(\mathcal{Q}, \mathcal{P})$. We'll write r.b.m. as shorthand for reduced bipartite monoid.

$(\mathcal{Q}, \mathcal{P})$ is a *subbipartite monoid* of $(\mathcal{S}, \mathcal{R})$ if $\mathcal{Q}$ is a submonoid of $\mathcal{S}$ and $\mathcal{R} \cap \mathcal{Q} = \mathcal{P}$. In this case we write $(\mathcal{Q}, \mathcal{P}) < (\mathcal{S}, \mathcal{R})$.

For a closed set $\mathscr{A}$ of impartial combinatorial games, the *indistinguishability relation* on $\mathscr{A}$ is given by

$$G \equiv H \pmod{\mathscr{A}} \quad \text{if and only if} \quad o^-(G+X) = o^-(H+X) \quad \text{for all } X \in \mathscr{A},$$

where $o^-(G)$ denotes the misère-play outcome of G. We write $[G]$ for the equivalence class of G modulo $\equiv$. Then the *misère quotient* of $\mathscr{A}$ is the bipartite monoid $(\mathcal{Q}, \mathcal{P})$, where $\mathcal{Q}$ is the quotient monoid of $\mathscr{A}$ modulo $\equiv$, and $\mathcal{P}$ is the corresponding *$\mathscr{P}$-portion*:

$$\mathcal{Q} = \{[G] : G \in \mathscr{A}\}; \quad \mathcal{P} = \{[G] : G \in \mathscr{A},\ o^-(G) = \mathscr{P}\}.$$

We denote the misère quotient of $\mathscr{A}$ by $\mathcal{Q}(\mathscr{A})$. It is necessarily reduced, and therefore isomorphic to the reduction of $\mathscr{A}$. The *quotient map* $\Phi : \mathscr{A} \to \mathcal{Q}$ is given by $\Phi(G) = [G]$.

If $\mathcal{Q}$ is a finite commutative monoid, then the intersection of all its ideals is called the *kernel* of $\mathcal{Q}$. The kernel $\mathcal{K}$ is itself an ideal; it is the smallest ideal of $\mathcal{Q}$, and is also the largest group that can be written as a homomorphic image of $\mathcal{Q}$. A misère quotient $(\mathcal{Q}, \mathcal{P})$ is said to be *normal* if $\mathcal{K} \cap \mathcal{P} = \{z\}$, where z is the group identity of $\mathcal{K}$.

The basic theory of misère quotients is introduced in [Plambeck 2005; Plambeck and Siegel 2008; Siegel 2015], and some familiarity with it is assumed throughout this paper.

Tame and restive quotients. The following facts were also established in [Plambeck 2005; Plambeck and Siegel 2008], and are discussed in [Siegel 2015] in this volume. When every element of $\mathscr{A}$ is tame (in the sense of [ONAG]), then its quotient is isomorphic to one of the *tame quotients* $\mathcal{T}_n$. Here $\mathcal{T}_0 = \{1\}$ (with empty $\mathscr{P}$-portion); $\mathcal{T}_1 = \{1, a\}$ (with the structure of $\mathbb{Z}_2$ and $\mathcal{P} = \{a\}$); and

$$\mathcal{T}_n \cong \langle a, b_1, b_2, \ldots, b_{n-1} \mid a^2 = 1, b_1^3 = b_1, b_2^3 = b_2, \ \ldots, \ b_{n-1}^3 = b_{n-1}, \\ b_1^2 = b_2^2 = \cdots = b_{n-1}^2 \rangle$$

for $n \geq 2$, with $\mathcal{P} = \{a, b_1^2\}$. (The condition that $(\mathcal{Q}, \mathcal{P}) \cong \mathcal{T}_n$, for some n, can be taken as an equivalent definition of tame.) The structure of $\mathcal{T}_n$, for $n \geq 2$, is best described as:

- a *kernel* $\mathcal{K}_n$ isomorphic to $\mathbb{Z}_2^n$, with identity element $z = b_1^2$, and generators $az, b_1, b_2, \ldots, b_{n-1}$;
- a separate component $\{1, a\}$ isomorphic to $\mathbb{Z}_2$, which maps onto $\{z, az\}$ under multiplication by z.

Thus the cardinality of $\mathcal{T}_n$ ($n \geq 2$) is $2^n + 2$.

Related to the tame quotients is the sequence of *restive quotients* $\mathcal{R}_{2^n+4}$, for $n \geq 2$, obtained by adjoining a new element t to $\mathcal{T}_n$ with $tz = t^2 = z$ (so that t is nilpotent to $\mathcal{K}_n$). We have $\mathcal{R}_{2^n+4} = \mathcal{T}_n \cup \{t, at\}$, so that $|\mathcal{R}_{2^n+4}| = 2^n + 4$; and $\mathcal{P} = \{a, b^2\}$, so that $\mathcal{R}_{2^n+4}$ also has just two $\mathscr{P}$-positions.

Tame extensions. The first result of this paper is a complete classification of misère quotients whose $\mathscr{P}$-portion has cardinality 2. If $(\mathcal{Q}, \mathcal{P})$ is a misère quotient, then the *tame extension* $\mathcal{T}(\mathcal{Q}, \mathcal{P})$ is a certain extension of $(\mathcal{Q}, \mathcal{P})$ (to be defined in Section 3) that adds no new $\mathscr{P}$-positions. It is defined in such a way that

$$\mathcal{T}_3 = \mathcal{T}(\mathcal{T}_2), \quad \mathcal{T}_4 = \mathcal{T}(\mathcal{T}_3), \quad \mathcal{T}_5 = \mathcal{T}(\mathcal{T}_4), \ldots$$

If we replace the "base" $\mathcal{T}_2$ by $\mathcal{R}_8$, then we recover the family of restive quotients:

$$\mathcal{R}_{12} = \mathcal{T}(\mathcal{R}_8), \quad \mathcal{R}_{20} = \mathcal{T}(\mathcal{R}_{12}), \quad \mathcal{R}_{36} = \mathcal{T}(\mathcal{R}_{20}), \ldots$$

The main result is that *every* finite quotient with $|\mathcal{P}| = 2$ is isomorphic to a quotient in one of these two families. It will follow that every finite quotient with $|\mathcal{P}| = 2$ has order $2^n + 2$ or $2^n + 4$, for some $n \geq 2$. Furthermore, if $(\mathcal{Q}, \mathcal{P})$ is an *infinite* quotient with $|\mathcal{P}| = 2$, then $(\mathcal{Q}, \mathcal{P}) \cong$ either $\mathcal{T}_\infty$ or $\mathcal{R}_\infty$, the limits of the two families (in a sense to be defined in Section 2).

"Almost tame" octal games. Tame extensions also have a useful application to octal games. Fix a (finite-length) octal game Γ and an integer M, and consider the partial quotient $(\mathcal{Q}, \mathcal{P}) = \mathcal{Q}_M(\Gamma)$ and quotient map Φ_M. Assume that $(\mathcal{Q}, \mathcal{P})$ is normal and $\mathcal{Q}$ is finite, and let $\mathcal{K} \subset \mathcal{Q}$ be the kernel of $\mathcal{Q}$. In Section 3, we will show that if $\Phi_M(H_n) \in \mathcal{K}$ for sufficiently many heaps H_n, then $\mathcal{Q}_{M+1}(\Gamma)$ is either $(\mathcal{Q}, \mathcal{P})$ or $\mathcal{T}(\mathcal{Q}, \mathcal{P})$, and $\Phi_{M+1}(H_{M+1}) \in \mathcal{K}$. "Sufficiently many" will be in the sense of [Guy and Smith 1956]: there exists n_0 such that $n_0 \leq n < 2n_0 + d$, where d is the octal length of Γ.

This theorem can be iterated, with strong consequences. In particular, if we determine that $\Phi_M(H_n) \in \mathcal{K}$ for sufficiently many n, then we can conclude that $\mathcal{Q}(\Gamma)$ is one of

$$(\mathcal{Q}, \mathcal{P}), \; \mathcal{T}(\mathcal{Q}, \mathcal{P}), \; \mathcal{T}(\mathcal{T}(\mathcal{Q}, \mathcal{P})), \ldots,$$

n	2	4	6	8	10	12	14	16	18
Quotients of order n	1	0	1	1	1	6	9	50	211

Table 1. The number of misère quotients of order $n \leq 18$ (up to isomorphism).

or possibly the limit $\mathcal{T}^\infty(\mathcal{Q}, \mathcal{P})$ of this sequence. Furthermore, $\Phi(H_n) \in \mathcal{K}$ for all but finitely many n. Since $\mathcal{Q}(\Gamma)$ is normal, normal and misère play coincide on $\mathcal{K}$; so we conclude that misère play reduces to normal play unless all the heaps are small. In practice, this means that once we have computed $\mathcal{Q}_M(\Gamma)$, then we have completely characterized the "misère-play divergence" of Γ; and its misère-play solution now depends only on finding a normal-play solution.

An example is the game **0.414**, which we mentioned in [Plambeck and Siegel 2008]. Its normal-play solution is unknown, despite the computation of at least 2^{24} $\mathcal{G}$-values by Flammenkamp [2012]. However, it is easy enough to compute $\mathcal{Q}_{18}(\mathbf{0.414})$, and to verify using the above logic that $\Phi(H_n) \in \mathcal{K}$ for all $n > 18$. Thus we know $\mathcal{Q}(\mathbf{0.414}) \cong \mathcal{T}^k(\mathcal{Q}_{18})$, for some k, and we need invest no further worry in the misère play of **0.414**: we may sit back and await a normal-play solution.

Recall the misère-play strategy for NIM: play normal NIM unless your move would leave only heaps of size 1. In that case, play to leave an odd number of heaps of size 1. We can now state an analogous strategy for **0.414**: play normal **0.414** unless your move would leave only heaps of size ≤ 18. In that case, consult the fine structure of $\mathcal{Q}_{18}$. We can state this reduction with confidence, despite the fact that the fullnormal-play strategy for **0.414** remains unknown.

Quotients of small order. We'll prove in Section 6 that $\mathcal{R}_8$ is the only quotient of order 8 (up to isomorphism). The succeeding sections focus instead on developing computational techniques for classifying quotients of small order.

In Section 7, we show that an arbitrary r.b.m. $(\mathcal{Q}, \mathcal{P})$ is a misère quotient if and only if there exists a *valid transition table* for $(\mathcal{Q}, \mathcal{P})$ — a certain combinatorial structure superimposed on $(\mathcal{Q}, \mathcal{P})$. This yields a computational method for testing whether $(\mathcal{Q}, \mathcal{P})$ is a misère quotient, which is optimized and applied in Section 8. The fruits of this effort are summarized in Table 1.

Preliminaries. We recall a few more key facts from [Plambeck and Siegel 2008].

Definition 1.1. Let $(\mathcal{Q}, \mathcal{P})$ be a bipartite monoid and fix $x \in \mathcal{Q}$. The *meximal set of* x *in* $(\mathcal{Q}, \mathcal{P})$, denoted $\mathcal{M}_x$, is defined by

$$\mathcal{M}_x = \{y \in \mathcal{Q} : \text{there is no } z \in \mathcal{Q} \text{ such that } xz \in \mathcal{P} \text{ and } yz \in \mathcal{P}\}.$$

The following statement is slightly more general than the rule given in [Plambeck and Siegel 2008], but the proof is identical.

Fact 1.2 (generalized mex rule). *Let* $(\mathcal{Q}, \mathcal{P}) = \mathcal{Q}(\mathscr{A})$, *and let* $(\mathcal{Q}, \mathcal{P}) < (\mathcal{S}, \mathcal{R})$. *Fix* G *with* $\mathrm{opts}(G) \subset \mathscr{A}$ *and fix* $x \in \mathcal{S}$. *The following are equivalent.*

(a) $\mathcal{Q}(\mathscr{A} \cup \{G\}) \cong (\mathcal{S}, \mathcal{R})$ *and* $\Phi(G) = x$.

(b) $\mathcal{S}$ *is generated by* $\mathcal{Q} \cup \{x\}$, *and the following two conditions hold.*

(i) $\Phi''G \subset \mathcal{M}_x$.

(ii) *For each* $Y \in \mathscr{A}$ *and* $n \geq 0$ *such that* $x^{n+1}\Phi(Y) \notin \mathcal{P}$, *we have either:* $x^{n+1}\Phi(Y') \in \mathcal{P}$ *for some option* Y' *of* Y; *or else* $x^n x' \Phi(Y) \in \mathcal{P}$ *for some* $x' \in \Phi''G$.

In [Plambeck and Siegel 2008] we stated Fact 1.2 for the special case $\mathcal{S} = \mathcal{Q}$. This will often be the case of greatest interest, but we shall have several occasions to use the more general form.

2. Limits and one-stage extensions

In this section we show that every misère quotient is the limit of a sequence of finitely generated quotients. Furthermore, each term of this sequence is an extension of the previous term, in a way we now make precise.

Definition 2.1. Let $(\mathcal{Q}, \mathcal{P})$, $(\mathcal{Q}^+, \mathcal{P}^+)$ be reduced bipartite monoids. We say that $(\mathcal{Q}^+, \mathcal{P}^+)$ is an *extension* of $(\mathcal{Q}, \mathcal{P})$ if there is some submonoid $(\mathcal{S}, \mathcal{R}) < (\mathcal{Q}^+, \mathcal{P}^+)$ such that $(\mathcal{Q}, \mathcal{P})$ is (isomorphic to) the reduction of $(\mathcal{S}, \mathcal{R})$. If $\mathcal{Q}^+$ is generated by $\mathcal{S} \cup \{x\}$ for some single element $x \in \mathcal{Q}^+ \setminus \mathcal{S}$, then we say that $(\mathcal{Q}^+, \mathcal{P}^+)$ is a *one-stage extension* of $(\mathcal{Q}, \mathcal{P})$.

Note that Definition 2.1 does *not* require that $\mathcal{Q}$ be a submonoid of $\mathcal{Q}^+$: it is only required that $(\mathcal{Q}, \mathcal{P})$ be the *reduction* of a subbipartite monoid of $(\mathcal{Q}^+, \mathcal{P}^+)$.

Lemma 2.2. *Let* $(\mathcal{Q}, \mathcal{P})$ *be a finitely generated misère quotient. Then there is a sequence of misère quotients*

$$0 = (\mathcal{Q}_0, \mathcal{P}_0), (\mathcal{Q}_1, \mathcal{P}_1), \ldots, (\mathcal{Q}_n, \mathcal{P}_n) = (\mathcal{Q}, \mathcal{P})$$

such that each $(\mathcal{Q}_{i+1}, \mathcal{P}_{i+1})$ *is a one-stage extension of* $(\mathcal{Q}_i, \mathcal{P}_i)$.

Proof. Write $(\mathcal{Q}, \mathcal{P}) = \mathcal{Q}(\mathscr{A})$ and choose a finite set $\mathscr{H} \subset \mathscr{A}$ so that $\Phi''\mathscr{H}$ generates $\mathcal{Q}$. (Here and throughout the sequel, $\Phi''\mathscr{H} = \{\Phi(H) : H \in \mathscr{H}\}$.) Since the hereditary closure of a finite set is finite, we may assume that $\mathscr{H}$ is hereditarily closed. Enumerate

$$\mathscr{H} = \{H_0, H_1, \ldots, H_m\}$$

so that the successive H_i's have nondecreasing birthdays, and put

$$(\mathcal{Q}_i, \mathcal{P}_i) = \mathcal{Q}(H_0, \ldots, H_i).$$

It is easily seen that either $(\mathcal{Q}_{i+1}, \mathcal{P}_{i+1}) = (\mathcal{Q}_i, \mathcal{P}_i)$, or else it is a one-stage extension of $(\mathcal{Q}_i, \mathcal{P}_i)$. A suitable reindexing gives the lemma. □

Now let $(\mathcal{Q}_n, \mathcal{P}_n)$ be a sequence of bipartite monoids, and suppose that for each n, there exists $(\mathcal{Q}_n^+, \mathcal{P}_n^+) < (\mathcal{Q}_{n+1}, \mathcal{P}_{n+1})$ such that $(\mathcal{Q}_n, \mathcal{P}_n)$ is the reduction of $(\mathcal{Q}_n^+, \mathcal{P}_n^+)$, with quotient map $\pi_n : \mathcal{Q}_n^+ \to \mathcal{Q}_n$. We call $(\mathcal{Q}_n, \mathcal{P}_n, \pi_n)$ a *partial inverse system*.

Let $\overleftarrow{\mathcal{Q}} = (\mathcal{Q}_n, \mathcal{P}_n, \pi_n)$ be a partial inverse system. It is convenient to regard the underlying sets of the $\mathcal{Q}_n$ as formally disjoint. A *thread of* $\overleftarrow{\mathcal{Q}}$ *starting at* n is a sequence $(x_n, x_{n+1}, x_{n+2}, \ldots)$, where $x_n \in \mathcal{Q}_n$ and for each $i > n$ we have $x_{i+1} \in \mathcal{Q}_i^+$ and $\pi_i(x_{i+1}) = x_i$. We say two threads $\overleftarrow{x}$ and $\overleftarrow{y}$ are equivalent, and write $\overleftarrow{x} \sim \overleftarrow{y}$, if one is a terminal segment of the other.

If $\overleftarrow{x} = (x_m, x_{m+1}, x_{m+2}, \ldots)$ and $\overleftarrow{y} = (y_n, y_{n+1}, y_{n+2}, \ldots)$ are threads, we can define their product as follows. Without loss of generality, assume that $m \le n$, and put

$$\overleftarrow{x} \cdot \overleftarrow{y} = (x_n y_n, x_{n+1} y_{n+1}, x_{n+2} y_{n+2}, \ldots).$$

It is easy to check that $\overleftarrow{x} \cdot \overleftarrow{y}$ is a thread and that the product respects the equivalence $\sim$. Further, $\overleftarrow{1} \cdot \overleftarrow{x} = \overleftarrow{x}$, where $\overleftarrow{1} = (1, 1, 1, \ldots)$ is a list of the identity elements of each $\mathcal{Q}_n$. Thus the threads modulo $\sim$ form a commutative monoid $\mathcal{Q}$. We can define a subset $\mathcal{P} \subset \mathcal{Q}$ by

$$\mathcal{P} = \{(x_n, x_{n+1}, x_{n+2}, \ldots) \in \mathcal{Q} : \text{some } x_i \in \mathcal{P}_i\}.$$

(Note that the condition that *some* $x_i \in \mathcal{P}_i$ is equivalent to *all* $x_i \in \mathcal{P}_i$, since each π_i is a bipartite monoid homomorphism.) This makes $(\mathcal{Q}, \mathcal{P})$ into a bipartite monoid, which we call the *partial inverse limit* of the system $\overleftarrow{\mathcal{Q}}$. We write $(\mathcal{Q}, \mathcal{P}) = \lim \overleftarrow{\mathcal{Q}} = \lim_n(\mathcal{Q}_n, \mathcal{P}_n)$.

The following lemma is an easy exercise.

Lemma 2.3. *If* $(\mathcal{Q}_n, \mathcal{P}_n)$ *is reduced for infinitely many values of* n, *then so is* $\lim_n(\mathcal{Q}_n, \mathcal{P}_n)$.

Theorem 2.4. *Suppose that* $\mathscr{A}_0 \subset \mathscr{A}_1 \subset \mathscr{A}_2 \subset \cdots$ *is a chain of closed sets of games. Then the quotients* $\mathcal{Q}(\mathscr{A}_n)$ *form a partial inverse system, and we have*

$$\mathcal{Q}\left(\bigcup_n \mathscr{A}_n\right) \cong \lim_n \mathcal{Q}(\mathscr{A}_n).$$

Proof. Let $\Phi_n : \mathscr{A}_n \to \mathcal{Q}(\mathscr{A}_n)$ be the quotient maps and put $\mathcal{Q}_n^+ = \Phi''_{n+1}\mathscr{A}_n$. Define $\pi_n : \mathcal{Q}_n^+ \to \mathcal{Q}(\mathscr{A}_n)$ by $\pi_n(\Phi_{n+1}(X)) = \Phi_n(X)$. Now if $X \equiv Y \pmod{\mathscr{A}_{n+1}}$, then necessarily $X \equiv Y \pmod{\mathscr{A}_n}$, so π_n is well-defined.

Now by Lemma 2.3, $\lim_n \mathcal{Q}(\mathscr{A}_n)$ is reduced. To complete the proof, it suffices to exhibit a surjective homomorphism $\Phi : \bigcup_n \mathscr{A}_n \to \lim_n \mathcal{Q}(\mathscr{A}_n)$. Let n be least so that $X \in \mathscr{A}_n$, and put

$$\Phi(X) = (\Phi_n(X), \Phi_{n+1}(X), \Phi_{n+2}(X), \ldots).$$

It is easily verified that Φ has the desired properties. □

An easy corollary of Theorem 2.4 will be central to the classification theory.

Corollary 2.5. *Suppose that* $(\mathcal{Q}, \mathcal{P})$ *is a misère quotient that is* not *finitely generated. Then there is some partial inverse system* $(\mathcal{Q}_n, \mathcal{P}_n, \pi_n)$ *of finitely generated misère quotients such that*:

(i) $(\mathcal{Q}_0, \mathcal{P}_0) = 0$;

(ii) *each* $(\mathcal{Q}_{n+1}, \mathcal{P}_{n+1})$ *is a one-stage extension of* $(\mathcal{Q}_n, \mathcal{P}_n)$; *and*

(iii) $(\mathcal{Q}, \mathcal{P}) = \lim_n (\mathcal{Q}_n, \mathcal{P}_n)$.

Proof. Write $(\mathcal{Q}, \mathcal{P}) = \mathcal{Q}(\mathscr{A})$ with $\mathscr{A}$ closed. Enumerate $\mathscr{A} = \{H_0, H_1, H_2, \ldots\}$ so that the birthdays of the H_n are nondecreasing. (This can always be done, since there are only finitely many games of each fixed birthday.) Then for each n, we have $\mathrm{opts}(H_n) \subset \{H_0, \ldots, H_{n-1}\}$. Put

$$(\mathcal{Q}_n, \mathcal{P}_n) = \mathcal{Q}(H_0, \ldots, H_n),$$

with quotient map Φ_n. Let $\mathcal{Q}_n^+$ be the submonoid of $\mathcal{Q}_{n+1}$ generated by

$$\{\Phi_{n+1}(H_0), \ldots, \Phi_{n+1}(H_n)\}.$$

The map $\pi_n : \mathcal{Q}_n^+ \to \mathcal{Q}_n$ defined by

$$\pi_n(\Phi_{n+1}(H)) = \Phi_n(H)$$

is well-defined, since each $G \equiv G' \pmod{\mathscr{A}_{n+1}}$ implies $G \equiv G' \pmod{\mathscr{A}_n}$. Now (i) is immediate, since necessarily $H_0 = 0$, and (ii) follows easily (after reindexing to eliminate cases where $\mathcal{Q}_{n+1} = \mathcal{Q}_n$). Now by Lemma 2.3, we know that $\lim_n (\mathcal{Q}_n, \mathcal{P}_n)$ is reduced. To prove (iii), it therefore suffices (by uniqueness of reductions) to show that $\lim_n (\mathcal{Q}_n, \mathcal{P}_n)$ is a quotient of $\mathscr{A}$.

Let $\Phi_n : \mathrm{cl}(\{H_0, \ldots, H_n\}) \to \mathcal{Q}_n$ be the usual quotient map, and define $\Phi : \mathscr{A} \to \mathcal{Q}$ by

$$\Phi(H_n) = (\Phi_n(H_n), \Phi_{n+1}(H_n), \Phi_{n+2}(H_n), \ldots).$$

It is easily checked that Φ is a surjective homomorphism of bipartite monoids. □

3. Normal quotients and tame extensions

In this section we introduce a certain algebraic property known as *faithful normality*, and we study one-stage extensions of faithfully normal quotients. In particular, we show that certain one-stage extensions of faithfully normal quotients behave just like extensions in normal play. The vast majority of quotients encountered in practice are faithfully normal, so this work has useful applications to octal games.

Definition 3.1. Let $(\mathcal{Q}, \mathcal{P})$ be a misère quotient with kernel $\mathcal{K}$, and let $z \in \mathcal{K}$ be the kernel identity. We say that $(\mathcal{Q}, \mathcal{P})$ is *regular* if $|\mathcal{K} \cap \mathcal{P}| = 1$, and *normal* if $\mathcal{K} \cap \mathcal{P} = \{z\}$.

Definition 3.2. Let $(\mathcal{Q}, \mathcal{P}) = \mathcal{Q}(\mathscr{A})$ and let $\Phi : \mathscr{A} \to \mathcal{Q}$ be the quotient map. Suppose that

$$\Phi(G) = \Phi(H) \Longrightarrow \mathscr{G}(G) = \mathscr{G}(H) \quad \text{for all } G, H \in \mathscr{A}.$$

Then we say that Φ is *faithful*. If in addition $(\mathcal{Q}, \mathcal{P})$ is normal, then we say that Φ is *faithfully normal*.

(Here and throughout, we write $\mathscr{G}(G)$ for the normal-play nim value of G.) Often we will abuse terminology and refer to the *quotient* as being faithful (or faithfully normal), rather than the quotient map. We recall the following fact from [Plambeck and Siegel 2008].

Fact 3.3. *Suppose $\Phi : \mathscr{A} \to \mathcal{Q}$ is faithfully normal. Then $\mathcal{K}$ is isomorphic to the normal quotient of $\mathscr{A}$.*

Roughly speaking, therefore, faithful normality asserts that normal and misère play coincide on $\mathcal{K}$. We have $\mathcal{K} \cong \mathbb{Z}_2^n$ for some n, and for each $i < 2^n$ there is a unique $z_i \in \mathcal{K}$ representing games of nim value i. For convenience, when $\mathcal{E} \subset \mathcal{K}$, we write $z_m = \operatorname{mex}(\mathcal{E})$ to mean $m = \operatorname{mex}\{i : z_i \in \mathcal{E}\}$.

Now fix a faithfully normal quotient $\mathcal{Q}(\mathscr{A})$ with kernel $\mathcal{K}$, and let $G \neq 0$ be a game such that $\operatorname{opts}(G) \subset \mathscr{A}$. Then $\mathcal{Q}(\mathscr{A} \cup \{G\})$ is necessarily a one-stage extension of $\mathcal{Q}(\mathscr{A})$. For the remainder of this section, we will focus on the special case where $\Phi''G \subset \mathcal{K}$. We will show that in this case, one-stage extensions behave *exactly* like normal-play extensions. In particular:

- Extensions by a *proper* subset of the kernel follow the mex rule. Formally, if $\Phi''G \subsetneq \mathcal{K}$, then $\mathcal{Q}(\mathscr{A} \cup \{G\}) \cong \mathcal{Q}(\mathscr{A})$ and $\Phi(G) = \operatorname{mex}(\Phi''G)$.
- Extensions by the *entire* kernel cause the kernel to grow (from $\mathbb{Z}_2^n$ to $\mathbb{Z}_2^{n+1}$). They behave like normal-play extensions whose nim values are new powers of 2. Formally, if $\Phi''G = \mathcal{K}$, then $\mathcal{Q}(\mathscr{A} \cup \{G\}) \cong \mathcal{T}(\mathcal{Q}(\mathscr{A}))$, where $\mathcal{T}(\mathcal{Q}(\mathscr{A}))$ is a certain "tame extension" of $\mathcal{Q}(\mathscr{A})$ generalizing the extension $\mathbb{Z}_2^n < \mathbb{Z}_2^{n+1}$.

We begin with the case $\Phi''G \subsetneq \mathcal{K}$.

Lemma 3.4. *Suppose $(\mathcal{Q}, \mathcal{P}) = \mathcal{Q}(\mathscr{A})$ is faithfully normal with kernel $\mathcal{K}$. Let G be a game with* $\operatorname{opts}(G) \subset \mathscr{A}$ *and suppose $\Phi''G \subsetneq \mathcal{K}$. Then $\mathcal{Q}(\mathscr{A} \cup \{G\}) \cong \mathcal{Q}(\mathscr{A})$ and* $\Phi(G) = \operatorname{mex}(\Phi''G)$.

Proof. We verify conditions (i) and (ii) of the generalized mex rule, with $(\mathcal{S}, \mathcal{R}) = (\mathcal{Q}, \mathcal{P})$ and $x = \operatorname{mex}(\Phi''G)$. Note that $x = z_m$, where $m = \mathscr{G}(G)$.

For (i), normality implies that $\mathcal{K} \setminus \{z_m\} \subset \mathcal{M}_x$. Since $\Phi''G \subset \mathcal{K} \setminus \{z_m\}$, this suffices. For (ii), fix $Y \in \mathscr{A}$ and $n \geq 0$, and suppose $x^{n+1}\Phi(Y) \notin \mathcal{P}$. If n is odd, then since $x \in \mathcal{K}$ and the quotient is faithfully normal, we have $\mathscr{G}(Y) > 0$. Thus $\mathscr{G}(Y') = 0$ for some Y', whence $x^{n+1}\Phi(Y') \in \mathcal{P}$.

Conversely, suppose that n is even. Then $\mathscr{G}(Y) \neq m$. If $\mathscr{G}(Y) > m$, then $\mathscr{G}(Y') = m$ for some Y', whence $x^{n+1}\Phi(Y') \in \mathcal{P}$. Otherwise, let $i = \mathscr{G}(Y)$. Since $\Phi''G \subset \mathcal{K}$ and $z_m = \operatorname{mex}(\Phi''G)$, we necessarily have $z_i \in \Phi''G$. But $z_i\Phi(Y) = z$, so $x^n z_i \Phi(Y) = z \in \mathcal{P}$. □

Tame extensions. We now consider the case where $\Phi''G = \mathcal{K}$. Let $(\mathcal{Q}, \mathcal{P})$ be a bipartite monoid with kernel $\mathcal{K}$, and define

$$\overline{\mathcal{K}} = \{\overline{x} : x \in \mathcal{K}\},$$

where each $\overline{x}$ is taken to be a formal symbol.

Definition 3.5. The *first tame extension* $\mathcal{T}(\mathcal{Q}, \mathcal{P}) = (\mathcal{Q}^+, \mathcal{P}^+)$ is defined as follows. $\mathcal{Q}^+ = \mathcal{Q} \cup \overline{\mathcal{K}}$, $\mathcal{P}^+ = \mathcal{P}$, and multiplication is extended by:

$$x \cdot \overline{y} = \overline{xy} \ (x \in \mathcal{Q},\ y \in \mathcal{K}); \quad \overline{x} \cdot \overline{y} = xy \ (x, y \in \mathcal{K}).$$

The *n-th tame extension* $\mathcal{T}^n(\mathcal{Q}, \mathcal{P})$ is defined by

$$\mathcal{T}^0(\mathcal{Q}, \mathcal{P}) = (\mathcal{Q}, \mathcal{P}); \quad \mathcal{T}^{n+1}(\mathcal{Q}, \mathcal{P}) = \mathcal{T}(\mathcal{T}^n(\mathcal{Q}, \mathcal{P})).$$

Finally, we define

$$\mathcal{T}^\infty(\mathcal{Q}, \mathcal{P}) = \lim_n \mathcal{T}^n(\mathcal{Q}, \mathcal{P}).$$

Observe that the sequence of normal quotients

$$0, \mathbb{Z}_2, \mathbb{Z}_2^2, \mathbb{Z}_2^3, \ldots, \mathbb{Z}_2^{\mathbb{N}}$$

can be written

$$\mathcal{T}^0(0), \mathcal{T}^1(0), \mathcal{T}^2(0), \mathcal{T}^3(0), \ldots, \mathcal{T}^\infty(0)$$

while the sequence of tame misère quotients

$$\mathcal{T}_0, \mathcal{T}_1, \mathcal{T}_2, \mathcal{T}_3, \mathcal{T}_4, \ldots, \mathcal{T}_\infty$$

can be written

$$\mathcal{T}_0, \mathcal{T}_1, \mathcal{T}^0(\mathcal{T}_2), \mathcal{T}^1(\mathcal{T}_2), \mathcal{T}^2(\mathcal{T}_2), \ldots, \mathcal{T}^\infty(\mathcal{T}_2).$$

Thus the normal quotients can be viewed as a tame sequence with base 0, and the tame misère quotients can be viewed as a tame sequence with base $\mathcal{T}_2$.

If $(\mathcal{Q}, \mathcal{P})$ is a misère quotient, then so is $\mathcal{T}(\mathcal{Q}, \mathcal{P})$, as the following lemma establishes (cf. Lemma 3.4). (Here we say that $(\mathcal{Q}, \mathcal{P})$ *is a misère quotient* if it is isomorphic to $\mathcal{Q}(\mathscr{A})$, for some set $\mathscr{A}$ of impartial games.)

Lemma 3.6. *Suppose* $(\mathcal{Q}, \mathcal{P}) = \mathcal{Q}(\mathscr{A})$ *is faithfully normal with kernel* $\mathcal{K}$. *Let* G *be a game with* $\text{opts}(G) \subset \mathscr{A}$ *and suppose* $\Phi'' G = \mathcal{K}$. *Then* $\mathcal{Q}(\mathscr{A} \cup \{G\}) \cong \mathcal{T}(\mathcal{Q}(\mathscr{A}))$ *and* $\Phi(G) = \bar{z}$.

Proof. Identical to the proof of Lemma 3.4. □

Corollary 3.7. *Suppose* $\mathcal{Q}(\mathscr{A})$ *is faithfully normal with kernel* $\mathcal{K}$. *Then for all* $n \in \mathbb{N} \cup \{\infty\}$, $\mathcal{T}^n(\mathcal{Q}(\mathscr{A}))$ *is a misère quotient.*

Proof. Let $G_0 = 0$, $\mathscr{A}_0 = \mathscr{A}$. Recursively choose G_{n+1} so that $\text{opts}(G_{n+1}) \subset \mathscr{A}_n$ and $\Phi'' G_{n+1} = \ker \mathcal{Q}(\mathscr{A}_n)$. Put $\mathscr{A}_{n+1} = \text{cl}(\mathscr{A}_n \cup \{G_{n+1}\})$.

By repeated application of Lemma 3.6, we have $\mathcal{Q}(\mathscr{A}_n) \cong \mathcal{T}^n(\mathcal{Q}(\mathscr{A}))$, and Theorem 2.4 therefore gives $\mathcal{Q}\big(\bigcup_n \mathscr{A}_n\big) = \mathcal{T}^\infty(\mathcal{Q}(\mathscr{A}))$. □

The quotients $\mathcal{R}_{2^n+4}$. If we start with a different base $(\mathcal{Q}, \mathcal{P})$, we obtain another sequence of quotients $\mathcal{T}^n(\mathcal{Q}, \mathcal{P})$. For example, if $(\mathcal{Q}, \mathcal{P}) = \mathcal{R}_8$, then for all $n \geq 2$, $\mathcal{T}^{n-2}(\mathcal{Q}, \mathcal{P})$ is a quotient of order $2^n + 4$, which we denote by $\mathcal{R}_{2^n+4}$. Likewise, we define $\mathcal{R}_\infty = \mathcal{T}^\infty(\mathcal{R}_8)$. Since $|\mathcal{P}| = 2$, all the $\mathcal{R}_n$'s have $\mathscr{P}$-portions of size 2. A major goal of this paper is to prove the following theorem.

Theorem 3.8. *Suppose* $(\mathcal{Q}, \mathcal{P})$ *is a misère quotient with* $|\mathcal{P}| = 2$. *Then either* $(\mathcal{Q}, \mathcal{P}) \cong \mathcal{T}_n$ *or* $(\mathcal{Q}, \mathcal{P}) \cong \mathcal{R}_n$, *for some* $n \in \mathbb{N} \cup \{\infty\}$.

Thus if $(\mathcal{Q}, \mathcal{P})$ is a misère quotient with $|\mathcal{P}| = 2$, it follows that either $|\mathcal{Q}| = \infty$, or $|\mathcal{Q}| = 2^n + 2$ or $2^n + 4$ for some $n \geq 2$. Furthermore, there is exactly one such quotient of each permissible finite order, and exactly two such infinite quotients.

"Almost tame" octal games. Lemmas 3.4 and 3.6 have useful implications for octal games, as summarized by the following theorem.

Theorem 3.9. *Let* Γ *be an octal game with last nonzero code digit* d. *Fix* n_0, *and suppose that* $(\mathcal{Q}, \mathcal{P}) = \mathcal{Q}_{2n_0+d-1}(\Gamma)$ *is faithfully normal with kernel* $\mathcal{K}$. *Suppose furthermore that*

$$\Phi(H_n) \in \mathcal{K} \text{ for all } n \text{ such that } n_0 \leq n < 2n_0 + d.$$

Then:

(i) $\mathcal{Q}(\Gamma)$ *is a faithfully normal quotient.*

(ii) $\mathcal{Q}(\Gamma) \cong \mathcal{T}^k(\mathcal{Q}, \mathcal{P})$ *for some* $k \in \mathbb{N} \cup \{\infty\}$.

(iii) $\Phi(H_n) \in \ker \mathcal{Q}(\Gamma)$, *for all* $n \geq n_0$.

Proof. We first show that (i)–(iii) hold for $\mathcal{Q}_n(\Gamma)$, for all n. By hypothesis we may assume that $n \geq 2n_0 + d$. Then a typical option of H_n is a position $H_a + H_b$, with $a + b \geq 2n_0$. Without loss of generality, we have $a \geq n_0$, so that $\Phi(H_a) \in \mathcal{K}$. Thus $\Phi(H_a)x \in \mathcal{K}$ for all x, and in particular $\Phi(H_a + H_b) \in \mathcal{K}$. This shows that $\Phi'' H_n \subset \mathcal{K}$, and Lemmas 3.4 and 3.6 immediately imply (i)–(iii).

If the partial quotients $\mathcal{Q}_n(\Gamma)$ eventually converge to some $\mathcal{T}^k(\mathcal{Q}, \mathcal{P})$, then $\mathcal{Q}(\Gamma) \cong \mathcal{T}^k(\mathcal{Q}, \mathcal{P})$. Otherwise $\mathcal{Q}(\Gamma) \cong \mathcal{T}^\infty(\mathcal{Q}, \mathcal{P})$; and in either case (i)–(iii) are immediate. □

Thus when the hypotheses of Theorem 3.9 are satisfied, we know that beyond heap n_0, the misère-play analysis of Γ is no harder than its normal-play analysis. It follows that we can stop computing partial quotients of Γ and revert to the much easier task of calculating nim values. We say that Γ is *tame relative to heap* n_0.

The hypotheses of Theorem 3.9 may seem rather restrictive, but there are several three-digit octal games that satisfy them; for example, **0.414**, **0.776**, and **4.76**. The misère-play solutions to these games now depend only on finding normal-play solutions, and we can regard them as "relatively solved."

The hexadecimal game **0.9092** is another interesting case. It is known to be arithmetic periodic in normal play (using methods due to Austin [1976] and Howse and Nowakowski [2004]). Furthermore, in misère play we can show that it is tame relative to heap 12 (using *MisereSolver*, say; see [Plambeck and Siegel 2007]). Now $\mathcal{Q}_{12}(\mathbf{0.9092}) \cong \mathcal{R}_8$, so by Theorem 3.9 (suitably generalized to hexadecimal games) we have $\mathcal{Q}(\mathbf{0.9092}) \cong \mathcal{T}^k(\mathcal{R}_8)$ for some k. Since the $\mathcal{G}$-values of **0.9092** are unbounded, k is necessarily ∞. Therefore $\mathcal{Q}(\mathbf{0.9092})$ is exactly $\mathcal{R}_\infty$.

4. One-stage extensions of $\mathcal{T}_n$

We next focus our attention on proving Theorem 3.8. The crux of the proof is an analysis of one-stage extensions of $\mathcal{T}_n$ and $\mathcal{R}_{2^n+4}$. This analysis also yields a useful structure theory for these quotients. In particular, we will prove the following two theorems.

Theorem 4.1. *If* $(\mathcal{Q}, \mathcal{P})$ *is a one-stage extension of* $\mathcal{T}_n$ *and* $|\mathcal{P}| = 2$, *then either* $(\mathcal{Q}, \mathcal{P}) \cong \mathcal{T}_n$, *or* $(\mathcal{Q}, \mathcal{P}) \cong \mathcal{T}_{n+1}$, *or else* $(\mathcal{Q}, \mathcal{P}) \cong \mathcal{R}_{2^n+4}$.

Theorem 4.2. *If* $(\mathcal{Q}, \mathcal{P})$ *is a one-stage extension of* $\mathcal{R}_{2^n+4}$ *and* $|\mathcal{P}| = 2$, *then either* $(\mathcal{Q}, \mathcal{P}) \cong \mathcal{R}_{2^n+4}$, *or else* $(\mathcal{Q}, \mathcal{P}) \cong \mathcal{R}_{2^{n+1}+4}$.

In this section we focus on Theorem 4.1, and we prove Theorem 4.2 in the following section.

Throughout the discussion there will be the implicit assumption that all quotients encountered are faithful. This is a slightly suspicious assumption, since it is unknown whether there exists an unfaithful quotient. However, since the argument proceeds "ground-up" by one-stage extensions, we are safe: a careful check of the proofs reveals that every extension under consideration preserves faithfulness. Therefore, if there exists an unfaithful quotient, it must necessarily satisfy $|\mathcal{P}| > 2$, and so will not interfere with the present argument. We will not be too careful about stating and restating this assumption of faithfulness, but in all cases the checks are routine.

The structure of $\mathcal{T}_n$. For the remainder of this section, fix a set of games $\mathscr{A}$, and suppose that $\mathcal{Q}(\mathscr{A}) \cong \mathcal{T}_n$, where $n \geq 2$. The structure of $(\mathcal{Q}, \mathcal{P}) = \mathcal{Q}(\mathscr{A})$ is described as follows. $\mathcal{Q} = \mathcal{K} \cup \{1, a\}$, where $\mathcal{K} \cong \mathbb{Z}_2^n$ and $a^2 = 1$. We write $\mathcal{K} = \{z_0, z_1, \ldots, z_{2^n-1}\}$, where z_0 is the identity, $z_1 = az_0$, and z_i corresponds to nim value i.

Now fix a game $G \not\cong 0$ with $\mathrm{opts}(G) \subset \mathscr{A}$, and write $m = \mathscr{G}(G)$, $\mathscr{B} = \mathrm{cl}(\mathscr{A} \cup \{G\})$, and $(\mathcal{Q}^+, \mathcal{P}^+) = \mathcal{Q}(\mathscr{B})$.

Definition 4.3. Let $\mathcal{E} \subset \mathcal{Q}$. We say that $\mathcal{E}$ is *complemented* if $\mathcal{E} \cap \{a, z\} \neq \varnothing$ and $\mathcal{E} \cap \{1, az\} \neq \varnothing$.

Lemma 4.4. *If $\Phi''G$ is complemented, then $2n \cdot G$ is a $\mathscr{P}$-position for all $n \geq 1$.*

Proof. Write the copies of G in pairs, as $n \cdot (G + G)$. Second player follows the mirror-image strategy on each pair *until* her move would remove the last copy of G. If that is the case, then the position must be

$$G + G' + Y, \qquad \text{with } Y \in \mathscr{A},$$

and since second player has been following the mirror-image strategy, we necessarily have $\mathscr{G}(Y) = 0$.

Case 1: $\Phi(G' + Y) \in \mathcal{K}$. Then second player moves to $G' + G' + Y$. Now

$$\mathscr{G}(G' + G' + Y) = \mathscr{G}(G' + G') + \mathscr{G}(Y) = 0 + 0.$$

Since also $\Phi(G' + G' + Y) \in \mathcal{K}$, faithful normality implies that

$$\Phi(G' + G' + Y) = z \in \mathcal{P}.$$

Case 2: $\Phi(G' + Y) = 1$. Then second player chooses an $H \in \mathrm{opts}(G)$ with $\Phi(H) \in \{a, z\}$, as guaranteed by complementarity, and we have

$$\Phi(H + G' + Y) = \Phi(H) \cdot \Phi(G' + Y) = \Phi(H) \cdot 1 \in \mathcal{P}.$$

Case 3: $\Phi(G'+Y) = a$. Then second player chooses H with $\Phi(H) \in \{1, az\}$, to the same effect.

Since $\mathcal{Q} = \mathcal{K} \cup \{1, a\}$, this exhausts all possibilities for $\Phi(G'+Y)$. □

Lemma 4.5. *Assume that* $|\mathcal{P}^+| \geq 2$, $\Phi''G$ *is complemented, and* $m = 0$ *or* 1. *Fix* $Y \in \mathscr{A}$ *with* $\Phi(Y) \in \mathcal{K}$ *and* $\mathscr{G}(Y) \neq \mathscr{G}(G)$. *Then* $G+Y$ *is an* $\mathscr{N}$*-position.*

Proof. The $m = 0$ and 1 cases are similar, so suppose $m = 0$. By Lemma 4.4, $G+G$ is a $\mathscr{P}$-position, so that $\Phi(G+G) \in \{a, z\}$. But $4 \cdot G$ is also a $\mathscr{P}$-position (again by Lemma 4.4), and $a^2 = 1$, so necessarily $\Phi(G+G) \neq a$. Therefore $\Phi(G+G) = z$. Moreover, since $\Phi(Y) \in \mathcal{K}$, we necessarily have $\Phi(Y+Y) = z$, so that $G+G \equiv Y+Y \pmod{\mathscr{B}}$.

Now consider $G+G+G$. A typical option is $G'+G+G \equiv G'+Y+Y \pmod{\mathscr{B}}$; but $\mathscr{G}(G') \neq 0$, so

$$\Phi(G'+Y+Y) = \Phi(G')z \notin \mathcal{P}.$$

Therefore $G+G+G$ is also a $\mathscr{P}$-position. Assume (for contradiction) that $G+Y$ is also a $\mathscr{P}$-position. Then either $G+Y \equiv * \pmod{\mathscr{B}}$ or $G+Y \equiv Y+Y \pmod{\mathscr{B}}$. But $G+G+Y+Y$ is also a $\mathscr{P}$-position, since it is equivalent to $4 \cdot Y$, so necessarily $G+Y \equiv Y+Y \pmod{\mathscr{B}}$. But now

$$G+G+G \equiv G+Y+Y \equiv Y+Y+Y \pmod{\mathscr{B}},$$

a contradiction, since $Y+Y+Y$ is an $\mathscr{N}$-position. □

Definition 4.6. Fix $\mathcal{E} \subset \mathcal{Q}$. The *discriminant* $\Delta = \Delta(\mathcal{E})$ is given by

$$\Delta = \mathcal{E} \cap \{1, a, z, az\}.$$

We say that $\mathcal{E}$ is *restive* if $\Delta = \{1, z\}$ or $\{a, az\}$, *restless* if $\Delta = \{a, z\}$ or $\{1, az\}$, and *tame* otherwise. We say that $\mathcal{E}$ is *wild* if it is restive or restless.

Lemma 4.7. *Assume that* $\Phi''G$ *is tame. If* $m < 2^n$, *then* $\mathcal{Q}(\mathscr{B}) \cong \mathcal{T}_n$; *if* $m = 2^n$, *then* $\mathcal{Q}(\mathscr{B}) \cong \mathcal{T}_{n+1}$. *In either case, we have*

$$\Phi(G) = \begin{cases} 1 & \text{if } \Delta = \{a\}, \\ a & \text{if } \Delta = \{1\}, \\ z_m & \text{otherwise}, \end{cases}$$

where $\Delta = \Delta(\Phi''G)$.

Proof. Let $\mathcal{E} = \Phi''G$. In each of the three possibilities for Δ, it is easily seen that $\mathcal{E}$ satisfies condition (i) of the generalized mex rule. We now verify condition (ii).

Case 1: $\Delta = \{a\}$. With $x = 1$, condition (ii) is equivalent to: for every $\mathscr{N}$-position $Y \in \mathscr{A}$, either $\Phi(Y') \in \mathcal{P}$ for some Y', or else $x'\Phi(Y) \in \mathcal{P}$ for some $x' \in \mathcal{E}$. But if $Y \neq 0$, then the first of these two conditions is satisfied a priori; while if $Y = 0$, then $x' = a$ suffices for the second.

Case 2: $\Delta = \{1\}$. We must verify (ii) with $x = a$. Fix $Y \in \mathscr{A}$ and $n \geq 0$ and suppose $a^{n+1}\Phi(Y) \notin \mathcal{P}$. If n is odd, then Y is an $\mathscr{N}$-position, so either $Y = 0$ or some Y' is a $\mathscr{P}$-position. If $Y = 0$, then we have $a^n \cdot 1 \cdot \Phi(Y) = a \in \mathcal{P}$; if Y' is a $\mathscr{P}$-position, then $a^{n+1}\Phi(Y') \in \mathcal{P}$. Finally, if n is even, then $Y + *$ is an $\mathscr{N}$-position. So either Y is a $\mathscr{P}$-position, in which case $a^n \cdot 1 \cdot \Phi(Y) = \Phi(Y) \in \mathcal{P}$; or else $Y' + *$ is a $\mathscr{P}$-position, in which case $a^{n+1}\Phi(Y') = a\Phi(Y') \in \mathcal{P}$.

Case 3: $\Delta \neq \{a\}, \{1\}$. Fix $Y \in \mathscr{A}$ and $n \geq 0$ and suppose $x^{n+1}\Phi(Y) \notin \mathcal{P}$. If n is odd, then $x^{n+1} = z$, so $\Phi(Y) \neq 1, z$. Therefore $\mathscr{G}(Y) \neq 0$, and Y has some option Y' with $\mathscr{G}(Y') = 0$. Therefore $x^{n+1}\Phi(Y') = z \in \mathcal{P}$.

If n is even, then $x^{n+1} = z_k$, so $\mathscr{G}(Y) \neq k$. If $\mathscr{G}(Y) > k$, then there is some option Y' with $\mathscr{G}(Y') = m$; hence $x^{n+1}\Phi(Y') = z \in \mathcal{P}$. So suppose $\mathscr{G}(Y) < m$. Then there is some option G' of G with $\mathscr{G}(G') = \mathscr{G}(Y)$. There are three subcases.

Subcase 3a: $n > 0$ or $\Phi(G') \in \mathcal{K}$ or $\Phi(Y) \in \mathcal{K}$. Then we have immediately that $x^n\Phi(G')\Phi(Y) = z \in \mathcal{P}$.

Subcase 3b: $n = 0$ and $\Phi(G') = \Phi(Y) = 1$. Then $1 \in \Delta$. Now $\Delta \neq \{1\}$ (since we are in Case 3), and furthermore $\Delta \neq \{1, az\}$ (since $\Phi''G$ is tame). So either $a \in \Phi''G$ or $z \in \Phi''G$. But if $x' = a$ or z, then $x'\Phi(Y) \in \mathcal{P}$, as needed.

Subcase 3c: $n = 0$ and $\Phi(G') = \Phi(Y) = a$. Then $a \in \Delta$. Now $\Delta \neq \{a\}$ (since we are in Case 3), and furthermore $\Delta \neq \{a, z\}$ (since $\Phi''G$ is tame). So either $1 \in \Phi''G$ or $az \in \Phi''G$. But if $x' = 1$ or az, then $x'\Phi(Y) \in \mathcal{P}$, as needed. □

Lemma 4.8. *Assume that $\Phi''G$ is restless. Then $|\mathcal{P}^+| \geq 3$.*

Proof. As in Lemma 4.7, we write $\mathcal{E} = \Phi''(G)$ and $\Delta = \Delta(\mathcal{E})$.

Case 1: $\Delta = \{1, az\}$. Then $\{a, z\} \cap \mathcal{E} = \varnothing$, so G is a $\mathscr{P}$-position. Furthermore, if G' is an option with $\Phi(G') = 1$ (resp. az), then $\Phi(G')z \in \mathcal{P}$ (resp. $\Phi(G')az \in \mathcal{P}$). This shows that $G + *2_2$ (resp. $G + *2_3$) is an $\mathscr{N}$-position. Therefore $G \not\equiv *2_2$ (mod $\mathscr{B}$) and $G \not\equiv *$ (mod $\mathscr{B}$); since G is a $\mathscr{P}$-position, this implies $|\mathcal{P}^+| \geq 3$.

Case 2: $\Delta = \{a, z\}$. Then $\{1, az\} \cap \mathcal{E} = \varnothing$, so $\{a, z\} \cap a\mathcal{E} = \varnothing$, and hence $G + *$ is a $\mathscr{P}$-position. Just as in Case 1, we see that $G + * \not\equiv *$ (mod $\mathscr{B}$) and $G + * \not\equiv *2_2$ (mod $\mathscr{B}$), so again $|\mathcal{P}^+| \geq 3$. □

Lemma 4.9. *Assume that $\Phi''G$ is restive and $|\mathcal{P}^+| = 2$. Then $\mathcal{Q}(\mathscr{B}) \cong \mathcal{R}_{2^n+4}$ and*

$$\Phi(G) = \begin{cases} t & \text{if } \Delta = \{a, az\}, \\ at & \text{if } \Delta = \{1, z\}. \end{cases}$$

(Here t denotes the same element as in the discussion of $\mathcal{R}_{2^n+4}$ in Section 1, and $\Delta = \Delta(\Phi''G)$ as in Lemma 4.7.)

Proof. As before, let $\mathcal{E} = \Phi''(G)$. The argument is similar in both cases, so suppose that $\Delta = \{a, az\}$. Now in $\mathcal{R}_{2^n+4}$ it is easy to compute $\mathcal{M}_t = \mathcal{Q} \setminus \{1, t, z\}$. Since $\mathcal{E} \cap \{1, z\} = \varnothing$, condition (i) of the generalized mex rule is therefore trivially satisfied.

For (ii), fix $Y \in \mathscr{A}$ and $n \geq 0$ and suppose that $t^{n+1}\Phi(Y) \notin \mathcal{P}$. There are three cases.

Case 1: $n > 0$. Then $t^{n+1} = z$, so necessarily $\mathscr{G}(Y) > 0$. Therefore $t^{n+1}\Phi(Y') \in \mathcal{P}$, where Y' is any option with $\mathscr{G}(Y') = 0$.

Case 2: $n = 0$ and $\Phi(Y) \notin \mathcal{K}$. If $\Phi(Y) = 1$, then we have $a\Phi(Y) \in \mathcal{P}$; if $\Phi(Y) = a$, then $az\Phi(Y) \in \mathcal{P}$. Since $a, az \in \mathcal{E}$, this suffices.

Case 3: $n = 0$ and $\Phi(Y) \in \mathcal{K}$. Then $\mathscr{G}(Y) \neq 0$. If $\Phi(Y') = z$ for some Y', then we are done, since $t\Phi(Y') \in \mathcal{P}$, so assume $\Phi(Y') \neq z$ for all Y'.

Now since G is restive, it is complemented, so by Lemma 4.5 $G + Y$ is an $\mathscr{N}$-position. Consider a typical $G + Y'$. By assumption, $\Phi(Y') \neq z$. If $\Phi(Y') = 1$, then $G' + Y'$ is a $\mathscr{P}$-position, where $\Phi(G') = a$. If $\Phi(Y') = a$ or az, then $G' + Y'$ is a $\mathscr{P}$-position, where $\Phi(G') = az$. If $\mathscr{G}(Y') \geq 2$, then by Lemma 4.5 $G + Y'$ is *a priori* an $\mathscr{N}$-position. (We automatically have $m = 0$ or 1, since G is restive; cf. [ONAG].) So in all cases, $G + Y'$ is an $\mathscr{N}$-position.

But $G + Y$ is an $\mathscr{N}$-position, so we must have $G' + Y$ a $\mathscr{P}$-position, for some G'. Then $x'\Phi(Y) \in \mathcal{P}$, where $x' = \Phi(G')$, completing the proof. □

Proof of Theorem 4.1. Immediate from the preceding lemmas. □

5. One-stage extensions of $\mathcal{R}_{2^n+4}$

In this section we generalize much of the machinery of Section 4. Recall (from Section 1) that $\mathcal{R}_{2^n+4} = \mathcal{T}_n \cup \{t, at\}$, where $t^2 = tz = z$, and $\mathcal{P} = \{a, z\}$.

For the rest of this section, assume that $(\mathcal{Q}, \mathcal{P}) = \mathcal{Q}(\mathscr{A})$ is faithful, with $(\mathcal{Q}, \mathcal{P}) \cong \mathcal{R}_{2^n+4}$. Fix G with $\text{opts}(G) \subset \mathscr{A}$, and write $\mathscr{B} = \text{cl}(\mathscr{A} \cup \{G\})$, $(\mathcal{Q}^+, \mathcal{P}^+) = \mathcal{Q}(\mathscr{B})$, $\mathcal{E} = \Phi''G$, and $m = \mathscr{G}(G)$. Assume throughout this section that $|\mathcal{P}^+| = 2$.

We can very quickly reduce to the case where $\mathcal{E}$ is complemented (in the sense of Definition 4.3: $\mathcal{E} \cap \{a, z\} \neq \varnothing$ and $\mathcal{E} \cap \{1, az\} \neq \varnothing$).

Lemma 5.1. *Assume that $\Phi''G$ is* not *complemented. Then $\mathcal{Q}(\mathscr{B}) \cong \mathcal{Q}(\mathscr{A})$.*

Proof. Case 1: $\{a, z\} \cap \mathcal{E} = \varnothing$. Since $\mathcal{P} = \{a, z\}$, this immediately implies that G is a $\mathscr{P}$-position, so since $|\mathcal{P}^+| = |\mathcal{P}| = 2$, we must have $G \equiv Y \pmod{\mathscr{B}}$ for some $Y \in \mathscr{A}$. Therefore $\mathcal{Q}(\mathscr{B}) \cong \mathcal{Q}(\mathscr{A})$.

Case 2: $\{1, az\} \cap \mathcal{E} = \varnothing$. If G is a $\mathscr{P}$-position, then the argument is just as in Case 1. Otherwise, consider $G + *$. Since $\{1, az\} \cap \mathcal{E} = \varnothing$, we have $\mathcal{P} \cap a\mathcal{E} = \varnothing$, so every $G' + *$ is an $\mathscr{N}$-position. Since $G + 0$ is also an $\mathscr{N}$-position, we conclude that $G + *$ is a $\mathscr{P}$-position.

But this implies $G + * \equiv Y \pmod{\mathscr{B}}$ for some $Y \in \mathscr{A}$, whence $G \equiv Y + *$ (mod $\mathscr{B}$), and again we have $\mathcal{Q}(\mathscr{B}) \cong \mathcal{Q}(\mathscr{A})$. □

We now consider the case when $\mathcal{E}$ is complemented. The key fact about complementarity is the following (cf. Lemma 4.4).

Lemma 5.2. *If $\Phi''G$ is complemented, then $2n \cdot G$ is a $\mathscr{P}$-position for all $n \geq 1$.*

Proof. Identical to the proof of Lemma 4.4. □

Lemma 5.3. *Assume that $\Phi''G$ is complemented and $m = 0$ or 1, and fix $Y \in \mathscr{A}$ with $\mathscr{G}(Y) \geq 2$. Then $G + Y$ is an $\mathscr{N}$-position.*

Proof. Identical to the proof of Lemma 4.5. □

Lemma 5.4. *Assume that $m \geq 2$, and fix $Y \in \mathscr{A}$ with $\Phi(Y) \in \{t, at\}$. Then $G+Y$ is an $\mathscr{N}$-position.*

Proof. Since $m \geq 2$ and $\mathscr{G}(Y) \leq 1$, there must exist an option G' such that $\mathscr{G}(G') = \mathscr{G}(Y)$. Then $\mathscr{G}(G' + Y) = 0$, so that $\Phi(G' + Y) \in \{1, t, z\}$. In all cases, $\Phi(G' + Y)z \in \mathcal{P}$. Fix $Z \in \mathscr{A}$ with $\Phi(Z) = z$; then $G + Y + Z$ is an $\mathscr{N}$-position and $Z + Z$ a $\mathscr{P}$-position, so $G + Y \not\equiv Z \pmod{\mathscr{B}}$.

Next choose an option G' with $\mathscr{G}(G') = \mathscr{G}(Y) \oplus 1$. Then $\mathscr{G}(G' + Y) = 1$, so $\Phi(G' + Y) \in \{a, at, az\}$. In all cases, $\Phi(G' + Y)az \in \mathcal{P}$. Fix $W \in \mathscr{A}$ with $\Phi(W) = az$; then $G + Y + W$ is an $\mathscr{N}$-position and $* + W$ a $\mathscr{P}$-position, so $G + Y \not\equiv * \pmod{\mathscr{B}}$.

Thus $G + Y \not\equiv Z \pmod{\mathscr{B}}$, and $G + Y \not\equiv * \pmod{\mathscr{B}}$. But $|\mathcal{P}^+| = 2$, so these are the only two classes of $\mathscr{P}$-position in $\mathscr{B}$, and it follows that $G + Y$ is an $\mathscr{N}$-position. □

Lemma 5.5. *Assume that $\Phi''G$ is complemented and $m \geq 2$, and fix $Y \in \mathscr{A}$. Then $G + Y$ is a $\mathscr{P}$-position if and only if $\Phi(Y) = z_m$.*

Proof. If $\Phi(Y) = z_i$, for some $i < m$, then $G' + Y$ (with $\mathscr{G}(G') = i$) is *a priori* a $\mathscr{P}$-position, so $G + Y$ is an $\mathscr{N}$-position. If $\Phi(Y) = z_i$ for some $i > m$, then $\mathscr{G}(Y') = m$ for some Y', so $\Phi(Y') = z_m$. By induction on the birthday of Y, we have that $G + Y'$ is a $\mathscr{P}$-position, so again $G + Y$ is an $\mathscr{N}$-position.

If $\Phi(Y) = z_m$, then by induction every $G + Y'$ is an $\mathscr{N}$-position. Likewise, for every G' we have $\mathscr{G}(G' + Y) \neq 0$ and $\Phi(G' + Y) \in \mathcal{K}$, so every $G' + Y$ is also an $\mathscr{N}$-position. Therefore $G + Y$ is a $\mathscr{P}$-position.

This leaves only the cases $\Phi(Y) \in \{1, a, t, at\}$. But if $\Phi(Y) = 1$ (resp. a), then $G' + Y$ is a $\mathscr{P}$-position, where $\Phi(G') \in \{a, z\}$ (resp. $\{1, az\}$), as guaranteed by complementarity. Therefore $G + Y$ is an $\mathscr{N}$-position. Conversely, if $\Phi(Y) \in \{t, at\}$, then Lemma 5.4 guarantees that $G + Y$ is an $\mathscr{N}$-position. □

We now proceed with the proof of Theorem 4.2. There are two fundamental cases, each stated as a separate lemma: $m \geq 2$, and $m \in \{0, 1\}$.

Lemma 5.6. *Assume that $\Phi''G$ is complemented and $m \geq 2$. Then $\mathcal{Q}(\mathscr{B}) \cong \mathcal{R}_{2^n+4}$ or $\mathcal{R}_{2^{n+1}+4}$.*

Proof. This is much like Lemma 3.4. It suffices to verify conditions (i) and (ii) in the generalized mex rule. Since $m \geq 2$, we have $\mathcal{M}_{z_m} = \mathcal{Q} \setminus \{z_m\}$. Since $\mathscr{G}(G) = m$, this suffices for (i). For (ii), fix $Y \in \mathscr{A}$ and $n \geq 0$, and suppose $z_m^{n+1}\Phi(Y) \notin \mathcal{P}$.

If n is odd, then $\mathscr{G}(Y) > 0$, so $z_m^{n+1}\Phi(Y') \in \mathcal{P}$, where $\mathscr{G}(Y') = 0$.

If n is even, then $\mathscr{G}(Y) \neq m$. If $\mathscr{G}(Y) > m$, then $z_m^{n+1}\Phi(Y') \in \mathcal{P}$, where $\mathscr{G}(Y') = m$. If $2 \leq \mathscr{G}(Y) < m$, then let $i = \mathscr{G}(Y)$. In this case z_i is the unique element of $\mathcal{Q}$ with $\mathscr{G}$-value i, so necessarily $z_i \in \mathcal{E}$. Since $z_i \cdot z_i \in \mathcal{P}$, this suffices.

If $\mathscr{G}(Y) = 0$, then we have $\Phi(Y) \in \{1, t, z\}$. If $\Phi(Y) = z$, then $x'\Phi(Y) \in \mathcal{P}$ for any $x' \in \mathcal{E} \cap \{1, t, z\}$. (This intersection must be nonempty, since $1, t, z$ are the only elements of $\mathcal{Q}$ corresponding to $\mathscr{G}$-value 0.) If $\Phi(Y) = 1$, then since $\mathcal{E}$ is complemented, we have $\mathcal{E} \cap \{a, z\} \neq \varnothing$; and either choice suffices. This leaves only the case $\Phi(Y) = t$. If $x't \in \mathcal{P}$ for some $x' \in \mathcal{E}$, then we are done. Otherwise, $G' + Y$ is an $\mathscr{N}$-position for every G'. But by Lemma 5.4 (and the assumption $|\mathcal{P}^+| = 2$), we know that $G + Y$ is an $\mathscr{N}$-position, so some $G + Y'$ must be a $\mathscr{P}$-position. By Lemma 5.5, we have specifically $\Phi(Y') = z_m$, whence $z_m^{n+1}\Phi(Y') = z \in \mathcal{P}$, as needed.

Finally, if $\mathscr{G}(Y) = 1$, then $\Phi(Y) \in \{a, at, az\}$, and the proof proceeds just as in the $\mathscr{G}(Y) = 0$ case. □

Lemma 5.7. *Assume that $\Phi''G$ is complemented and $m = 0$ (resp. 1). Then $\mathcal{Q}(\mathscr{B}) \cong \mathcal{R}_{2^n+4}$, and $\Phi(G) = t$ (resp. at).*

Proof. The two cases are essentially identical, so assume $m = 0$. As always, we use the generalized mex rule. Note that

$$\mathcal{M}_t = \mathcal{Q} \setminus \{1, t, z\} = \{x : \mathscr{G}(x) \neq 0\},$$

and since $\mathscr{G}(G) = 0$, this suffices for (i). For (ii), fix $Y \in \mathscr{A}$ and $n \geq 0$, and suppose $t^{n+1}\Phi(Y) \notin \mathcal{P}$. There are four cases.

Case 1: $n \geq 1$. Then $t^{n+1} = z$, so $z\Phi(Y) \notin \mathcal{P}$. Thus $\Phi(Y) \neq 1, t, z$, so $\mathscr{G}(Y) \neq 0$. We conclude that $t^{n+1}\Phi(Y') \in \mathcal{P}$, where Y' is any option with $\mathscr{G}(Y') = 0$.

Case 2: $n = 0$ and $\mathscr{G}(Y) = 0$. Then $\Phi(Y) \in \{1, t, z\}$, and since $t\Phi(Y) \notin \mathcal{P}$, necessarily $\Phi(Y) = 1$. But since $\mathcal{E}$ is complemented, $\mathcal{E} \cap \{a, z\} \neq \varnothing$, so $x'\Phi(Y) \in \mathcal{P}$, where $x' = a$ or z.

Case 3: $n = 0$ and $\mathscr{G}(Y) = 1$. Since $\mathcal{E}$ is complemented, we have $\mathcal{E} \cap \{1, az\} \neq \varnothing$. Since $m = 0$, we know that $1 \notin \mathcal{E}$, so necessarily $az \in \mathcal{E}$. Since $\mathscr{G}(Y) = 1$, we always have $az\Phi(Y) \in \mathcal{P}$, so this suffices.

Case 4: $n = 0$ and $\mathscr{G}(Y) \geq 2$. If $t \in \Phi''Y$ or $z \in \Phi''Y$, then $t\Phi(Y') = z$ and there is nothing to prove. Otherwise, put $i = \mathscr{G}(Y)$; to complete the proof, it suffices to show that $z_i \in \mathcal{E}$, because $z_i\Phi(Y) = z \in \mathcal{P}$. So consider $G + Y$. We first show that every $G + Y'$ is an $\mathscr{N}$-position. If $\mathscr{G}(Y') = 0$, then $\Phi(Y') = 1$ (since we are assuming $t, z \notin \Phi''Y$). Since G is complemented and $\mathscr{G}(G) = 0$, we necessarily have $a \in \Phi''G$, so $a\Phi(Y') \in \mathcal{P}$ and hence $G + Y'$ is an $\mathscr{N}$-position. If $\mathscr{G}(Y') = 1$, then since G is complemented and $\mathscr{G}(G) = 0$, we necessarily have $az \in \Phi''G$, so $az\Phi(Y') \in \mathcal{P}$ and again $G + Y'$ is an $\mathscr{N}$-position. Finally, if $\mathscr{G}(Y') \geq 2$, then the desired conclusion follows from Lemma 5.3.

This shows that every $G + Y'$ is an $\mathscr{N}$-position. But by Lemma 5.3, $G + Y$ itself is an $\mathscr{N}$-position. Therefore some $G' + Y$ is necessarily a $\mathscr{P}$-position. Since $\Phi(Y) = z_i$, we conclude that $\Phi(G') = z_i$ as well, completing the proof. □

6. Uniqueness of $\mathcal{R}_8$

The following theorem emerges readily from previous work.

Theorem 6.1. *$\mathcal{R}_8$ is the only misère quotient of order* 8 (*up to isomorphism*).

Proof. Let $(\mathcal{Q}, \mathcal{P})$ be a misère quotient of order 8. By Lemma 2.2, $(\mathcal{Q}, \mathcal{P})$ must arise as a one-stage extension of $\mathcal{T}_2$ (since $|\mathcal{T}_2|$ has order 6, and all nontrivial quotients must have even order, as shown in [Plambeck and Siegel 2008]). So there is some closed set $\mathscr{A}$, and some G with $\mathrm{opts}(G) \subset \mathscr{A}$, such that

$$\mathcal{Q}(\mathscr{A}) \cong \mathcal{T}_2 \quad \text{and} \quad \mathcal{Q}(\mathscr{A} \cup \{G\}) \cong (\mathcal{Q}, \mathcal{P}).$$

Let $\Phi : \mathrm{cl}(\mathscr{A} \cup \{G\}) \to \mathcal{Q}$ be the quotient map, and write

$$a = \Phi(*), \quad b = \Phi(*2), \quad t = \Phi(G).$$

Since $\mathcal{Q}(\mathscr{A}) \cong \mathcal{T}_2$ and $\mathcal{Q}(\mathscr{A} \cup \{G\}) \ncong \mathcal{T}_2$, t is not in the submonoid generated by a, b. Thus neither is at (since $a^2 = 1$), and it follows immediately that

$$\mathcal{Q} = \mathcal{T}_2 \cup \{t, at\}.$$

Now put $\mathcal{E} = \Phi''G$. $\mathcal{E}$ cannot be tame, since then Lemma 4.7 would imply that $(\mathcal{Q}, \mathcal{P}) \cong \mathcal{T}_2$ or $\mathcal{T}_3$, neither of which has order 8.

If $\mathcal{E}$ is restive, then either $\{1, z\} \subset \mathcal{E}$ or $\{a, az\} \subset \mathcal{E}$, and it follows that G and $G + *$ are both $\mathscr{N}$-positions (since $\Phi''G$ contains either z or a directly, and $\Phi''(G + *)$ contains either $1 \cdot a$ or $az \cdot a$, equal to a or z respectively). Therefore $t, at \notin \mathcal{P}$, so $|\mathcal{P}| = 2$. By Theorem 4.1, we have $(\mathcal{Q}, \mathcal{P}) \cong \mathcal{R}_8$.

We complete the proof by assuming $\mathcal{E}$ to be restless and obtaining a contradiction. There are two cases.

Case 1: $\Delta = \{1, az\}$. Then $a, z \notin \mathcal{E}$, so G is a $\mathscr{P}$-position. Therefore $*2 + G$ is an $\mathscr{N}$-position; and since $\Phi(*2 + G) = bt$, we have $bt \notin \mathcal{P}$, so that $bt \in \{1, b, ab, az, at\}$. To obtain a contradiction, we show that $*2 + G$ is distinguishable from some representative of each of these possibilities.

The table below summarizes. The first column of each row lists one possibility for bt, along with an inequality $x \neq y$ that rules out this possibility. In each case, x is known to be in $\mathcal{P}$, and the second column exhibits an $\mathscr{N}$-position Y that witnesses $y \notin \mathcal{P}$. The winning move Y' is shown in the third column; the notation $\Phi^{-1}(x)$ is used to represent a typical option of G with $\Phi(G) = x$.

Distinction(s)	Typical $\mathscr{N}$-position	Winning Move
$1 \neq bt \Leftarrow a \neq abt$ $az \neq bt \Leftarrow z \neq abt$ $at \neq bt \Leftarrow t \neq abt$	$* + *2 + G$	$* + * + G$
$b \neq bt \Leftarrow z \neq zt$	$*2 + *2 + G$	$*2 + *2 + \Phi^{-1}(1)$
$ab \neq bt \Leftarrow z \neq azt$	$* + *2 + *2 + G$	$* + *2 + *2 + \Phi^{-1}(az)$

Case 2: $\Delta = \{a, z\}$. This is similar. Clearly G is an $\mathscr{N}$-position, so since $1, az \notin \mathcal{E}$, we have that $* + G$ is a $\mathscr{P}$-position. As before, this implies that $*2 + G$ is an $\mathscr{N}$-position. The following table parallels the table from Case 1.

Distinction(s)	Typical $\mathscr{N}$-position	Winning Move
$1 \neq bt \Leftarrow a \neq abt$ $az \neq bt \Leftarrow z \neq abt$ $t \neq bt \Leftarrow at \neq abt$	$* + *2 + G$	$* + G$
$b \neq bt \Leftarrow z \neq zt$	$*2 + *2 + G$	$*2 + *2 + \Phi^{-1}(z)$
$ab \neq bt \Leftarrow z \neq azt$	$* + *2 + *2 + G$	$* + *2 + *2 + \Phi^{-1}(a)$

This exhausts all possibilities and completes the proof. □

Theorem 6.1 can be extended: for example, $\mathcal{T}_3$ is the unique misère quotient of order 10. But the proof of Theorem 6.1 gives us pause. The uniqueness of $\mathcal{R}_8$ takes shape through a somewhat subtle combinatorial analysis. To prove the uniqueness of $\mathcal{T}_3$ by hand, we would need to sharpen the restless cases of Theorem 6.1, and then show that every one-stage extension of $\mathcal{R}_8$ has order ≥ 12. This appears to be quite a lot of work, so we now refocus our efforts on automating this sort of analysis.

7. Valid transition tables

Transition algebras were introduced in [Plambeck and Siegel 2008], and there they proved to be useful in the study of mex functions. We now abstract out some of their structure.

Definition 7.1. Let $\mathcal{Q}$ be a commutative monoid. A *transition table on* $\mathcal{Q}$ is a subset $T \subset \mathcal{Q} \times Pow(\mathcal{Q})$.

Note that if $\mathscr{A}$ is a closed set of games, then $T(\mathscr{A})$ is a transition table on $\mathcal{Q}(\mathscr{A})$.

Definition 7.2. Let T be a transition table on a bipartite monoid $(\mathcal{Q}, \mathcal{P})$. T is said to be *valid* if and only if the following four conditions hold.

(i) (parity) For each $(x, \mathcal{E}) \in T$, we have
$$x \in \mathcal{P} \Longleftrightarrow \mathcal{E} \neq \varnothing \text{ and } \mathcal{E} \cap \mathcal{P} = \varnothing.$$

(ii) (completeness) For each $x \in \mathcal{Q}$, there is some set $\mathcal{E}$ such that $(x, \mathcal{E}) \in T$.

(iii) (closure) If $(x, \mathcal{E}), (y, \mathcal{F}) \in T$, then $(xy, x\mathcal{F} \cup y\mathcal{E}) \in T$.

(iv) (well-foundedness) There exists a map $R : \mathcal{Q} \to \mathbb{N}$ (a *rank function* for $\mathcal{Q}$) with the following property. $R(1) = 0$, and for each $x \in \mathcal{Q}$, there is some $(x, \mathcal{E}) \in T$ such that $R(y) < R(x)$ for all $y \in \mathcal{E}$.

We note that condition (iv) implies (ii), but nonetheless we include (ii) for clarity. Note also that condition (iii) implies a monoid structure.

Definition 7.3. A transition table T is a *transition algebra* if it is closed (in the sense of Definition 7.2(iii)).

We will use the terms "valid transition table" and "valid transition algebra" interchangeably. The main result is the following.

Theorem 7.4. *Let $(\mathcal{Q}, \mathcal{P})$ be a r.b.m. with $1 \notin \mathcal{P}$. The following are equivalent.*

(i) *There exists a closed set of games $\mathscr{A}$ with $\mathcal{Q}(\mathscr{A}) = (\mathcal{Q}, \mathcal{P})$.*

(ii) *There exists a valid transition table T on $(\mathcal{Q}, \mathcal{P})$.*

Proof. (i) $\Rightarrow$ (ii): Put $T = T(\mathscr{A})$. It is straightforward to check that T is valid. A suitable rank function is given by $R(x) = \min\{\text{birthday}(G) : \Phi(G) = x\}$.

(ii) $\Rightarrow$ (i): First define, for each $x \in \mathcal{Q}$, a game H_x as follows. The definition is by induction on $R(x)$. Let $(x, \mathcal{E}) \in T$ be such that $R(y) < R(x)$ for each $y \in \mathcal{E}$, and put
$$H_x = \{H_y : y \in \mathcal{E}\}.$$

Now define a game H_t for each $t \in T$:
$$H_{(x,\mathcal{E})} = \{H_y : y \in \mathcal{E}\}.$$

Let

$$\mathscr{A} = \mathrm{cl}(\{H_t : t \in T\}).$$

We claim that $\mathcal{Q}(\mathscr{A}) = (\mathcal{Q}, \mathcal{P})$.

Since $(\mathcal{Q}, \mathcal{P})$ is a r.b.m., it suffices (by [Plambeck and Siegel 2008, Proposition 4.7]) to exhibit a surjective homomorphism $\Phi : \mathscr{A} \to \mathcal{Q}$. Regarding $\mathscr{A}$ as a free commutative monoid on the generators H_t, we define Φ as a monoid homomorphism by

$$\Phi(H_{(x,\mathcal{E})}) = x.$$

By completeness (condition (ii) in the definition of validity), Φ is surjective. To complete the proof, we need to show that, for all $G \in \mathscr{A}$,

$$\Phi(G) \in \mathcal{P} \Longleftrightarrow G \neq 0 \quad \text{and} \quad \Phi(G') \notin \mathcal{P} \quad \text{for any option } G'.$$

So fix $G = H_{t_1} + \cdots + H_{t_k}$, and write $t_i = (x_i, \mathcal{E}_i)$. Write $x = x_1 x_2 \cdots x_k$, and denote by x/x_i the product $x_1 x_2 \cdots x_{i-1} x_{i+1} \cdots x_k$. Put

$$\mathcal{E} = \bigcup_{1 \le i \le k} \frac{x}{x_i} \mathcal{E}_i,$$

and let $t = (x, \mathcal{E})$. By closure (condition (iii) in the definition of validity), $t \in T$. By parity (condition (i)), we have

$$x \in \mathcal{P} \Longleftrightarrow \mathcal{E} \neq \varnothing \text{ and } \mathcal{E} \cap \mathcal{P} = \varnothing.$$

But clearly $\Phi(G) = x$, and $\mathcal{E} = \Phi'' G$. This suffices except for the case when $\mathcal{E} = \varnothing$; but then G has no options, so $\Phi(G) = 1$. Since we assumed that $1 \notin \mathcal{P}$, this completes the proof. □

Theorem 7.4 yields an algorithm for counting the number of misère quotients of order n: for each r.b.m. of order n, iterate over all transition tables and check whether any are valid. This is an atrociously poor algorithm, however; even if one could effectively enumerate the r.b.m.'s of order n, each one admits 2^{n2^n} transition tables! Theorem 7.4 is still important, however, since it reduces the search for misère quotients to a finite problem.

8. Enumerating quotients of small order

We now show how the techniques of the previous section can be made (reasonably) efficient. We first show that every misère quotient can be represented by a certain restricted type of transition algebra.

Definition 8.1. Let $(\mathcal{Q}, \mathcal{P})$ be a bipartite monoid. Fix $x_1, \ldots, x_k \in \mathcal{Q}$, and for $0 \le i \le k$ let $\mathcal{S}_i$ be the submonoid of $\mathcal{Q}$ generated by $x_1, \ldots, x_i$. We say that $x_1, \ldots, x_k$ is a *construction sequence* for $(\mathcal{Q}, \mathcal{P})$ if:

(i) $\mathcal{S}_k = \mathcal{Q}$;

(ii) for each i, $x_i \notin \mathcal{S}_{i-1}$;

(iii) for each $i < k$, the reduction of $(\mathcal{S}_i, \mathcal{P} \cap \mathcal{S}_i)$ is a misère quotient.

Definition 8.2. Let $(\mathcal{Q}, \mathcal{P})$ be a bipartite monoid. A transition algebra T on $(\mathcal{Q}, \mathcal{P})$ is said to be a *minimex algebra* if there exists a construction sequence $x_1, \ldots, x_k \in \mathcal{Q}$ that generates T in the following sense. Write $\mathcal{E}_i = \mathcal{M}_{x_i} \cap \mathcal{S}_{i-1}$, where the $\mathcal{S}_i$'s are as in the previous definition. Then T is generated by

$$(x_1, \mathcal{E}_1), \ldots, (x_k, \mathcal{E}_k).$$

We say that T is the minimex algebra *constructed by* $x_1, \ldots, x_k$.

Lemma 8.3. *Suppose T is a transition algebra on a finite r.b.m. $(\mathcal{Q}, \mathcal{P})$. Fix generators $x_1, \ldots, x_k \in \mathcal{Q}$ and suppose that, for each i, there is an $\mathcal{E}_i \subset \mathcal{S}_{i-1}$ such that $(x_i, \mathcal{E}_i) \in T$. Then T admits a rank function.*

Proof. Define a map $R^* : \mathcal{Q} \to \mathbb{N}^k$ as follows. For each $x \in \mathcal{Q}$, write

$$x = x_1^{n_1} x_2^{n_2} \cdots x_k^{n_k},$$

choosing the *lexicographically least* expression on the generators $x_1, \ldots, x_k$. Put $R^*(x) = (n_1, \ldots, n_k)$.

Now order the elements of $\mathbb{N}^k$ lexicographically. We claim that R^* is a "rank function" under this ordering. For if $R^*(x) = (n_1, \ldots, n_k)$, then let

$$(x, \mathcal{E}) = (x_1, \mathcal{E}_1)^{n_1} (x_2, \mathcal{E}_2)^{n_2} \cdots (x_k, \mathcal{E}_k)^{n_k}.$$

By the assumptions on the $\mathcal{E}_i$, we know that $R^*(y) < R^*(x_i)$ for each $y \in \mathcal{E}_i$. Therefore $R^*(y) < R^*(x)$ for each $y \in \mathcal{E}$.

Finally, R^* can be converted into a suitable rank function $R : \mathcal{Q} \to \mathbb{N}$ by enumerating the finite range of R^*. □

Theorem 8.4. *Let $(\mathcal{Q}, \mathcal{P})$ be a finitely generated r.b.m. with $1 \notin \mathcal{P}$. The following are equivalent.*

(i) *There exists a closed set of games $\mathscr{A}$ with $\mathcal{Q}(\mathscr{A}) = (\mathcal{Q}, \mathcal{P})$.*

(ii) *There exists a valid minimex algebra on $(\mathcal{Q}, \mathcal{P})$.*

Proof. (ii) $\Rightarrow$ (i) is immediate from Theorem 7.4, since every minimex algebra is automatically a valid transition table. So we must prove (i) $\Rightarrow$ (ii).

Since $\mathcal{Q}$ is finitely generated, we may assume that $\mathscr{A}$ is also finitely generated (passing, if necessary, to a suitable finitely generated subset of $\mathscr{A}$, and noting that the closure of a finitely generated set is finitely generated). Choose generators $H_1, \ldots, H_l$ for $\mathscr{A}$ such that $\text{opts}(H_i) \subset \langle H_1, \ldots, H_{i-1} \rangle$ for each i. (Here $\langle H_1, \ldots, H_{i-1} \rangle$ denotes the submonoid of $\mathscr{A}$ generated by $H_1, \ldots, H_{i-1}$.)

Put $y_i = \Phi(H_i)$ and consider the sequence $y_1, \ldots, y_l \in \mathcal{Q}$. Define a subsequence $y_{j_1}, \ldots, y_{j_k}$ inductively: let j_i be the *least* index such that

$$y_{j_i} \notin \mathcal{S}_{i-1} = \langle y_{j_1}, \ldots, y_{j_{i-1}} \rangle,$$

and stop when the subsequence $y_{j_1}, \ldots, y_{j_k}$ generates $\mathcal{Q}$. To avoid excessive use of nested subscripts, put $x_i = y_{j_i}$.

We claim that $x_1, \ldots, x_k$ is a construction sequence. Conditions (i) and (ii) are immediate from the inductive definition, and for (iii) note that

$$(\mathcal{S}_i, \mathcal{P} \cap \mathcal{S}_i) \text{ reduces to } \mathcal{Q}(H_1, H_2, H_3, \ldots, H_{j_i}).$$

Next let $\mathcal{E}_i = \Phi'' H_{j_i}$ and let U be the submonoid of $T(\mathscr{A})$ generated by $(x_i, \mathcal{E}_i)$. We claim that U is valid. Conditions (i) and (iii) (in the definition of "valid") are immediate, since U is a submonoid of a valid transition table; and condition (ii) follows because the x_i's generate $\mathcal{Q}$. Finally, the choice of x_i's guarantees that $\mathcal{E}_i \subset \mathcal{S}_{i-1}$, so (iv) is a consequence of Lemma 8.3.

Finally, let $\mathcal{E}'_i = \mathcal{M}_{x_i} \cap \mathcal{S}_{i-1}$. Let U' be generated by $(x_i, \mathcal{E}'_i)$. To complete the proof, we show that U' is valid; then U' will satisfy all the requirements of a minimex algebra. Conditions (ii), (iii) and (iv) follow as before. It remains to prove (i). Now for each i, we know that $\mathcal{E}_i \subset \mathcal{S}_{i-1}$. Since U is valid, we have furthermore that $\mathcal{E}_i \subset \mathcal{M}_{x_i}$. Therefore $\mathcal{E}_i \subset \mathcal{E}'_i$. It follows that, whenever $(x, \mathcal{E}') \in U'$, then there is some $\mathcal{E} \subset \mathcal{E}'$ with $(x, \mathcal{E}) \in U$.

To conclude, fix any $(x, \mathcal{E}') \in U'$. If $x \in \mathcal{P}$, then $\mathcal{E} \cap \mathcal{P} = \varnothing$ because each $\mathcal{E}'_i \subset \mathcal{M}_{x_i}$. If $x \notin \mathcal{P}$, then choose $\mathcal{E} \subset \mathcal{E}'$ with $(x, \mathcal{E}) \in U$. Since U is valid, we know that $\mathcal{E} \cap \mathcal{P} \neq \varnothing$. Therefore $\mathcal{E}' \cap \mathcal{P} \neq \varnothing$. This proves (i), showing that U' is a minimex algebra. □

We now describe the algorithm for enumerating quotients of order n. Define a *construction scheme* to be a tuple $(\mathcal{Q}, \mathcal{P}, x_1, \ldots, x_k)$, such that $(\mathcal{Q}, \mathcal{P})$ is a bipartite monoid and $x_1, \ldots, x_k$ is a construction sequence for $\mathcal{Q}$. A *simple extension* of $(\mathcal{Q}, \mathcal{P}, x_1, \ldots, x_k)$ is a construction scheme $(\mathcal{Q}^+, \mathcal{P}^+, x_1, \ldots, x_{k+1})$ such that $\mathcal{Q} \subset \mathcal{Q}^+$ and $\mathcal{P}^+ \cap \mathcal{Q} = \mathcal{P}$.

It is worth emphasizing a subtle, but crucial, technicality in the definition of construction scheme. No restrictions are placed on the b.m. $(\mathcal{Q}, \mathcal{P})$. However, it is required that every *proper* initial segment $(\mathcal{S}_i, \mathcal{P} \cap \mathcal{S}_i)$ reduce to a genuine misère quotient. Therefore, simple extensions are meaningful only in the special case where $(\mathcal{Q}, \mathcal{P})$ is indeed a misère quotient.

By the above theorems, $(\mathcal{Q}, \mathcal{P})$ is a misère quotient if and only if there is a construction scheme $(\mathcal{Q}, \mathcal{P}, x_1, \ldots, x_k)$ such that the minimex algebra constructed by $x_1, \ldots, x_k$ is valid. To find all misère quotients of order n, we can therefore enumerate all construction schemes of order n and check which ones generate valid minimex algebras.

	$\mathcal{Q}$	$\mathcal{P}$	G
$\mathcal{S}_{12}$	$\langle a, b, c \mid a^2 = 1, b^4 = b^2, b^2c = b^3, c^2 = 1\rangle$	$\{a, b^2, ac\}$	$*2_{\#}1$
$\mathcal{S}'_{12}$	$\langle a, b, c \mid a^2 = 1, b^3 = b, c^2 = 1\rangle$	$\{a, b^2, c\}$	$*2_{\#}321$
$\mathcal{R}_{12}$	$\langle a, b, c, d \mid a^2 = 1, b^3 = b, b^2c = c, c^2 = b^2, bd = b, cd = c, d^2 = b^2\rangle$	$\{a, b^2\}$	$*2_{\#\#}54321$
	$\langle a, b, c \mid a^2 = 1, b^4 = b^2, b^2c = b^3, c^2 = b^2\rangle$	$\{a, b^2, c\}$	$*H2_{\#\#}$
	$\langle a, b, c, d \mid a^2 = 1, b^3 = b, bc = b, c^2 = b^2, bd = ab, d^2 = b^2\rangle$	$\{a, b^2, d\}$	$*H_{\#}$
	$\langle a, b, c, d \mid a^2 = 1, b^4 = b^2, b^2c = ab^3, c^2 = abc\rangle$	$\{a, b^2, c\}$	$*HK2_{\#\#}0$

Table 2. The six misère quotients of order 12. $H = *2_{\#\#}321$, $K = *2_{\#\#}2_{\#}$.

$$*\,2_{\#}0$$
$$*\,G2_{\#}32$$
$$*\,H_{\#G}320$$
$$*\,(G2_{\#})(G2_{\#2}2_{\#})$$
$$*\,(G2_{\#})(G2_{\#2}2_{\#}1)$$
$$*\,(G2_{\#})(G2_{\#3}2_{\#}1)$$
$$*\,(G2_{\#2}2_{\#}1)(G2_{\#3}2_{\#}1)$$
$$*\,2_{\#\#}4_2 54320$$
$$*\,K_2K_1KG2_{\#}321$$

Table 3. Nine games that generate nonisomorphic quotients of order 14. $G = *2_{\#}320$, $H = *2_{\#\#}321$, $K = *2_{\#\#}2_{\#}32$.

This method is made useful by a crucial optimization. Built into the definition of construction sequence is the assumption that *each proper initial segment reduces to a known misère quotient.* We can therefore use the following strategy. First, recursively compute *all* misère quotients of order $< n$. Now start with the trivial construction scheme $(\{1\}, \varnothing)$. Given a construction scheme $\Sigma = (\mathcal{Q}, \mathcal{P}, x_1, \ldots, x_k)$, consider every possible simple extension $\Sigma^+ = (\mathcal{Q}^+, \mathcal{P}^+, x_1, \ldots, x_{k+1})$ such that $|\mathcal{Q}^+| \leq n$. The key is that if $|\mathcal{Q}^+| < n$, then $(\mathcal{Q}^+, \mathcal{P}^+)$ *must* reduce to a known quotient. If it does not, then we can discard Σ^+ from further consideration.

We have therefore reduced the search space to small simple extensions of known quotients. Since a simple extension is just a monoid extension by a single generator, there are relatively few possibilities, and the algorithm is tractable. It is summarized as Algorithm 1 on the next page.

Recursively compute all quotients of size $< n$
$\mathscr{X} \leftarrow \varnothing$
Put the trivial construction scheme $(\{1\}, \varnothing)$ into $\mathscr{X}$
for all $\Sigma = (\mathcal{Q}, \mathcal{P}, x_1, \ldots, x_k)$ **in** $\mathscr{X}$ **do**
  $\mathscr{Y} \leftarrow$ the set of all simple extensions of Σ of order $\leq n$
  for all $(\mathcal{Q}^+, \mathcal{P}^+, x_1, \ldots, x_{k+1})$ **in** $\mathscr{Y}$ **do**
    if $|\mathcal{Q}^+| = n$ **then**
      $T \leftarrow$ the minimex algebra on $(\mathcal{Q}^+, \mathcal{P}^+)$ constructed by $x_1, \ldots, x_{k+1}$
      if $(\mathcal{Q}^+, \mathcal{P}^+)$ is reduced and T is valid **then**
        Output $(\mathcal{Q}^+, \mathcal{P}^+)$ ▷ It's a misère quotient
      end if
    else ▷ $|\mathcal{Q}^+| \leq n - 2$
      $(\mathcal{S}, \mathcal{R}) \leftarrow$ the reduction of $(\mathcal{Q}^+, \mathcal{P}^+)$
      if $(\mathcal{S}, \mathcal{R})$ is a misère quotient **then**
        Put $(\mathcal{Q}^+, \mathcal{P}^+, x_1, \ldots, x_{k+1})$ into $\mathscr{X}$
      end if
    end if
  end for
end for

Algorithm 1. Classification algorithm.

Acknowledgement

I wish to thank the referee for his unusually extensive and helpful comments.

References

[Austin 1976] R. B. Austin, *Impartial and partisan games*, Masters thesis, University of Calgary, 1976.

[Flammenkamp 2012] A. Flammenkamp, "Sprague–Grundy values of octal games", 2012, http://wwwhomes.uni-bielefeld.de/achim/octal.html.

[Guy and Smith 1956] R. K. Guy and C. A. B. Smith, "The G-values of various games", *Proc. Cambridge Philos. Soc.* **52** (1956), 514–526.

[Howse and Nowakowski 2004] S. Howse and R. J. Nowakowski, "Periodicity and arithmetic-periodicity in hexadecimal games: Algorithmic cominatorial game theory", *Theoret. Comput. Sci.* **313**:3 (2004), 463–472.

[ONAG] J. H. Conway, *On numbers and games*, 2nd ed., A K Peters, Natick, MA, 2001.

[Plambeck 2005] T. E. Plambeck, "Taming the wild in impartial combinatorial games", *Integers* **5**:1 (2005), G05.

[Plambeck 2009] T. E. Plambeck, "Advances in losing", pp. 57–89 in *Games of no chance 3* (Banff, AB, 2005), edited by M. H. Albert and R. J. Nowakowski, Math. Sci. Res. Inst. Publ. **56**, Cambridge University Press, 2009.

[Plambeck and Siegel 2007] T. E. Plambeck and A. N. Siegel, "Misère quotients for impartial games: Supplementary material", preprint, 2007. arXiv 0705.2404

[Plambeck and Siegel 2008] T. E. Plambeck and A. N. Siegel, "Misère quotients for impartial games", *J. Combin. Theory Ser. A* **115**:4 (2008), 593–622.

[Siegel 2015] A. N. Siegel, "The structure and classification of misère quotients", pp. 241–266 in *Games of no chance 4*, Math. Sci. Res. Inst. Publ. **63**, Cambridge, New York, 2015.

aaron.n.siegel@gmail.com *San Francisco, CA, United States*

Games of No Chance 4
MSRI Publications
Volume **63**, 2015

An algorithm for computing indistinguishability quotients in misére impartial combinatorial games

MIKE WEIMERSKIRCH

This paper advances the theory of impartial misère octal games by developing an algorithm for finding certain infinite quotient monoids. The notion of a misère quotient monoid was introduced by Thane Plambeck, who also, together with Aaron Siegel, gave an algorithm for finding finite misère quotients. This paper examines the periodicity of outcomes when changing the number of heaps of various sizes. The quotient monoid for misère 0.3122 up to heaps of size 7 is found. It is the first example of an infinite misère quotient monoid.

1. Introduction

This paper gives an algorithm for computing certain misère indistinguishability quotient monoids. The approach employed here is not the genus theory of [Berlekamp et al. 2003, Chapter 13], but rather the quotient monoid approach introduced by Thane Plambeck [2005].

The algorithm described here was initially designed to analyze octal games, but is also valid for a broader class of games which will be called "heap rulesets".

2. Heap rulesets

The notion of a heap ruleset comes from Nim, which is played with heaps of beans. The rules of Nim, and its variations, specify how a player may remove beans from a heap. The terminology below is influenced greatly by play of Nim, and readers may wish to keep games like Nim in mind when reading this paper. However, the collection of heap rulesets includes many other impartial games whose standard descriptions do not involve heaps. Chomp and Cram are examples.

We take a different perspective than is customary in describing our games. We will begin with the set $\mathcal{H} = \{H_1, H_2, \dots\}$ called the *heap alphabet*. The positions that can occur during the course of play are arbitrary finite multisets of $\mathcal{H}$. The

Keywords: none.

set of all such positions is called the *free heap monoid* (terminology developed in private conversations with Ezra Miller), and is denoted $\mathcal{A}$. Elements of $\mathcal{A}$ will be represented as vectors $(x_1, x_2, \dots)$ where x_i denotes the number of copies of H_i. Note that $\mathcal{A}$ is isomorphic to $\mathbb{N}^\infty = \bigoplus_{n \in \mathbb{Z}^+} \mathbb{N}^n$, which is a commutative monoid, with addition being the standard vector addition. $\mathcal{H}$ is a generating set for $\mathcal{A}$. If we take the set consisting of only the first n elements of $\mathcal{H}$, we have a *partial heap alphabet* which we will denote $\mathcal{H}_n = \{H_1, \dots, H_n\}$. The *size* of a heap is its index, that is H_i is the heap of size i. A *move* will change a heap of size i into a finite (perhaps empty) collection of heaps, all of which have size $< i$. The collection of moves allowed for a particular contest will be denoted Γ. A heap alphabet together with a set of legal moves Γ, will be called a *ruleset*. In essence, the ruleset specifies "how the game is played". The term *game* (misused several times above), is synonymous with position, that is, a game is a particular element of $\mathcal{A}$. Moves will be represented as vectors in $\mathbb{Z}^\infty = \bigoplus_{n \in \mathbb{Z}^+} \mathbb{Z}^n$, the condition stating that the rightmost non-zero coordinate of a move must be -1 and the prior coordinates are all non-negative.

3. Quotient maps and outcome functions

The goal is to be able to look at a game $G = (x_1, x_2, \dots) \in \mathcal{A}$ and determine whether there is a winning strategy for the **N**ext player to move, (an N-position) or not, in which case it is a game you would like to move to as the **P**revious player, (a P-position). We define the game outcome function $o^- : \mathcal{A} \to \{N, P\}$ accordingly.

The outcome of all games under a particular ruleset $(\mathcal{H}, \Gamma)$ can be determined recursively as follows:

Let G be a game and let $G + \Gamma$ be the set of options of G. If for all options $H \in G + \Gamma$, $o^-(H) = N$, then G is a P-position. If there exists any option $H \in G + \Gamma$ with $o^-(H) = P$, then G is an N-position. That is,

$$o^-(G) = \begin{cases} N & \text{if there exists an option } H \text{ of } G \text{ with } o^-(H) = P, \\ P & \text{if for all options } H \text{ of } G,\ o^-(H) = N \end{cases}$$

Note that we have a slight problem with this algorithm. Some elements of $G+\Gamma$ may have negative coordinates and therefore are not in $\mathcal{A}$. They correspond to moves which change a heap of size i into some smaller heaps, but no heap of size i was present in G. These moves do not exist in the actual course of play. We get around this difficulty by defining $o^-(\gamma) = N$ if any coordinate of γ is negative. The addition of extra moves to N-positions does not alter the above algorithm, which searches for the existence of a P-position amongst the options.

Remembering the outcome of every game in $\mathcal{A}$ is cumbersome, and so some simplifications are necessary.

Definition 1. Two games $G, H \in \mathcal{A}$ are *indistinguishable* if $o^-(G + X) = o^-(H + X)$ for all games $X \in \mathcal{A}$.

Indistinguishablility is a congruence relation on $\mathcal{A}$ and the congruence classes form the quotient monoid $\mathcal{Q}$. (When the free heap monoid $\mathcal{A}$ corresponds to a partial heap alphabet $\mathcal{H}_n$, it is customary to call the quotient a *partial quotient* and denote it $\mathcal{Q}_n$.) The induced map from $\mathcal{A}$ to $\mathcal{Q}$ is denoted Φ and is called the *quotient map*. The quotient map induces a quotient outcome function $\mathcal{O} : \mathcal{Q} \to \{N, P\}$ making the following diagram commute.

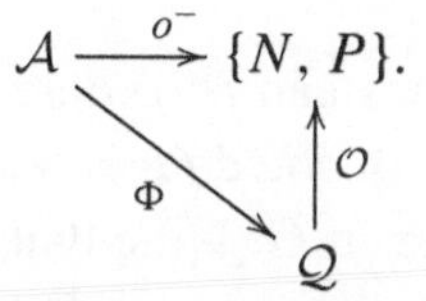

This represents a notational change from earlier work. What Plambeck and Siegel [2008] call the "P-portion of $\mathcal{Q}$" and denote by $\mathcal{P}$ is here the inverse image of P under the quotient outcome map and is denoted $\mathcal{O}^{-1}(P)$.

It is customary to use multiplicative notation in $\mathcal{Q}$, which can be achieved by first passing to the multiplicative monoid $\mathcal{M}$ with generators $\{a, b, c, \dots\}$ replacing $\{H_1, H_2, H_3, \dots\}$.

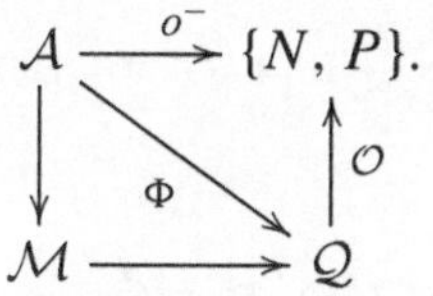

4. Computation of periodicity of outcomes and the corresponding quotients for finite heap alphabets

For a finite heap alphabet $\mathcal{H}_n$, when the quotient $\mathcal{Q}$ is finite, we are able to find relations of the form $\alpha^r = \alpha^{r+d}$ with $d > 0$ for each free heap monoid generator α according to the following algorithm.

4.1. ***Algorithm for computing periodicity.*** We wish to find for all $x_i \geq 0$, $1 \leq i \leq n-1$, the smallest values of the preperiod, r_n, and the period, d_n, so that for all $k \geq 0$, $o^-(x_1, x_2, \dots, x_{n-1}, r_n + k) = o^-(x_1, x_2, \dots, x_{n-1}, r_n + d_n + k)$. We achieve this in the following manner.

To find the periodicity for the n-th heap size:

(1) Begin with the preperiods $(r_1, \dots, r_{n-1})$ and periods $(d_1, \dots, d_{n-1})$ calculated for the previous heap sizes so that $o^-(r_1 + k_1, r_2 + k_2, \dots, r_{n-1} + k_{n-1}) =$

$o^-(r_1+d_1+k_1, r_2+d_2+k_2, \ldots, r_{n-1}+d_{n-1}+k_{n-1})$ for all $k_1, \ldots, k_{n-1} \geq 0$. Set $m = 1$.

(2) Calculate outcomes for games with $x_n = m$ up to the periodicity of the previous heap sizes, that is, for $x_i \leq r_i + d_i$. Verify for each heap size, that the previous periodicity still holds. If not, calculate additional outcomes in order to update the preperiods and periods.

(3) Check to see if the outcomes for $x_n = m$ agree with the outcomes for $x_n = l$, for some $l < m$. If so, then $r_n = l$ and $d_n = m - l$, if not, increase m by 1 and repeat.

4.2. *Periodicity theorem.* Games can be ordered using the *colexicographical order*, in which G precedes H, denoted $G < H$, if the rightmost coordinate in which G and H differ is smaller in G. Note that if $G < H$, then starting at the game G, no sequence of moves can arrive at H.

The following theorem is central to the algorithm. An example which shows the use of this theorem in the algorithm appears in the following subsection.

Theorem 2. *Fix a heap ruleset $(\mathcal{H}_n, \Gamma)$, under misère play, and fix values for $i, y_{i+1}, y_{i+2}, \ldots, y_n$. Suppose that the following assumptions are satisfied for some r_i, d_i:*

(i) *The outcomes for games of the form*

$$G = (x_1, x_2, \ldots, x_{i-1}, r_i, y_{i+1}, y_{i+2}, \ldots, y_n)$$

agree with the outcomes of

$$G^* = (x_1, x_2, \ldots, x_{i-1}, r_i + d_i, y_{i+1}, y_{i+2}, \ldots, y_n)$$

for all $x_1, x_2, \ldots, x_{i-1}$.

(ii) *The outcomes for games of the form*

$$K = (x_1, x_2, \ldots, x_{i-1}, r_i + u, x_{i+1}, x_{i+2}, \ldots, x_n)$$

agree with the outcomes of

$$K^* = (x_1, x_2, \ldots, x_{i-1}, r_i + d_i + u, x_{i+1}, x_{i+2}, \ldots, x_n)$$

for all $x_1, x_2, \ldots, x_{i-1}$, for $(x_{i+1}, x_{i+2}, \ldots, x_n)$ preceding $(y_{i+1}, y_{i+2}, \ldots, y_n)$ in the colexicographic order and all $u \geq 0$.

Then $o^-(G + vH_i) = o^-(G^* + vH_i)$ for all $v, x_1, x_2, \ldots, x_{i-1} \geq 0$.

Proof by induction. The statement is true for $v = 0$ by supposition. Assume the statement is true for v. All moves from the heaps $H_{i+1}, \ldots H_n$ correspond to options with the same outcomes for $G + (v+1)H_i$ and $G^* + (v+1)H_i$ by

(ii) and all moves from H_i correspond to options with the same outcomes for $G+(v+1)H_i$ and $G^*+(v+1)H_i$ by the induction hypothesis. These are the only moves available when $x_1=0, x_2=0, \dots x_{i-1}=0$. Since the options have the same outcomes, the games themselves have the same outcome. We then induct on x_1, leaving $x_2=0, x_3=0, \dots, x_{i-1}=0$. Since the only new options are to move from H_1, the outcomes of the new options will be the same by induction, and the games themselves again have the same outcome. Now we have the statement true for arbitrary x_1, $x_2=0$ and $x_3=0, x_4=0, \dots x_{i-1}=0$. We next induct on x_2. The new options make use of H_2 which are known to have the same value by a similar argument. Continue inducting on each heap size up to $i-1$ to complete the proof. □

4.3. *Example.* As a first example, we will compute several partial quotients for the quaternary ruleset 0.3122.

First partial quotient of 0.3122. This first quotient shows the fundamental nature of the periodicity of outcomes.

For the first partial heap alphabet $\mathcal{H}_1$, there is only one legal move, namely $\Gamma=\{(-1)\}$. The outcome of the game (0) is defined to be N in misère play. $o^-(1)=P$, since the only move is to the N-position (0) and $o^-(2)=N$, since there is a move to the P-position (1).

What about the sum of several copies of H_1? That is, what is the outcome of (x_1) for arbitrary x_1. Since the only move is (-1), the outcome of (x_1) will be P if $o^-(x_1-1)=N$ and vice versa. The outcomes are periodic with period two.

$$\begin{array}{rcccccccc} x_1 = & 0 & 1 & 2 & 3 & 4 & 5 & \dots \\ o^-(x_1) = & N & P & N & P & N & P & \dots \end{array}$$

The key fact is that the outcome of (3) is a function of $o^-(2)$, just as the outcome of (1) is a function of $o^-(0)$. In fact, $o^-(j+1)$ is a function of $o^-(j)$ for all $j \geq 0$. Moreover, the outcomes of $(j+k)$ for $j, k > 0$ are all determined by a single outcome, $o^-(j)$. Since the outcome of (2) is the same as the outcome of (0), then the outcome of $(2+k)$ must be the same as the outcome of $(0+k)$ for all $k \geq 0$.

We will record this periodicity for the partial heap alphabet $\mathcal{H}_n$ as follows: R_n and D_n are n-tuples where R_n records the preperiod in each component and D_n records the period. That is, when $R_n=(r_1, r_2, \dots, r_n)$ and $D_n=(d_1, d_2, \dots, d_n)$, if you have at least r_i heaps of size i, adding d_i more heaps of size i won't change the outcome.

Thus in the current example, $R_1=(0)$ and $D_1=(2)$

This periodicity gives us a candidate for the quotient monoid, namely $\mathcal{Q}^* = \langle a \mid a^2=1 \rangle$. The relation $a^2=a^0$ is clearly valid from the periodicity. The candidate quotient is therefore a refinement of the true quotient $\mathcal{Q}$ in the sense that every game associated to a particular element of $\mathcal{Q}^*$ belongs to a single indistinguishability congruence class in $\mathcal{Q}$. However, $\mathcal{Q}^*$ may have too many elements by having further partitioned the congruence classes in $\mathcal{Q}$. In general, we need to check that there do not exist distinct elements $p, q \in \mathcal{Q}^*$ where for all $x \in \mathcal{Q}^*$, $\mathcal{O}(px) = \mathcal{O}(qx)$. If we find two such elements, then we add the relation $p = q$ to create a new candidate quotient. When the full periodicity R_n, D_n has been found for a partial heap alphabet, the initial candidate quotient is finite, as will be the list of relations to check. Each successive quotient will have strictly smaller cardinality than the previous quotient, and therefore this verification will converge to the true quotient $\mathcal{Q}$ in a finite number of steps.

In this case, since the N-position (0) is mapped to 1 in $\mathcal{Q}^*$ and the P-position (1) is mapped to a, then $\mathcal{O}(1) = N$ and $\mathcal{O}(a) = P$. Thus 1 and a are distinguishable and the candidate $\mathcal{Q}^*$ is as small as possible, that is $\mathcal{Q}^* = \mathcal{Q}_1$.

Second partial quotient of 0.3122. This second step shows the recursive nature of the algorithm, and provides an example where the periodicity of outcomes occurs after a preperiod.

For the second partial quotient $\mathcal{Q}_2$, the legal moves are

$$\Gamma = \{(-1, 0),\ (0, -1),\ (1, -1)\}.$$

We already know the outcomes of games of the form $(x_1, 0)$ and notice that $o^-(x_1, 1) = N$ for all $x_1 \geq 0$ since $(x_1, 1)$ has options to $(x_1, 0)$ and $(x_1 + 1, 0)$, one of which is a P-position. Moving on to the next row, we compute the first three outcomes.

$$\begin{array}{rllllllll} x_1 = & 0 & 1 & 2 & 3 & 4 & 5 & \ldots \\ o^-(x_1, 0) = & N & P & N & P & N & P & \ldots \\ o^-(x_1, 1) = & N & N & N & N & N & N & \ldots \\ o^-(x_1, 2) = & P & N & P \end{array}$$

We use the theorem to claim that this row must also be periodic. Compare the options of $(3, 2)$ versus the options of $(1, 2)$. The moves from $(3, 2)$ which make use of H_2, that is, the moves $(1, -1)$ and $(0, -1)$ to the positions $(4, 1)$ and $(3, 1)$ respectively, are moves with the same outcomes as the options of $(1, 2)$ to $(2, 0)$ and $(1, 0)$, since the previous row was periodic. (Part (ii) of the theorem.) Also, $o^-(2, 2) = o^-(0, 2)$. Since the outcomes of the options of $(3, 2)$ agree with the outcomes of the options of $(1, 2)$, then $o^-(3, 2) = o^-(1, 2)$. We then proceed inductively using part (i) of the theorem. For all $k > 0$, $o^-(3 + k, 2) = o^-(1 + k, 2)$ since using the move $(1, -1)$ from each game has $o^-(4+k, 1) = o^-(2+k, 1)$ by the previous row's periodicity. Likewise, using the move $(0, -1)$ has $o^-(3 + k, 1) = o^-(1 + k, 1)$ by the previous row's periodicity.

Finally, using the move $(-1,0)$ has $o^-(3+k-1,1) = o^-(1+k-1,1)$ by induction.

We then repeat this process to calculate the outcome of further rows.

$$\begin{array}{rccccccc} x_1 = & 0 & 1 & 2 & 3 & 4 & 5 & \cdots \\ o^-(x_1,0) = & N & P & N & P & N & P & \cdots \\ o^-(x_1,1) = & N & N & N & N & N & N & \cdots \\ o^-(x_1,2) = & P & N & P & N & P & N & \cdots \\ o^-(x_1,3) = & N & N & N & N & N & N & \cdots \end{array}$$

Notice that for *all* $x_1 \geq 0$ we have $o^-(x_1,3) = o^-(x_1,1)$. We now claim to have established periodicity in the second dimension. The outcomes of $(x_1,4)$ for all $x_1 \geq 0$ are determined by the outcomes of the previous row $(x_2 = 3)$, just as the outcomes of $(x_1,2)$ for all $x_1 \geq 0$ were determined by the previous row $(x_2 = 1)$. Since $o^-(x_1,3) = o^-(x_1,1)$, then $o^-(x_1,4) = o^-(x_1,2)$. By induction, $o^-(x_1,3+k) = o^-(x_1,1+k)$ for all $x_1 \geq 0$ and $k \geq 0$.

Computing the quotient monoid $\mathcal{Q}_2$, the periodicity $R_2 = (0,1)$ $D_2 = (2,2)$ gives the candidate quotient $\mathcal{Q}^* = \langle a, b \mid a^2 = 1, b^3 = b\rangle$.

Again, we check to see if we can find $p, q \in \mathcal{Q}^*$ with $\mathcal{O}(px) = \mathcal{O}(qx)$ for all $x \in \mathcal{Q}^*$. In fact, all six elements are distinguishable (for instance, a and b^2 are distinguishable, since $\mathcal{O}[a(b^2)] = N$, but $\mathcal{O}[b^2(b^2)] = \mathcal{O}(b^4) = \mathcal{O}(b^2) = P$.) Therefore $\mathcal{Q}^*$ is the true quotient $\mathcal{Q}_2$

In the second row, we verified periodicity by comparing the options of $(3,2)$ versus the options of $(1,2)$, but the periodicity occurs from the beginning of the row. The difficulty in comparing $(2,2)$ and $(0,2)$ is that from $(0,2)$, the move $(-1,0)$ takes us off the board to $(-1,2) \notin \mathcal{A}$. We can fix this problem by extending the domain of outcome function o^- to include vectors in $\mathbb{Z}^n$ by declaring that $o^-(x_1, x_2, \ldots x_n) = N$ if any $x_i < 0$. Adding in extra illegal moves to N-positions will not alter the existence or non-existence of legal options to a P-position.

$$\begin{array}{rcccccccc} x_1 = & -1 & 0 & 1 & 2 & 3 & 4 & 5 & \cdots \\ o^-(x_1,-1) = & \mathbf{N} & \mathbf{N} & \mathbf{N} & \mathbf{N} & \mathbf{N} & \mathbf{N} & \mathbf{N} & \cdots \\ o^-(x_1,0) = & \mathbf{N} & N & P & N & P & N & P & \cdots \\ o^-(x_1,1) = & \mathbf{N} & N & N & N & N & N & N & \cdots \\ o^-(x_1,2) = & \mathbf{N} & P & N & P & N & P & N & \cdots \\ o^-(x_1,3) = & \mathbf{N} & N & N & N & N & N & N & \cdots \\ & & & & \vdots & & & & \end{array}$$

Third and fourth partial quotients of 0.3122. Calculation of the third partial quotient is similar to the second. The fourth partial quotient gives an example of step 2) of the algorithm updating the prior periodicity.

We can continue to calculate outcomes for $x_1 = 0, 1, 2$, and the reader unfamiliar with *reversibility* may wish to continue to include these outcomes. However, in order to decrease the amount of data we need to keep track of, we will employ the following simplification. It is a global condition on any ruleset $(\mathcal{H}, \Gamma)$, for any game whose only option is to the identity, (H_1 in this case), that two copies of that game are reversible to, and therefore indistinguishable from, the identity. (See [Berlekamp et al. 2003, Chapter 13].) Therefore we only need to keep track of columns 0, 1 and do not need to verify the periodicity for heaps of size 1. Column 2 will always agree with column 0.

For $\mathcal{H}_4$, the legal moves are $\Gamma = \{(-1, 0, 0, 0),\ (0, -1, 0, 0),\ (1, -1, 0, 0),\ (0, 1, -1, 0),\ (1, 0, 0, -1),\ (0, 0, 1, -1)\}$

We calculate the outcomes for $x_3 = 1$ up to the previous periodicity on x_2 which requires us to calculate rows 0, 1, 2, 3. The periodicity for heaps of size 2 checks, that is $o^-(x_1, 3, 1) = o^-(x_1, 1, 1)$, but the outcomes for $(x_1, x_2, 1)$ do not agree with the outcomes of $(x_1, x_2, 0)$, so we continue to $x_3 = 2$.

$\mathbf{x_3 = 0}$ $x_1 =$	0	1	$\mathbf{x_3 = 1}$ $x_1 =$	0	1	$\mathbf{x_3 = 2}$ $x_1 =$	0	1
$x_2 = 0$	N	P		P	N		P	N
$x_2 = 1$	N	N		N	N		N	N
$x_2 = 2$	P	N		P	N		P	N
$x_2 = 3$	N	N		N	N		N	N

Now the outcomes for $(x_1, x_2, 2)$ agree with those for $(x_1, x_2, 1)$ so $R_3 = (0, 1, 1)$; $D_3 = (2, 2, 1)$.

When $x_4 = 1$ and $x_3 = 0$, the outcomes for $x_2 = 3$ no longer agree with the outcomes of $x_2 = 1$, so we need to calculate additional outcomes. The new preperiod for x_2 must be greater than the old preperiod and the new period must be a multiple of the old period. In this case the new preperiod is 2 and the period remains 2.

$\mathbf{x_4 = 1,\ x_3 = 0}$ $x_1 =$	0	1
$x_2 = 0$	N	P
$x_2 = 1$	N	N
$x_2 = 2$	N	N
$x_2 = 3$	P	N
$x_2 = 4$	N	N

We continue calculating outcomes until the full periodicity $R_4 = (0, 2, 1, 0)$, $D_4 = (2, 2, 1, 2)$ is reached. (See figure on the next page.)

The candidate quotient arrived at from the periodicity is

$$\mathcal{Q}^* = \langle a, b, c, d \mid a^2 = 1, b^4 = b^2, c^2 = c, d^2 = 1 \rangle.$$

$\boldsymbol{x_4 = 0}$

	$\boldsymbol{x_3 = 0}$		$\boldsymbol{x_3 = 1}$		$\boldsymbol{x_3 = 2}$	
$x_1 =$	0	1	0	1	0	1
$x_2 = 0$	N	P	P	N	P	N
$x_2 = 1$	N	N	N	N	N	N
$x_2 = 2$	P	N	P	N	P	N
$x_2 = 3$	N	N	N	N	N	N
$x_2 = 4$	P	N	P	N	P	N

$\boldsymbol{x_4 = 1}$

	$\boldsymbol{x_3 = 0}$		$\boldsymbol{x_3 = 1}$		$\boldsymbol{x_3 = 2}$	
$x_1 =$	0	1	0	1	0	1
$x_2 = 0$	N	P	N	N	N	N
$x_2 = 1$	N	N	P	N	P	N
$x_2 = 2$	N	N	N	N	N	N
$x_2 = 3$	P	N	P	N	P	N
$x_2 = 4$	N	N	N	N	N	N

$\boldsymbol{x_4 = 2}$

	$\boldsymbol{x_3 = 0}$		$\boldsymbol{x_3 = 1}$		$\boldsymbol{x_3 = 2}$	
$x_1 =$	0	1	0	1	0	1
$x_2 = 0$	N	P	P	N	P	N
$x_2 = 1$	N	N	N	N	N	N
$x_2 = 2$	P	N	P	N	P	N
$x_2 = 3$	N	N	N	N	N	N
$x_2 = 4$	P	N	P	N	P	N

A check of outcomes reveals that $c = b^2$ and $b^2 d = b^3$, so that

$$\mathcal{Q}_4 = \langle a, b, d \mid a^2 = 1, b^4 = b^2, b^2 d = b^3, d^2 = 1 \rangle,$$

with outcome map $\Phi(H_1) = a$, $\Phi(H_2) = b$, $\Phi(H_3) = b^2$, $\Phi(H_4) = d$ and $\mathcal{O}^{-1}(P) = \{a, b^2, ad\}$.

5. Infinite quotients

If the quotient $\mathcal{Q}$ is infinite for a particular partial heap alphabet, the algorithm will fail to terminate. Similar techniques searching for periodicity in directions involving more than one heap size may be employed to discover these infinite monoids.

One example is the fifth partial quotient of the quaternary game 0.31011. The moves are given as follows:

$$\Gamma = \{(-1,0,0,0,0),\ (1,-1,0,0,0),\ (0,-1,0,0,0),\ (0,1,-1,0,0),\\ (0,0,1,-1,0),\ (0,0,0,-1,0),\ (0,0,0,1,-1),\ (0,0,0,0,-1)\}$$

For each value of x_5, the first three preperiods are $r_1 = 0$, $r_2 = 2$, $r_3 = 3$ and the first three periods are $d_1 = 2$, $d_2 = 2$, $d_3 = 1$. However, r_4 continues to increase with x_5. A "diagonal" periodicity is produced, so that for all games G with $x_4 + x_5 > 11$, $o^-(G) = o^-(G + 2H_4 + 2H_5)$. One verifies this by

(1) finding preperiods and periods in the first four dimensions for $x_5 = 0, 1, 2$;

(2) verifying that

(i) $o^-(x_1, x_2, x_3, 12, 0) = o^-(x_1, x_2, x_3, 14, 0) = o^-(x_1, x_2, x_3, 14, 2) = o^-(x_1, x_2, x_3, 16, 2)$,
(ii) $o^-(x_1, x_2, x_3, 13, 0) = o^-(x_1, x_2, x_3, 15, 2)$,
(iii) $o^-(x_1, x_2, x_3, 11, 1) = o^-(x_1, x_2, x_3, 13, 1)$;

(3) verifying for $a + b = 12$, $o^-(x_1, x_2, x_3, a, b) = o^-(x_1, x_2, x_3, a + 2, b + 2)$;

(4) verifying that

(i) $o^-(x_1, x_2, x_3, 0, 12) = o^-(x_1, x_2, x_3, 0, 14) = o^-(x_1, x_2, x_3, 2, 16)$,
(ii) $o^-(x_1, x_2, x_3, 1, 12) = o^-(x_1, x_2, x_3, 1, 14) = o^-(x_1, x_2, x_3, 1, 16)$,
(iii) $o^-(x_1, x_2, x_3, 1, 13) = o^-(x_1, x_2, x_3, 1, 15)$,
(iv) $o^-(x_1, x_2, x_3, 0, 13) = o^-(x_1, x_2, x_3, 2, 15)$

This yields the candidate quotient $\mathcal{Q}_5^* = \langle a, b, c, d, e \mid a^2 = 1;\ b^4 = b^2;\ c^4 = c^3;\ e^2d^{14} = d^{12};\ e^3d^{13} = ed^{11};\ e^4d^{12} = e^2d^{10};\ e^5d^{11} = e^3d^9;\ e^6d^{10} = e^4d^8;\ e^7d^9 = e^5d^7;\ e^8d^8 = e^6d^6;\ e^9d^7 = e^7d^5;\ e^{10}d^6 = e^8d^4;\ e^{11}d^5 = e^9d^3;\ e^{12}d^4 = e^{10}d^2;\ e^{13}d^3 = e^{11}d;\ e^{14}d^2 = e^{12}\rangle$.

Additional relations can be found, giving a new candidate quotient $\mathcal{Q}_5^{**} = \langle a, b, c, d, e \mid a^2 = 1;\ b^4 = b^2;\ cb^2 = c;\ c^4 = c^3;\ dc^3 = c^3a;\ d^2b^2 = d^2;\ ec^3 = c^3b;\ e^2b^2 = e^2;\ e^2d^6 = d^4;\ e^7d^5 = e^5d^3;\ e^8d^4 = e^6d^2;\ e^9d^3 = e^7d;\ e^{10}d^2 = d^8\rangle$.

Another example for which the verification has been performed is the seventh partial quotient of the quaternary game 0.3122: $\mathcal{Q}_7 = \langle a, b, c, d, e, f \mid a^2 = 1;\ b^5 = b^3;\ cb^3 = b^4;\ c^2b^2 = b^2;\ c^3b = cb;\ c^4 = c^2;\ d^2b^4 = d^2b^2;\ d^2cb^2 = d^2b^3;\ d^3b^2 = d^2b^2;\ eb^2 = b^4a;\ e^2 = b^4;\ fb^2 = d^2b^3a;\ fd^3cb = d^2c^2ba;\ fd^3c^2 = d^2c^3a;\ f^2ed^2c^2 = fedc^3a;\ f^3dcb = f^2c^2ba;\ f^3dc^2 = f^2c^3a;\ f^4d^6 = d^2;\ f^5ed^5 = fed;\ f^6d^4 = f^2\rangle$, with outcome map Φ having values

$$\Phi(H_1) = a, \quad \Phi(H_2) = b, \quad \Phi(H_3) = b^2d^2, \quad \Phi(H_4) = c,$$
$$\Phi(H_5) = d, \quad \Phi(H_6) = e, \quad \Phi(H_7) = f.$$

6. Further work

The two examples of infinite quotients were discovered on a case-by-case basis. Automation of the process to search for "diagonal" periodicities would greatly enhance our ability to find other infinite quotients. Included in finding such an algorithm is a determination of exactly which directions are possible directions for periodicity given a ruleset.

An algorithm which exhausts the list of possible relations in the infinite case is also needed.

References

[Berlekamp et al. 2003] E. R. Berlekamp, J. H. Conway, and R. K. Guy, *Winning ways for your mathematical plays, II*, 2nd ed., A K Peters, Natick, MA, 2003.

[Plambeck 2005] T. E. Plambeck, "Taming the wild in impartial combinatorial games", *Integers* **5**:1 (2005), G05.

[Plambeck and Siegel 2008] T. E. Plambeck and A. N. Siegel, "Misère quotients for impartial games", *J. Combin. Theory Ser. A* **115**:4 (2008), 593–622.

weim0024@math.umn.edu *School of Mathematics, University of Minnesota, 206 Church St. SE, Minneapolis, 55455, United States*

Games of No Chance 4
MSRI Publications
Volume **63**, 2015

Unsolved problems in combinatorial games

RICHARD J. NOWAKOWSKI

We have sorted the problems into sections:

- A. Taking and breaking
- B. Pushing and placing pieces
- C. Playing with pencil and paper
- D. Disturbing and destroying
- E. Theory of games

The numbers in parentheses are the old numbers used in each of the lists of unsolved problems given on pp. 183–189 of AMS *Proc. Sympos. Appl. Math.* **43** (1991), called PSAM **43** below; on pp. 475–491 of *Games of No Chance*, hereafter referred to as GONC; on pp. 457–473 of *More Games of No Chance* (MGONC); and on pp. 475–500 of *Games of No Chance 3* (GONC3). Some numbers have little more than the statement of the problem if there is nothing new to be added. References [year] may be found in Fraenkel's bibliography at the end of this volume. References [#] are at the end of this article. A useful reference for the rules and an introduction to many of the specific games mentioned below is M. Albert, R. J. Nowakowski and D. Wolfe, *Lessons in Play: An Introduction to the Combinatorial Theory of Games*, A. K. Peters, 2007 (LIP) or Berlekamp, Conway and Guy, *Winning Ways for your Mathematical Plays*, vol. 1–4, A. K. Peters, 2000–2004 (WW).

A. Taking and breaking games

A1. (1) Subtraction games with finite subtraction sets are known to have periodic nim-sequences. Investigate the relationship between the subtraction set and the length and structure of the period. The same question can be asked about **partizan** subtraction games, in which each player is assigned an individual subtraction set. See Fraenkel and Kotzig [1987].

(A move in the game $S(s_1, s_2, s_3, \ldots)$ is to take a number of beans from a heap, provided that number is a member of the **subtraction-set**, $\{s_1, s_2, s_3, \ldots\}$.

Keywords: none.

Analysis of such a game and of many other heap games is conveniently recorded by a **nim-sequence**,

$$n_0 n_1 n_2 n_3 \ldots,$$

meaning that the nim-value of a heap of h beans is n_h; i.e., that the value of a heap of h beans in this particular game is the **nimber** $*n_h$.)

For examples see Table 2 in Section **4** on p. 67 of "Impartial games" in GONC.

It would now seem feasible to give the complete analysis for games whose subtraction sets have just three members, though this has so far eluded us. Several people, including Mark Paulhus and Alex Fink, have given a complete analysis for all sets $\{1, b, c\}$ and for sets $\{a, b, c\}$ with $a < b < c < 32$.

In general, period lengths can be surprisingly long, and it has been suggested that they could be superpolynomial in terms of the size of the subtraction set. However, Guy conjectures that they are bounded by polynomials of degree at most $\binom{n}{2}$ in s_n, the largest member of a subtraction set of cardinality n. It would also be of interest to characterize the subtraction sets which yield a purely periodic nim-sequence, i.e., for which there is no preperiod. Carlos Santos [6] reduced this upper bound slightly by using a dynamical system approach.

Angela Siegel [20] considered infinite subtraction sets which are the complement of finite ones and showed that the nim-sequences are always arithmetic periodic. That is, the nim-values belong to a finite set of arithmetic progressions with the same common difference. The number of progressions is the period and their common difference is called the **saltus**. For instance, the game $S\{\hat{4}, \hat{9}, \widehat{26}, \widehat{30}\}$ (in which a player may take any number of beans except 4, 9, 26 or 30) has a preperiod of length 243, period-length 13014 and saltus 4702. Marla Clusky and Danny Sleator have proved her conjecture that the class of games $S\{\hat{a}, \hat{b}, \widehat{a+b}\}$ is purely periodic with period length $3(a+b)$.

For infinite subtraction games in general there are corresponding questions about the length and purity of the period. Suppose the elements of the finite subtraction sets are not constants. It was shown by Fraenkel [2011] that the game with subtraction set

$$\{1, 2, \ldots, t-1, \lfloor n/t \rfloor\},$$

where $t \geq 2$ is a fixed integer, n the pile size, has an aperiodic Sprague–Grundy sequence. See also Fraenkel [2012] and Guo [2012].

A2. **(2)** Are all finite **octal games** ultimately periodic? (If the binary expansion of the k-th code digit in the game with code $\mathbf{d}_0{\cdot}\mathbf{d}_1\mathbf{d}_2\mathbf{d}_3\ldots$ is

$$\mathbf{d}_k = 2^{a_k} + 2^{b_k} + 2^{c_k} + \cdots,$$

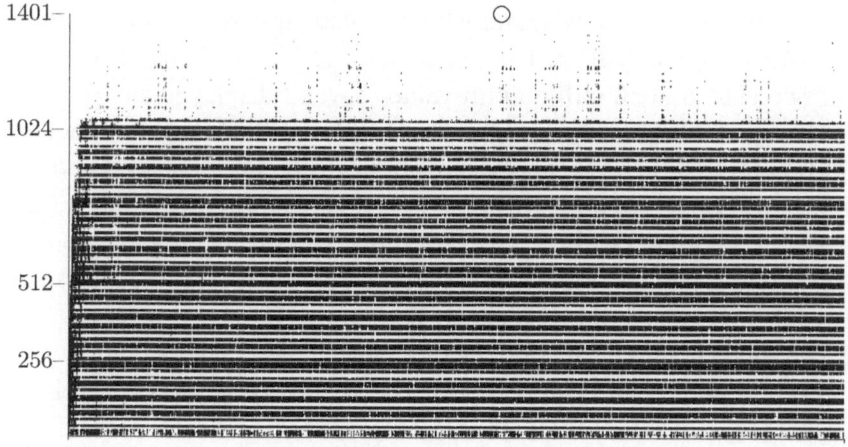

Figure 1. Plot of 11000000 nim-values of the octal game ·**007**.

where $0 \le a_k < b_k < c_k < \dots$, then it is legal to remove k beans from a heap, provided that the rest of the heap is left in exactly a_k or b_k or c_k or ... nonempty heaps. See WW, pp. 81–115. Some specimen games are exhibited in Table 3 of Section **5** of "Impartial games" in GONC.)

Resolve any number of outstanding particular cases, e.g., ·**6** (**Officers**), ·**04**, ·**06**, ·**14**, ·**36**, ·**37**, ·**64**, ·**74**, ·**76**, ·**004**, ·**005**, ·**006**, ·**007**, ·**014**, ·**015**, ·**016**, ·**024**, ·**026**, ·**034**, ·**064**, ·**114**, ·**125**, ·**126**, ·**135**, ·**136**, ·**142**, ·**143**, ·**146**, ·**162**, ·**163**, ·**164**, ·**166**, ·**167**, ·**172**, ·**174**, ·**204**, ·**205**, ·**206**, ·**207**, ·**224**, ·**244**, ·**245**, ·**264**, ·**324**, ·**334**, ·**336**, ·**342**, ·**344**, ·**346**, ·**362**, ·**364**, ·**366**, ·**371**, ·**374**, ·**404**, ·**414**, ·**416**, ·**444**, ·**564**, ·**604**, ·**606**, ·**744**, ·**764**, ·**774**, ·**776** and **Grundy's Game** (split a heap into two unequal heaps; WW, pp. 96–97, 111–112; LIP, p. 142), which has been analyzed, first by Dan Hoey, and later by Achim Flammenkamp, as far as heaps of 2^{35} beans.

J.P. Grossman (preprint) using a new approach based on the sparse space phenomenon, has analyzed ·**.6** up to heaps of size 2^{47} and has found no periodicity.

Perhaps the most notorious and deserving of attention is the game ·**007**, one-dimensional Tic-Tac-Toe, or **Treblecross**, which Flammenkamp has pushed to 2^{25}. Figure 1 shows the first 11 million nim-values, a small proportion of which are ≥ 1024; the largest, $\mathcal{G}(6193903) = 1401$ is shown circled. Will 2048 ever be reached?

Flammenkamp has settled ·**106**: it has the remarkable period and preperiod lengths of 328226140474 and 465384263797. For information on the current status of each of these games, we refer the read to Flammenkamp's web page at uni-bielefeld.de/ achim/octal.html.

A game similar to Grundy's, and which is also unsolved, is John Conway's **Couples-Are-Forever** (LIP, p. 142) where a move is to split any heap except a heap of two. The first 50 million nim-values haven't displayed any periodicity. See Caines et al. [1999]. More generally, Bill Pulleyblank suggests looking at splitting games in which you may only split heaps of size $> h$, so that $h = 1$ is She-Loves-Me-She-Loves-Me-Not and $h = 2$ is Couples-Are-Forever. David Singmaster suggested a similar generalization: you may split a heap provided that the resulting two heaps each contain at least k beans: $k = 1$ is the same as $h = 1$, while $k = 2$ is the third cousin of **Dawson's Chess**.

Explain the structure of the periods of games known to be periodic.

In *Discrete Math.*, **44** (1983), pp. 331–334, Problem 38, Fraenkel raised questions concerning the computational complexity (see Section E1 below) of octal games. In Problem 39, he and Kotzig define **partizan octal games** in which distinct octals are assigned to the two players. Mesdal [2009] show that in many cases, if the game is "all-small" (WW, pp. 229–262; LIP, pp. 183–207), then the atomic weights are arithmetic periodic. (See Section E13 for an explanation of the new name "dicot" games.) In Problem 40, Fraenkel introduces **poset games**, played on a partially ordered set of heaps, each player in turn selecting a heap and then removing a nonnegative number of beans from this heap and from each heap above it in the ordering, at least one heap being reduced in size. For posets of height one, new regularities in the nim-sequence can occur; see Horrocks and Nowakowski [2003].

Note that this includes, as particular cases, Subset Takeaway, Chomp or Divisors, and Green Hackenbush forests. Compare Problems A3, D1 and D2 below.

A3. (3) Hexadecimal games have code digits $\mathbf{d}_k$ in the interval from $\mathbf{0}$ to $\mathbf{f}$ ($= \mathbf{15}$), so that there are options splitting a heap into three heaps. See WW, pp. 116–117.

Such games may be arithmetically periodic. Nowakowski has calculated the first 100000 nim-values for each of the 1-, 2- and 3-digit games. Richard Austin's Theorem 6.8 in his thesis [1976] and the generalization by Howse and Nowakowski [2004] suffice to confirm the arithmetic periodicity of several of these games.

Some interesting specimens are ·**28** = ·**29**, which have period 53 and saltus 16, the only exceptional value being $\mathcal{G}(0) = 0$; ·**9c**, which has period 36, preperiod 28 and saltus 16; and ·**f6** with period 43 and saltus 32, but its apparent preperiod of 604 and failure to satisfy one of the conditions of the theorem prevent us from verifying the ultimate periodicity. The game ·**205200c** is arithmetic periodic with preperiod length of 4, period length of 40, saltus 16 except that $40k + 19$ has nim-value 6 and $40k+39$ has nim-value 14. This regularity, (which also seems to be exhibited by ·**660060008** with a period length of approximately 300,000), was first reported in Horrocks and Nowakowski [2003] (see Problem A2). Grossman and Nowakowski [this volume] have shown that the nim-sequences for ·**200...0048**, with an odd number of zero code digits, exhibit "ruler function" patterns. The game ·**9** has not so far yielded its complete analysis, but, as far as analyzed (to heaps of size 100000), exhibits a remarkable fractal-like set of nim-values. See Howse and Nowakowski [2004]. Also of special interest are ·**e**; ·**7f** (which has a strong tendency to period 8, saltus 4, but, for $n \le 100,000$, has 14 exceptional values, the largest being $\mathcal{G}(94156) = 26614$); ·**b6** (which "looks octal"); ·**b33b** (where a heap of size n has nim-value n except for 27 heap sizes which appear to be random).

Other unsolved hexadecimal games are

·**1x**,	$\mathbf{x} \in \{8, 9, c, d, e, f\}$,	·**2x**,	$a \le \mathbf{x} \le f$,
·**3x**,	$8 \le \mathbf{x} \le e$,	·**4x**,	$\mathbf{x} \in \{9, b, d, f\}$,
·**5x**,	$8 \le \mathbf{x} \le f$,	·**6x**,	$8 \le \mathbf{x} \le f$,
·**7x**,	$8 \le \mathbf{x} \le f$,	·**9x**,	$1 \le \mathbf{x} \le a$,
·**9d**,		·**bx**,	$\mathbf{x} \in \{6, 9, d\}$,
·**dx**,	$1 \le \mathbf{x} \le f$,	·**fx**,	$\mathbf{x} \in \{4, 6, 7\}$.

A4. (53) *N*-heap Wythoff game. Given $N \ge 2$ heaps of finitely many tokens, whose sizes are $p_1, \ldots, p_N$ with $p_1 \le \cdots \le p_N$. Players take turns removing any positive number of tokens from a *single* heap or removing $(a_1, \ldots, a_N)$ from *all* the heaps — a_i from the i-th heap — subject to the conditions: (i) $0 \le a_i \le p_i$ for each i, (ii) $\sum_{i=1}^{N} a_i > 0$, (iii) $a_1 \oplus \cdots \oplus a_N = 0$, where $\oplus$ is nim addition. The player making the last move wins and the opponent loses. Note that the classical Wythoff game is the case $N = 2$.

For $N \ge 3$, Fraenkel makes the following conjectures.

Conjecture 1. For every fixed set $K := (A^1, \ldots, A^{N-2})$ there exists an integer $m = m(K)$ (i.e, m depends only on K), such that

$$(A^1, \ldots, A^{N-2}, A_n^{N-1}, A_n^N), \quad A^{N-2} \le A_n^{N-1} \le A_n^N,$$

with $A_n^{N-1} < A_{n+1}^{N-1}$ for all $n \ge 1$, is the n-th $\mathcal{P}$-position, and

$$A_n^{N-1} = \operatorname{mex}\left(\{A_i^{N-1}, A_i^N : 0 \le i < n\} \cup T\right), \quad A_n^N = A_n^{N-1} + n,$$

for all $n \ge m$, where $T = T(K)$ is a (small) set of integers.

That is, if you fix $N-2$ of the heaps, the $\mathcal{P}$-positions resemble those for the classical Wythoff game. For example, for $N = 3$ and $A^1 = 1$, we have $T = \{2, 17, 22\}$, $m = 23$.

Conjecture 2. For every fixed K there exist integers $a = a(K)$ and $M = M(K)$ such that

$$A_n^{N-1} = \lfloor n\phi \rfloor + \varepsilon_n + a \quad \text{and} \quad A_n^N = A_n^{N-1} + n,$$

for all $n \geq M$, where $\phi = (1+\sqrt{5})/2$ is the golden section, and $\varepsilon_n \in \{-1, 0, 1\}$.

In Fraenkel and Krieger [2004] the following was shown, inter alia: let $t \in \mathbb{Z}_{\geq 1}$, $\alpha = (2 - t + \sqrt{t^2+4})/2$ ($\alpha = \phi$ for $t = 1$), $T \subset \mathbb{Z}_{\geq 0}$ a finite set, $A_n = (\text{mex}\,\{A_i, B_i : 0 \leq i < n\} \cup T)$, where $B_n = A_n + nt$. Let $s_n := \lfloor n\alpha \rfloor - A_n$. Then there exist $a \in \mathbb{Z}$ and $m \in \mathbb{Z}_{\geq 1}$, such that for all $n \geq m$, either $s_n = a$, or $s_n = a + \varepsilon_n$, $\varepsilon_n \in \{-1, 0, 1\}$. If $\varepsilon_n \neq 0$, then $\varepsilon_{n-1} = \varepsilon_{n+1} = 0$. Also the general structure of the ε_n was characterized succinctly.

This result was then applied to the N-heap Wythoff game. In particular, for $N = 3$ (such that $K = A^1$) it was proved that $A_n^2 = \text{mex}\left(\{A_i^2, A_i^3 : 0 \leq i < n\} \cup T\right)$, where

$$T = \{x \geq K : \text{there exists } 0 \leq k < K \text{ such that } (k, K, x) \text{ is a } P\text{-position}\} \cup \{0, \ldots, K-1\}.$$

The following upper bound for A_n^3 was established: $A_n^3 \leq (K+3)A_n^2 + 2K + 2$. It was also proved that Conjecture 1 implies Conjecture 2.

In Sun and Zeilberger [2004], a sufficient condition for the conjectures to hold was given. It was then proved that the conjectures are true for the case $N = 3$, where the first heap has up to 10 tokens. For those 10 cases, the parameter values m, M, a, T were listed in a table.

Sun [2005] obtained results similar to those in Fraenkel and Krieger [2004], but the proofs are different. It was also proved that Conjecture 1 implies Conjecture 2. A method was given to compute a in terms of certain indexes of the A_i and B_j.

A5. (23) Burning-the-Candle-at-Both-Ends. Conway and Fraenkel ask us to analyze Nim played with a row of heaps. A move may only be made in the leftmost or in the rightmost heap. When a heap becomes empty, then its neighbor becomes the end heap.

Albert and Nowakowski [2001] have determined the outcome classes in impartial and partizan versions (called **End-Nim**, LIP, pp. 210, 263) with finite heaps, and Duffy, Kolpin and Wolfe [2009] extend the partizan case to infinite ordinal heaps. Wolfe asks for the actual values.

Nowakowski suggested to analyze impartial and partizan **End-Wythoff**: take from either end-pile, or the *same* number from both ends. The impartial game is solved by Fraenkel and Reisner [2009]. Fraenkel [1982] asks a similar question about a generalized Wythoff game: take from either end-pile or take $k > 0$ from one end-pile and $\ell > 0$ from the other, subject to $|k - \ell| < a$, where a is a fixed integer parameter ($a = 1$ is **End-Wythoff**).

There is also **Hub-and-Spoke Nim**, proposed by Fraenkel. One heap is the hub and the others are arranged in rows forming spokes radiating from the hub. Albert notes that this game can be generalized to playing on a forest, i.e., a graph each of whose components is a tree. The most natural variant is that beans may only be taken from a leaf (valence 1) or isolated vertex (valence 0).

The partizan game of **Red-Blue Cherries** is played on an arbitrary graph. A player picks an appropriately colored cherry from a vertex of minimum degree, which disappears at the same time. Albert, Grossman, McCurdy, Nowakowski and Wolfe [1] show that if the graph has a leaf, then the value is an integer. They ask: Is every **Red-Blue Cherries** position an integer? See also [10].

A6. (17) Extend the analysis of **Kotzig's Nim** (WW, pp. 515–517). Is the game eventually periodic in terms of the length of the circle for every finite move set? Analyze the misère version of **Kotzig's Nim**.

Let $\Gamma(S; n)$ be the outcome of the game S lists all the moves and n is the size of the circle. Recently, Ward and Xin [2010] gave more evidence of the periodicity by showing: If $n \in \{1, 3, 5, 7, 15\}$, or if $n \equiv 3 \pmod 5$ and $n \geq 23$, then $\Gamma(1, 4; n) \in \mathcal{P}$; otherwise, $\Gamma(1, 4; n) \in \mathcal{N}$. They also give several conjectures including: If n is odd and $n \notin \{9, 11, 17\}$ then $\Gamma(3, 5; n) \in \mathcal{P}$ and is in $\mathcal{N}$ otherwise. Their evidence bolsters the periodicity conjecture but also indicates that there is likely to be a lot of noise when n is small.

A7. (18) Obtain asymptotic estimates for the proportions of $\mathcal{N}$-, $\mathcal{O}$- and $\mathcal{P}$-positions in Epstein's **Put-or-Take-a-Square** game (WW, pp. 518–520).

A8. Gale's nim. This is Nim played with four heaps, but the game ends when three of the heaps have vanished, so that there is a single heap left. Brouwer and Guy have independently given a partial analysis, but the situation where the four heaps have distinct sizes greater than 2 is open. An obvious generalization is to play with h heaps and play finishes when k of them have vanished.

A9. Euclid's Nim is played with a pair of positive integers, a move being to diminish the larger by any multiple of the smaller. The winner is the player who reduces a number to zero. Analyses have been given by Cole and Davie [1969], Spitznagel [1973], Lengyel [2003], Collins [2005], Fraenkel [2005] and Nivasch [2006]. Lengyel [2003] reports that Schwartz first found that (a, b) is the sequential sum of nim-heaps given by the normal continued fraction of a/b, the last number in the continued fraction depends on the presence or absence of fibonacci numbers. Gurvich [2007] shows that the nim-value, $g^+(a, b)$ for the pair (a, b) in normal play is the same as the misère nim-value, $g^-(a, b)$

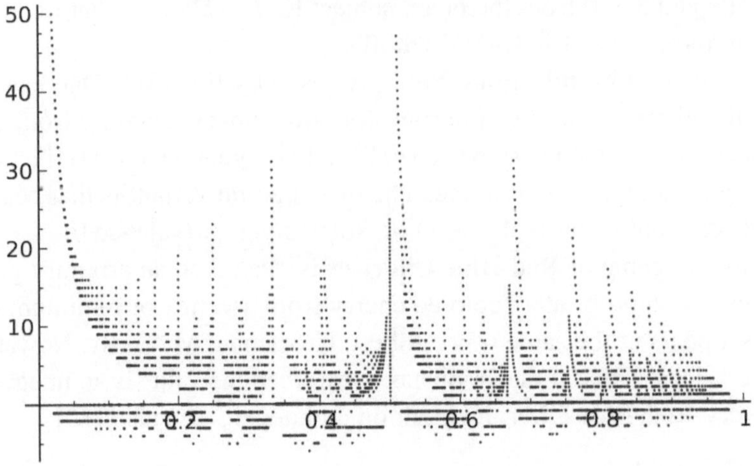

Figure 2. Partizan Euclid: mean atomic weights for $0 < q < p < 101$.

except for $(a, b) = (kF_i, kF_{i+1})$ where $k > 0$ and F_i is the i-th Fibonacci number. In this case, $g^+(kF_i, kF_{i+1}) = 0$ and $g^-(kF_i, kF_{i+1}) = 1$ if i is even and the values are reversed if i is odd. Aaron Siegel notes that this can be restated as: **Euclid** is tame, and the fickle positions have the form (kF_i, kF_{i+1}). The latter have genus 0^1 when i is even and 1^0 when i is odd.

Collins and Lengyel [392] play **Euclid's Nim** with three integers and solve some special cases.

In **Partizan Euclid**, from the position (p, q), where $p = kq + r$, $0 \leq r < q$, Left moves to (q, r) and Right to $(q, (k+1)q - p)$. The outcome classes are investigated in [12]. Determining the values seems hard. The atomic weights may be easier — see Figure 2, where the *mean atomic-weights* for $0 < q < p \leq 100$ are given, where the coordinates (x, y) are $x = q/p$ and y is mean atomic-weight of (p, q).

A10. (20) D.U.D.E.N.E.Y (WW, pp. 521–523) is **Nim**, but with an upper bound, Y, on the number of beans that may be taken, and with the restriction that a player may not repeat his opponent's last move. If Y is even, the analysis is easy. Some advance in the analysis, when Y is odd, has been made by Marc Wallace, Alex Fink and Kevin Saff.

We can, for example, extend the table of strings of pearls given in WW, p. 523, with the values of Y in Table 1, which have the pure periods shown, where D=$Y+2$, E=$Y+1$. The first entry corrects an error of $128r+31$ in WW.

It seems likely that the string for $Y = 2^{2k+1} + 2^{2k} - 1$ has the simple period E for all values of k. But there is some evidence to suggest that an analysis will

$256r+31$	DEE	$512r+153$	DEE	$1024r+415$	DEE
$512r+97$	DDEDDDE	$512r+159$	DEE	$512r+425$	DE
$1024r+103$	DE	$512r+225$	DDE	$512r+487$	DEE
$128r+119$	DEE	$512r+255$	E	$1024r+521$	DDDE
$1024r+127$	DEEE	$512r+257$	DDDDE	$1024r+607$	DDE
$512r+151$	DDDEE	$512r+297$	DDEDEDE	$1024r+735$	DEEE

Table 1. Strings of "pearls" in D.U.D.E.N.E.Y for values of Y of various forms.

never be complete. Indeed, consider the following table showing the fraction, among 2^k cases, that remain undetermined:

$k=$	3	5	6	7	8	9	10	11	12	13	14	15	16	17
fraction	$\frac{1}{2}$	$\frac{5}{16}$	$\frac{9}{32}$	$\frac{11}{64}$	$\frac{21}{128}$	$\frac{33}{256}$	$\frac{60}{512}$	$\frac{97}{1024}$	$\frac{177}{2048}$	$\frac{304}{4096}$	$\frac{556}{8192}$	$\frac{974}{16384}$	$\frac{1576}{32768}$	$\frac{2763}{65536}$

Moreover, the periods of the pearl-strings appear to become arbitrarily long.

A11. (21) Schuhstrings is the same as **D.U.D.E.N.E.Y**, except that a deduction of zero is also allowed, but cannot be immediately repeated (WW, pp. 523–524). In *Winning Ways* it was stated that it was not known whether there is any Schuhstring game in which three or more strings terminate simultaneously. Kevin Saff has found three such strings (when the maximum deduction is $Y = 3430$, the three strings of multiples of 2793, 3059, 3381 terminate simultaneously) and he conjectures that there can be arbitrarily many such simultaneous terminations.

A12. (22) Analyze **Dude**, i.e., unbounded Dudeney, or Nim in which you are not allowed to repeat your opponent's last move.

Let $[h_1, h_2, \ldots, h_k; m]$, $h_i \le h_{i+1}$, be the game with heaps of size h_1 through h_k, where m is the move just made and $m = 0$ denotes a starting position. Then [5], the $\mathcal{P}$-positions are

$$[(2s+1)2^{2j}; (2s+1)2^{2j}] \quad \text{for } k = 1,$$
$$[(2s+1)2^{2j}, (2s+1)2^{2j}; 1] \quad \text{for } k = 2,$$

and for $k \ge 3$ the heap sizes are arbitrary, the only condition being that the previous move was 1. The nim-values do not seem to show an easily described pattern.

A13. Nim with pass. David Gale suggested an analysis of Nim played with the option of a single pass by either of the players, which may be made at any time up to the penultimate move. It may not be made at the end of the game. Once a player has passed, the game is as in ordinary Nim. The game ends when all heaps have vanished. Morrison, Friedman and Landsberg [2011] have looked at this game with their renormalization techniques.

A14. Games with a Muller twist. In such games, each player specifies a condition on the set of options available to her opponent on his next move.

In **Odd-or-Even Nim**, for example, each player specifies the parity of the opponent's next move. This game was analyzed by Smith and Stănică [2002], who propose several other such games which are still open (see also Gavel and Strimling [2004]).

The game of **Blocking Nim** proceeds in exactly the same way as ordinary Nim with N heaps, except that before a given player takes his turn, his opponent is allowed to announce a **block**, $(a_1, \ldots a_N)$; i.e., for each pile of counters, he has the option of specifying a positive number of counters which may not be removed from that pile. Flammenkamp, Holshouser and Reiter [2003, 2004] give the $\mathcal{P}$-positions for three-heap Blocking Nim with an incomplete block containing only one number, and ask for an analysis of this game with a block on just two heaps, or on all three. There are corresponding questions for games with more than three heaps. Larsson [2011] shows with two piles of counters where at most $k-1$ moves, for some fixed k, may be blocked off at each stage. Then the $\mathcal{P}$-positions are of the form $\{x, y\}$, where either $|y-x| < k$ and $y - x \equiv k-1 \pmod 2$ or $x + y < k$.

Let S be a set of positive integers. The **complementary subtraction** game $\hat{S}$ is played on a heap where the last act of a move is say whether the next subtraction is to be a number from S or from its complement. Horrocks and Trenton [2008] introduced this variant. They analyzed heaps up to size 8000 the case where S is the Fibonacci numbers without finding periodicity although there appears to be regularity. They also ask about the case $S = \{x \mid x \equiv a \text{ or } b \pmod c\}$. They report that the nim-values appear chaotic. The set of $(a_n, a_n + n)$ where $a_1 = 1$ and $a_n = \text{mex}\{a_i : i < n\}$ form the $\mathcal{P}$-positions of Wythoff's game. We ask what happens in this game if $S = \{a_n : n + 1, 2, 3, \ldots\}$?

A15. (13) Misère caternary and octal games. Misère analysis has been revolutionized by Thane Plambeck and Aaron Siegel [2008] with their concept of the **misère quotient** of a game, though the number of unsolved problems continues to increase. See Section E13 for more theoretical questions and an explanation of some of the concepts mentioned here.

Plambeck and Siegel ask the specific questions:

(1) The misère quotient of **·07** (**Dawson's Kayles**) has order 638 at heap size 33. Is it infinite at heap size 34? Even if the misère quotient is infinite at heap 34 then, by Rédei's theorem [8, p. 142; 17], it must be isomorphic to a finitely presented commutative monoid. Call this monoid D_{34}. Exhibit a monoid presentation of D_{34}, and having done that, exhibit D_{35}, D_{36}, etc, and explain what is going on in general. Given a set of games $\mathcal{A}$, describe an algorithm

to determine whether the misère quotient of $\mathscr{A}$ is infinite. Much harder: if the quotient is infinite, give an algorithm to compute a presentation for it.

(2) Give complete misère analyses for any of the (normal-play periodic) octal games that show "algebraic-periodicity" in misère play. Some examples are **·54**, **·261**, **·355**, **·357**, **·516** and **·724**. Give a precise definition of algebraic periodicity and describe an algorithm for detecting and generalizing it. This is a huge question: if such an algorithm exists, it would likely instantly give solutions to at least a half-dozen unsolved 2- and 3-digit octals.

Plambeck also offers prizes of US $500.00 for a complete analysis of **Dawson's Chess, ·137** (alias **Dawson's Kayles, ·07**); US $200.00 for the "wild quaternary game", **·3102**; and US $25.00 each for **·3122**, **·3123** and **·3312**.

There is a website miseregames.org which contains thousands of misère quotients for octal games.

Siegel notes that Dawson first proposed his problem in 1935, making it perhaps the oldest open problem in combinatorial game theory. [Michael Albert offers the alternative "Is chess a first player win?"] It may be of historical interest to note that Dawson showed the problem to one of the present authors around 1947. Fortunately, he forgot that Dawson proposed it as a losing game, was able to analyze the normal play version, rediscover the Sprague–Grundy theory, and get Conway interested in games.

A16. A16 is now **E12**.

A17. Nem and mnem. Mem is played with heaps of tokens. Remove any number of tokens from any one heap. The number of tokens removed must be at least as large as the number that were removed on the previous move from that heap. Equivalently, a "heap" is a pair of integers (n, k), and a move is to any pair $(n-i, i)$, where $k \le i \le n$.

Mnem is exactly the same as **Mem**, with additional options: Either player may add tokens to a heap instead of removing them. If adding tokens, the number added must be strictly less than the number of tokens added or removed on the previous move. If removing tokens, the number must be at least as large as the number added or removed on the previous move. So a move from (n, k) is to $(n-i, i)$ for $k \le i \le n$, as in **Mem**, or to $(n+i, i)$ for $1 \le i < k$.

Conway and (Aaron) Siegel have investigated this game. They conjecture: *Every position in* **Mnem** *has finite nim value*. They verified this experimentally up to $n = 10,000$. They also conjecture: *For both* **Mem** *and* **Mnem**, *if* $k^2 \ge n$, *then* $(n, k) = \lfloor n/k \rfloor$. Siegel reports that they have no idea how to prove either so these may be very difficult problems.

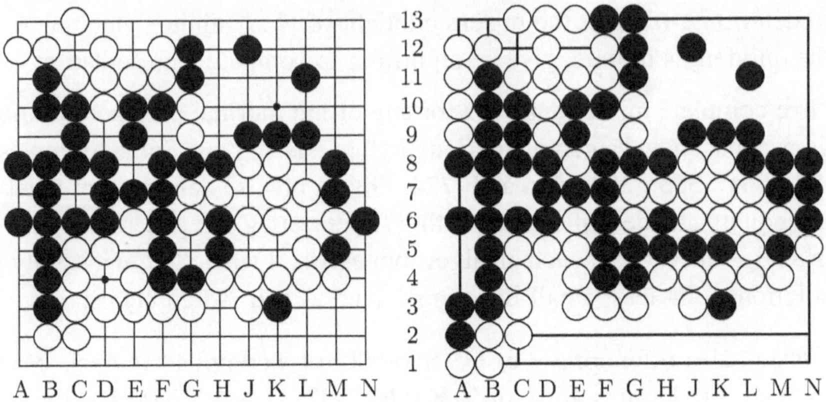

Figure 3. Jiang v. Rui, MSRI, July, 2000.

B. Pushing and placing pieces

B1. (5) Go is of particular interest, partly because of the loopiness induced by the "ko" rule, and many problems involve general theory; see E4 and E5.

Elwyn writes:

> I attach one region that has been studied intermittently over the past several years. The region occurs in the southeast corner of the board (Figure 3). At move 85 Black takes the ko at L6. What then is the temperature at N4? This position is copied from the game Jiang and Rui played at MSRI in July 2000. In 2001, Bill Spight and I worked out a purported solution by hand, assuming either Black komaster or White komaster. I've recently been trying to get that rather complicated solution confirmed by GoExplorer, which would then presumably also be able to calculate the dogmatic solution. I've been actively pursuing this off and on for the past couple weeks, and haven't gotten there yet.

Elwyn also writes:

> Nakamura (GONC3) has shown how capturing races in Go can be analyzed by treating liberties as combinatorial games. Like atomic weights, when the values are integers, each player's best move reduces his opponent's resources by one. The similarities between atomic weights and Nakamura's liberties are striking.

Theoretical problem: Either find a common formulation which includes much or all of atomic weight theory and Nakamura's theory of liberties, OR find some significant differences.

Important practical applied problem: Extend Nakamura's theory to include other complications which often arise in **Go**, such as simple kos, either internal and/or external.

B2. Woodpush (see LIP, pp. 214, 275) is a game that involves ko but that is simpler than **Go**. **Woodpush** is played on a finite strip of squares. Each square is empty or occupied by a black or white piece. A piece of the current player's color retreats: Left retreats to the left and Right to the right — to the next empty square, or off the board if there is no empty square; except, if there is a contiguous string containing an opponent's piece then it can move in the opposite direction *pushing* the string ahead of it. Pieces can be moved off the end of the strip. Immediate repetition of a global board positions is not allowed. A "ko" threat must be played first. For example:

$$LRR\square \xrightarrow{\text{Left}} \square LRR \xrightarrow{\text{Right}} LR\square R \xrightarrow{\text{Left}} \text{ko-threat} \xrightarrow{\text{Right}} R\square\square R$$

Note that Right's first move to $LRR\square$ is illegal because it repeats the immediately prior board position and Left's second move to $\square LRR$ is also illegal so he must play a ko-threat. Also note that in $\square LRR\square$, Right never has to play a ko-threat since he can always push with either of his two pieces — with Left moving first,

$$\begin{array}{ccccc} \square LRR\square & \xrightarrow{\text{Left}} & \square\square LRR & \xrightarrow{\text{Right}} & \square LR\square R \\ & \rightarrow & \text{ko-threat} & \rightarrow & \text{Right answers ko-threat} \\ & \rightarrow & \square\square LRR & \rightarrow & \square LRR\square \end{array}$$

Berlekamp, Plambeck, Ottaway, Aaron Siegel and Spight (work in progress) use top-down thermography to analyze the three piece positions. What about more pieces? Cazenave and Nowakowski (this volume) show that the position $L.L.R.R$ is ± 4 but that $L.L...R.R$ and $L.L.....R.R$ are draws by superko (repetition of the board position after more than 2 moves).

B3. (40) Chess. Noam Elkies [2002] has examined Dawson's chess, but played under usual Chess rules, so that capture is not obligatory.

He would still welcome progress with his conjecture that the value $*k$ occurs for all k in (ordinary chess) pawn endings on sufficiently large chessboards.

Thea van Roode has suggested **Impartial Chess**, in which the players may make moves of either color. Checks need not be responded to and Kings may be captured. The winner could be the first to promote a pawn.

B4. (30) Chess: King and Rook. Low and Stamp [2006] have given a strategy in which White wins the King and Rook vs. King problem within an 11×9 region. Kanungo and Low [2007] show that with the initial position WK(1, 1),

WR(x, y) and BK(a, b), where $1 < x < a$, White has a winning strategy on the $(a+b+3) \times (a+b+5)$ board.

B5. Nonattacking queens. Noon and Van Brummelen [2006] alternately place queens on an $n \times n$ chessboard so that no queen attacks another. The winner is the last queen placer. They give nim-values for boards of sizes $1 \leq n \leq 10$ as 1121312310 and ask for the values of larger boards.

B6. (55) Amazons. Müller [14] has shown that the 5×5 game is a first player win and asks about the 6×6 game.

B7. Phutball, more properly, Conway's **Philosopher's Football**, is usually played on a Go board with positions (i, j), $-9 \leq i, j \leq 9$ and the ball starting at (0,0). For the rules, see WW, pp. 752–755. The game is loopy (see Section E5 below), and Nowakowski, Ottaway and Siegel (see [19]) discovered positions that contained tame cycles, i.e., cycles with only two strings, one each of Left and Right moves. Aaron Siegel asks if there are positions in such combinatorial games which are stoppers but contain a **wild cycle**, i.e., one which contains more than one alternation between Left and Right moves. Demaine, Demaine and Eppstein [2002] show that it is NP-complete to decide if a player can win on the next move. Loosen (see **directional phutball**) raises the question of whether there is any $\mathcal{P}$-position.

Phlag Phutball is a variant played on an $n \times n$ board with the initial position of the ball at $(0, 0)$ except that now only the ball may occupy the positions $(2i, 2j)$ with both coordinates even. This eliminates "tackling", and is an extension of one-dimensional **Oddish Phutball**, analyzed in Grossman and Nowakowski [2002]. The $(3, 2n+1)$ board (i.e., (i, j), $i = 0, 1, 2$ and $-n \leq j \leq n$) is already interesting and requires a different strategy from that appropriate to **Oddish Phutball**.

Loosen [2008] introduces **Directional Phutball**, a nonloopy version, which is also played on a grid. The ball starts in the bottom left corner with Left's goal-line is the right edge, Right's is the top edge. Players can only jump toward their opponent's goal-line, and win by jumping on over that goal-line. Men can only be placed ahead of the ball (i.e., in the positive quadrant with origin on the ball). She notes that there is no $\mathcal{P}$-position in this game and that making an off-parity move ("poultry") can be good. She asks for a complete analysis of the $2 \times n$ board. She shows that the atomic weights of the $m \times n$ board for $2 \leq m, n, \leq 4$ is 0 and is 0, 1, 1, 2 for $2 \times n$ for $n = 5, 6, 7, 8$ respectively.

B8. Hex. (LIP, pp. 264–265) Nash's strategy stealing argument shows that Hex is a first player win but few quantitative results are known. Arneson et al. report that 8×8 is solved as are most openings on the 9×9.

Garikai Campbell [2004] asks:

(1) For each n, what is the shortest path on an $n \times n$ board with which the first player can guarantee a win?

(2) What is the least number of moves that guarantees the first player a win? Campbell showed that this is at least n on an $n \times n$ board. Peng et al. [2010] show that it is 7 on the 5×5 board.

B9. (54) Fox and geese. Berlekamp and Siegel [19, Chapter 2] and WW, pp. 669–710, "analysed the game fairly completely, relying in part on results obtained using CGSuite." On p. 710 of WW the following open problems are given.

(1) Define a position's **span** as the maximum occupied row-rank minus its minimum occupied row-rank. Then quantify and prove an assertion such as the following: If the backfield is sufficiently large, and the span is sufficiently large, and if the separation is sufficiently small, and if the Fox is neither already trapped in a daggered position along the side of the board, nor immediately about to be so trapped, then the Fox can escape and the value is **off**.

(2) Show that any formation of three Geese near the center of a very tall board has a "critical rank" with the following property: If the northern Goose is far above, and the Fox is far below, then the value of the position is either positive, HOT, or **off**, according as the northern Goose is closer, equidistant, or further from the critical rank than the Fox.

(3) Welton asks what happens if the Fox is empowered to retreat like a Bishop, going back several squares at a time in a straight line? More generally, suppose his straight-line retreating moves are confined to some specific set of sizes. Does $\{1, 3\}$, which maintains parity, give him more or less advantage than $\{1, 2\}$?

(4) What happens if the number of Geese and board widths are changed?

In Aaron Siegel's thesis there are several other questions:

(5) In the critical position, with Geese at (we use the algebraic Chess notation of a, b, c, d, ... for the files and $1, 2, 3, \ldots, n$ for the ranks) (b, n), (d, n), $(\mathrm{e}, n-1)$, $(\mathrm{h}, n-1)$, and Fox at $(\mathrm{c}, n-1)$, which has value $1 + 2^{-(n-8)}$ on an $n \times 8$ board with $n \geq 8$ in the usual game, is the value $-2n + 11$ for all $n \geq 6$ when played with "Ceylonese rules"? (Fox allowed two moves at each turn.)

(6) On an $n \times 4$ board with $n \geq 5$ and Geese at (b, n) and $(\mathrm{c}, n-1)$ do all Fox positions have value **over**? With the Geese on (b, n) and (d, n) are only other values 0 at $(\mathrm{c}, n-1)$ and $\{\textbf{over} \mid 0\}$ at $(\mathrm{b}, n-2)$ and $(\mathrm{d}, n-2)$?

(7) On an $n \times 6$ board with $n \geq 8$ and Geese at (b, n), (d, n) and $(\mathrm{e}, n-1)$ do the positions $(\mathrm{a}, n-2k+1)$, $(\mathrm{c}, n-2k+1)$, $(\mathrm{e}, n-2k+1)$, all have value 0,

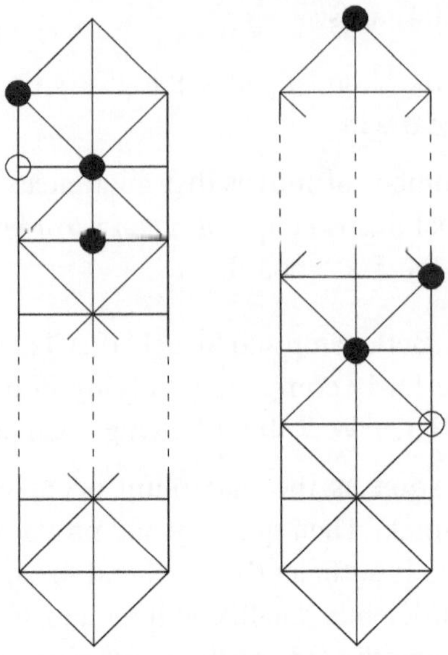

Figure 4. Sequences of hare and hounds positions.

and those at (b, $n-2k$), (d, $n-2k$), (f, $n-2k$) all have value Star? And if the Geese are at (b, n), (d, n) and (f, n) are the zeroes and Stars interchanged?

B10. Hare and Hounds. Aaron Siegel asks if the sequences of positions of increasing board length shown in Section B10 are, on the left, increasingly hot, and, on the right, have arbitrarily large negative atomic weight. He also conjectures that the starting position on a $6n+5\times 3$ board, for $n>0$, has value

$$-(n-1)+\left\{b, c\,|0\|0\,\Big|\,0\,\Big\|\,0\ldots\,\Bigg\|\,0\right\}$$

where there are $2n+4$ zeroes and slashes and $b=\left\{0, a\,\|\,0, \{0\mid \mathbf{off}\}\right\}$,

$$c=\left\{0\,\Big\|\,\downarrow_{[2]}*|0\|0\right\}\quad\text{and}\quad a=\{0, \downarrow_{[2]}*\mid 0, \downarrow_{[2]}*\}.$$

B11. (4) Domineering. (WW, pp. 119–122, 138–142; LIP pp. 1–7, 260). Extend the analysis.

(Left and Right take turns to place dominoes on a checker-board. Left orients her dominoes North–South and Right orients his East–West. Each domino exactly covers two squares of the board and no two dominoes overlap. A player unable to play loses.)

See Berlekamp [1988] and the second edition of WW, pp. 138–142, where some new values are given. For example David Wolfe and Dan Calistrate have

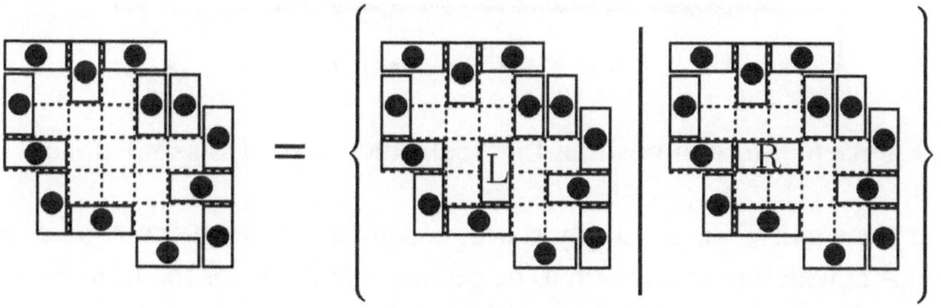

Figure 5. A **Domineering** position of value $\pm 2*$.

found the values (to within '-ish', i.e., infinitesimally shifted) of 4×8, 5×6 and 6×6 boards. The value for a 5×7 board is

$$\left\{ \tfrac{3}{2} \,\middle|\, \{\tfrac{5}{4} | -\tfrac{1}{2}\}, \left\{ \tfrac{3}{2} \,\middle|\, -\tfrac{1}{2}, \{\tfrac{3}{2} | -1\} \,\middle\|\, -1 | -3 \right\} \,\middle\|\, -1, \left\{ \tfrac{3}{2} \,\middle|\, -\tfrac{1}{2} \,\middle\|\, -1 \right\} \,\middle|\, -3 \right\}$$

Lachmann, Moore and Rapaport [2002] discover who wins on rectangular, toroidal and cylindrical boards of widths 2, 3, 5 and 7, but do not find their values. Bullock [3, p. 84] showed that 19×4, 21×4, 14×6 and 10×8 are wins for Left and that 10×10 is a first player win.

Berlekamp notes that the value of a $2 \times n$ board, for n even, is only known to within "ish", and that there are problems on $3 \times n$ and $4 \times n$ boards that are still open.

Berlekamp asks, as a hard problem, to characterize all hot **Domineering** positions to within "ish". As a possibly easier problem he asks for a **Domineering** position with a new temperature, i.e., one not occurring in Table 1 on GONC, p. 477. Gabriel Drummond-Cole [2002] found values with temperatures between 1.5 and 2. Figure 5 shows a position of value $\pm 2*$ and temperature 2.

Shankar and Sridharan [2005] have found many **Domineering** positions with temperatures other than those shown in Table 1 on p. 477 of GONC. Blanco and Fraenkel [2] have obtained partial results for the game of **Tromineering**, played with trominoes in place of (or, alternatively, in addition to) dominoes.

B12. NoGo can be found under the name "Anti-Atari Go" at the Sensei Library (see senseis.xmp.net/?AntiAtariGo) and was invented independently by Neil McKay. On a Go-board (or on any graph) pieces are placed as in Go, the only restriction is that every connected group, of each player, must be adjacent to at least one empty intersection. Kyle Burke has shown that **NoGo** is NP-hard on a graph. McKay, Nowakowski and (Angela) Siegel found positions of value 1, $\frac{1}{2}$, $\frac{1}{4}$, $\frac{1}{8}$ on a Go board. Using CGSuite, one can find the nimbers $*$, $*2$, $*4$, $*8$ on a one-dimensional board. The position $\bullet \cdot \circ \cdot \bullet \cdot \circ \ldots$, where $\bullet$ is a black piece, $\circ$ is

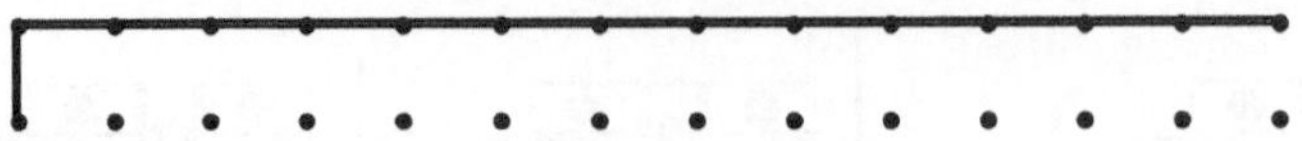

Figure 6. Starting position for "Icelandic" $1 \times n$ Dots-and-Boxes.

a white piece and '·' is an empty space, is equivalent to the octal game **.6** (see Section A2) which is not known to be periodic. We ask for the nim-dimension of both the one- and two-dimensional board. Is there a limit to the denominator of the fractions found on the two-dimensional board?

C. Playing with pencil and paper

C1. (51) Dots-and-Boxes. Elwyn Berlekamp asks for a complete theory of the "Icelandic" $1 \times n$, i.e., with starting position as in Figure 6.

See Berlekamp's book [2000] for more problems about this popular children's (and adults') game (see also WW, pp. 541–584; LIP, pp. 21–28, 260).

C2. (25) Sprouts. Extend the analysis of this Conway–Paterson game in either the normal or misère form. (WW, pp. 564–568).

(A move joins two spots, or a spot to itself by a curve which doesn't meet any other spot or previously drawn curve. When a curve is drawn, a new spot must be placed on it. The valence of any spot must not exceed three.)

C3. (26) Sylver coinage. (WW, pp. 575–597): Extend the analysis.

(Players alternately name distinct positive integers, but may not name a number which is a sum, with repetitions allowed, of previously named integers. Whoever names 1 loses.) Sicherman [2002] contains the most recent information.

C4. (28) Extend Úlehla's or Berlekamp's analysis of **von Neumann's Game** from directed forests to directed acyclic graphs (WW, pp. 570–572; Úlehla [1980]).

(Von Neumann's game, or Hackendot, is played on one or more rooted trees. The roots induce a direction, towards the root, on each edge. A move is to delete a node, together with all nodes on the path to the root, and all edges incident with those nodes. Any remaining subtrees are rooted by the nodes that were adjacent to deleted nodes.)

C5. (43) Inverting Hackenbush. Thea van Roode has written a thesis [24] investigating both this and **Reversing Hackenbush**, but there is plenty of room for further analysis of both games.

In **Inverting Hackenbush**, when a player deletes an edge from a component, the remainder of the component is replanted with the new root being the pruning point of the deleted edge. In **Reversing Hackenbush**, the colors of the edges are all changed after each deletion. Both games are hot, in contrast to **Blue-Red Hackenbush** (WW, pp. 1–7; LIP, pp. 82, 88, 111–112, 212, 266) which is cold, and **Green Hackenbush** (WW, pp. 189–196), which is tepid.

C6. (42) Beanstalk and **Beans-Don't-Talk** are games invented respectively by John Isbell and John Conway. See Guy [1986]. **Beanstalk** is played between Jack and the Giant. The Giant chooses a positive integer, n_0. Then Jack and the Giant play alternately $n_1, n_2, n_3, \ldots$, according to the rule

$$\begin{aligned} n_{i+1} &= n_i/2 \text{ if } n_i \text{ is even,} \\ &= 3n_i \pm 1 \text{if } n_i \text{ is odd;} \end{aligned}$$

that is, if n_i is even, there's only one option, while if n_i is odd there are just two. The winner is the person moving to 1.

We still don't know if there are any $\mathcal{O}$**-positions** (positions of infinite remoteness).

C7. (63) The Erdős–Szekeres game [7] (and see [18]) was introduced by Harary, Sagan and West [1985]. From a deck of cards labeled from 1 through n, Alexander and Bridget alternately choose a card and append it to a sequence of cards. The game ends when there is an ascending subsequence of a cards or a descending subsequence of d cards.

The game appears to have a strong bias towards the first player. Albert et al., (this volume) show that for $d = 2$ and $a \leq n$ the outcome is $\mathcal{N}$ or $\mathcal{P}$ according as n is odd or even, and is $\mathcal{O}$ (drawn) if $n < a$. They conjecture that for $a \geq d \geq 3$ and all sufficiently large n, it is $\mathcal{N}$ with both normal and misère play, and also with normal play when played with the rationals in place of the first n integers.

They also suggest investigating the form of the game in which players take turns naming pairs (i, π_i) subject to the constraint that the chosen values form part of the graph of some permutation of $\{1, 2, \ldots, n\}$.

D. Disturbing and destroying

D1. (27) Extend the analysis of **Chomp** (WW, pp. 598–599; LIP pp. 19, 46, 216).

David Gale offered \$300.00 for the solution of the infinite 3D version where the board is the set of all triples (x, y, z) of nonnegative integers, that is, the lattice points in the positive octant of $\mathbb{R}^3$. The problem is to decide whether it is a win for the first or second player.

Chomp (Gale [1974]) is equivalent to **Divisors** (Schuh [1952]). **Chomp** is easily solved for $2 \times n$ arrays, Sun [2002], and indeed a recent result by Steven

Byrnes [2003] shows that any poset game eventually displays periodic behavior if it has two rows, and a fixed finite number of other elements. See also the Fraenkel poset games mentioned near the end of Section A2.

Thus, most of the work in recent years has been on three-rowed **Chomp**. The situation becomes quite complicated when a third row is added, see Zeilberger [2001] and Brouwer et al. [2005]. A novel approach (renormalization) is taken by Friedman and Landsberg (GONC3). They demonstrate that three-rowed **Chomp** exhibits certain scaling and self-similarity patterns similar to chaotic systems. Is there a deterministic proof that there is a unique winning move from a $3 \times n$ rectangle? The renormalization approach is based on nonlinear dynamics techniques from physics; its results are highly suggestive but as of yet not fully mathematically rigorous.

Transfinite Chomp has been investigated by Huddleston and Shurman [2002]. An open question is to calculate the nim value of the position $\omega \times 4$—they conjecture this to be $\omega \cdot 2$, but it could be as low as 46, or even uncomputable! Perhaps the most fascinating open question in **Transfinite Chomp** is their *Stratification Conjecture*, which states that if the number of elements taken in a move is $< \omega^i$, then the change in the nim-value is also $< \omega^i$.

D2. (33) Subset Take-away. Given a finite set, players alternately choose proper subsets subject to the rule that once a subset has been chosen no proper subset can be removed. Last player to move wins.

Many people play the dual, that is, a nonempty subset must be chosen and no proper superset of this can be chosen. We discuss this version of the game which now can be considered a poset game with the sets ordered by inclusion.

The $(n; k)$ Subset Take-away game is played using all subsets of sizes 1 through k of a n-element set. In the $(n; n)$ game one has the whole set (i.e., the set of size n) as an option, so a strategy-stealing argument shows this must be a first player win.

(1) Gale and Neyman [1982], in their original paper on the game, conjectured that the winning move in the $(n; n)$ game is to remove just the whole set. This is equivalent to the statement that the $(n; n-1)$ game is a second-player win, which has been verified only for $n \leq 5$.

(2) A stronger conjecture states that $(n; k)$ is a second player win if and only if $k+1$ divides n. This was proved in the original paper only for $k = 1$ or 2.

See also Fraenkel and Scheinerman [1991].

D3. (39) Sowing or **Mancala** games. There appears to have been no advance on the papers mentioned in MGONC, although we feel that this should be a fruitful field of investigation at several different levels.

D4. Annihilation Games. **k-annihilation**. Initially place tokens on some of the vertices of a finite digraph. Denote by $\rho_{\text{out}}(u)$ the outvalence of a vertex u. A move consists of removing a token from some vertex u, and "complementing" $t := \min(k, \rho_{\text{out}}(u))$ (immediate) followers of u, say $v_1, \ldots, v_t$: if there is a token on v_j, remove it; if there is no token there, put one on it. The player making the last move wins. If there is no last move, the outcome is a draw. For $k = 1$, there is an $O(n^6)$ algorithm for deciding whether any given position is in $\mathcal{P}$, $\mathcal{N}$, or $\mathcal{D}$; and for computing an optimal next move in the last 2 cases (Fraenkel and Yesha [1982]). Fraenkel asks: Is there a polynomial algorithm for $k > 1$? For an application of k-annihilation games to lexicodes; see Fraenkel and Rahat [2003].

D5. Toppling Dominoes (LIP, pp. 110–112, 274) is played with a row of vertical dominoes each of which is either blue or red. A player topples one of his/her dominoes to the left or to the right.

See Fink et al [this volume] for the proofs that every number occurs exactly once and is a palindrome; there are exactly n positions with value $*n$. Several conjectures are listed but the most intriguing seems to be: *if G is a palindrome then G's value appears uniquely.*

There are several variants of **Toppling Dominoes**. If all the dominoes must be toppled in the same direction then this is a **Hackenbush** string. **Timber** [15] is an impartial version of **Toppling Dominoes** played on a directed graph. The dominoes are on the edges. A player chooses a domino which is toppled in the direction of the edge. The dominoes on incident edges are then toppled (regardless of the underlying edge) and the process is iterated. If only outcomes are required then only trees are interesting. In [15], an algorithm is given to determine the outcome class of a tree, which surprisingly requires nim-values. Values, however, appear to be difficult to determine. The normal and misère versions on a path are related to Dyck paths and Catalan numbers; see Section E15 for more details. The partizan version, imaginatively called **Partizan Timber**, has Left and Right dominoes placed on the edges of a directed graph. Any **Toppling Dominoes** position can be transformed into a **Partizan Timber** position by subdividing each edge and directing the edge toward the new vertex. A similar algorithm to that of **Timber** can be used to determine the outcome class of a tree. The uniqueness of numbers doesn't hold even on a path. In the other direction, are there any values that do not occur in the game?

D6. Hanoi Stick-Up is played with the disks of the Towers of Hanoi puzzle, starting with each disk in a separate stack. A move is to place one stack on top of another such that the size of the bottom of the first stack is less than the size of the top of the second; the two stacks then fuse (&) into one. The only relevant information about a stack are its top and bottom sizes, and it's often possible to collapse the labeling of positions: so, for instance, starting with 8 disks and fusing 1&7 and 2&5,

we have stacks	0	1&7	2&5	3	4	6
which can be relabeled	0	1&3	1&2	1	2	3

in which the legal moves are still the same. John Conway, Alex Fink and others have found that the $\mathcal{P}$-positions of height ≤ 3 in normal Hanoi Stickup are exactly those which, after collapsing, are of the form $0^a\ 01^b\ 1^c\ 12^d\ 2^e$ with $\min(a+b+c, c+d+e, a+e)$ even, except that if $a+e \leq a+b+c$ and $a+e \leq c+d+e$ then both a and e must be even (02 can't be involved in a legal move so can be dropped).

They also found the normal and misère outcomes of all positions with up to six stacks, but there is more to be discovered.

D7. (56) Are there any draws in **Beggar-My-Neighbor**? Marc Paulhus showed that there are no cycles when using a half-deck of two suits, but the problem for the whole deck (one of Conway's "antiHilbert" problems) is still open.

E. Theory of games

E1. (49) Fraenkel updates Berlekamp's questions on computational complexity as follows:

Demaine, Demaine and Eppstein [2002] proved that deciding whether a player can win in a *single* move in **Phutball** (WW, pp. 752–755; LIP, p. 212) is NP-complete. Grossman and Nowakowski [2002] gave constructive partial strategies for one-dimensional **Phutball**. Thus, these papers do not show that **Phutball** has the required properties.

Perhaps **Nimania** (Fraenkel and Nešetřil [1985]) and **Multivision** (Fraenkel [1998]) satisfy the requirements. **Nimania** begins with a single positive integer, but after a while there is a multiset of positive integers on the table. At move k, a copy of an existing integer m is selected, and 1 is subtracted from it. If $m = 1$, the copy is deleted. Otherwise, k copies of $m-1$ are adjoined to the copy $m-1$. The player first unable to move loses and the opponent wins. It was proved: (i) The game terminates. (ii) Player I can win. In Fraenkel, Loebl and Nešetřil [1988], it was shown that the max number of moves in **Nimania** is an Ackermann function, and the min number satisfies $2^{2^{n-2}} \leq \mathrm{Min}(n) \leq 2^{2^{n-1}}$.

The game is thus intractable simply because of the length of its play. This is a *provable* intractability, much stronger than NP-hardness, which is normally only a *conditional* intractability. One of the requirements for the tractability of a game is that a winner can consummate a win in at most $O(c^n)$ moves, where

$c > 1$ is a constant, and n a sufficiently succinct encoding of the input (this much is needed for nim on 2 equal heaps of size n).

To consummate a win in **Nimania**, player I can play randomly most of the time, but near the end of play, a winning strategy is needed, given explicitly. Whether or not this is an "intricate" solution, depends on the beholder. But it seems that it's of even greater interest to construct a game with a very *simple* strategy which still has high complexity!

Also every play of **Multivision** terminates, the winner can be determined in linear time, and the winning moves can be computed linearly. But the length of play can be arbitrarily long. So let's ask the following: Is there a game which has

1. simple, playable rules,
2. a simple explicit strategy,
3. length of play at most exponential, and
4. is NP-hard or harder.

Tung [1987] proved the following:

Theorem. *Given a polynomial $P(x, y) \in \mathbb{Z}[x, y]$, the problem of deciding whether for all x there exists $y[P(x, y) = 0]$ holds over $\mathbb{Z}_{\geq 0}$, is co-NP-complete.*

Define the following game of length 2: player I picks $x \in \mathbb{Z}_{\geq 0}$, player II picks $y \in \mathbb{Z}_{\geq 0}$. Player I wins if $P(x, y) \neq 0$, otherwise player II wins. For winning, player II has only to compute y such that $P(x, y) = 0$, given x, and there are many algorithms for doing so.

Also Jones and Fraenkel [1995] produced games, with small length of play, which satisfy these conditions.

So we are led to the following reformulation of Berlekamp's question: Is there a game which has

1. simple, playable rules,
2. a finite set of options at every move,
3. a simple explicit strategy,
4. length of play at most exponential,
5. and is NP-hard or harder.

E2. Complexity closure. Aviezri Fraenkel asks: Are there partizan games G_1, G_2, G_3 such that: (i) G_1, G_2, G_3, $G_1 + G_2$, $G_2 + G_3$ and all their options have polynomial-time strategies, (ii) $G_1 + G_3$ is NP-hard?

E3. Sums of switch games. David Wolfe considers a sum of games G, each of the form $a\|b|c$ or $a|b\|c$ where a, b, and c are integers specified in unary. Is there a polynomial time algorithm to determine who wins in G, or is the problem NP-hard?

E4. (52) How does one play **sums** of games with varied overheating operators? Sentestrat and Top-down thermography (LIP, p. 214):

David Wolfe would like to see a formal proof that sentestrat works, an algorithm for top-down thermography, and conditions for which top-down thermography is computationally efficient.

Aaron Siegel asks the following generalized thermography questions.

(1) Show that the Left scaffold of a dogmatic (neutral ko-threat environment; LIP, p. 215) thermograph is decreasing as function of t. (Note, this is NOT true for komaster thermographs.) [Dogmatic thermography was invented by Berlekamp and Spight. See [21] for a good introduction.]

(2) Develop the machinery for computing dogmatic thermographs of double kos (multiple alternating 2-cycles joined at a single node).

In the same vein as (2):

(3) Develop a temperature theory that applies to all loopy games.

Siegel thinks that (3) is among the most important open problems in combinatorial game theory. The temperature theory of Go appears radically different from the classical combinatorial theory of loopy games (where infinite plays are draws). It would be a huge step forward if these could be reconciled into a "grand unified temperature theory". Problem (2) seems to be the obvious next step toward (3).

Conway asks for a natural set of conditions under which the mapping $G \mapsto \int^* G$ is the *unique* homomorphism that annihilates all infinitesimals.

E5. Loopy games (WW, pp. 334–377; LIP, pp. 213–214) are partizan games that do not satisfy the ending condition. A **Stopper** is a game that, when played on its own, has no ultimately alternating, Left and Right, infinite sequence of legal moves. Aaron Siegel reminds us of WW, 2nd ed., p. 369, where the authors tried hard to prove that every loopy game had stoppers, until Clive Bach found the Carousel counterexample. Is there an alternative notion of simplest form that works for *all* finite loopy games, and, in particular, for the Carousel? The simplest form theorem for stoppers is at WW, p. 351.

Siegel conjectures that, if Q is an arbitrary cycle of Left and Right moves that contains at least two moves for each player, and is not strictly alternating, then there is a stopper consisting of a single cycle that matches Q, together

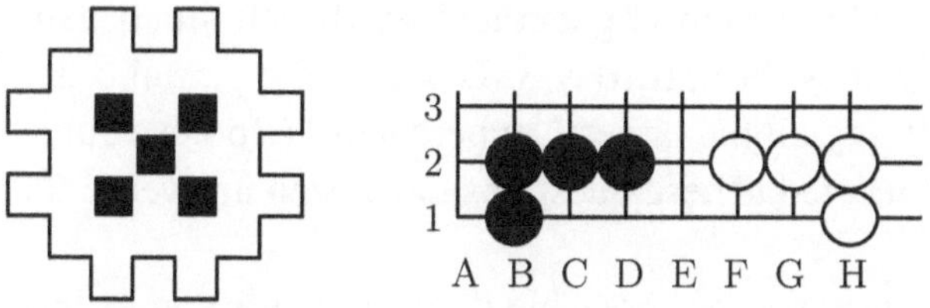

Figure 7. A **Domineering** position and a Chilled Go position of value $*2$.

with various exits to enders, i.e., games which end in a finite, though possibly unbounded, number of moves. (Note that games normally have Left and Right playing alternately, but if the game is a sum, then play in one component can have arbitrary sequences of Left and Right moves, not just alternating ones.)

A long cycle is *tame* if it alternates just once between Left and Right, otherwise it is *wild*. Aaron Siegel writes:

> I can produce wild cycles "in the laboratory," by specifying their game graphs explicitly. So the question is to detect one "in nature", i.e., in an actual game with (reasonably) playable rules such as **phutball** [Problem B7].

Siegel also asks under what conditions does a given infinitesimal have a well-defined atomic weight, and asks to specify an algorithm to calculate the atomic weight of an infinitesimal stopper g. The algorithm should succeed whenever the atomic weight is well-defined, i.e., whenever g can be sandwiched between loopfree all-smalls of equal atomic weight.

E6. (45) Elwyn Berlekamp asks for the **habitat** of $*2$, where $*2 = \{0, *|0, *\}$. Gabriel Drummond-Cole [2005] has found **Domineering** positions with this value. See, for example, Figure 7, which also shows a Go position, found by Nakamura and Berlekamp [2003], whose chilled value is $*2$. The Black and White groups are both connected to life via unshown connections emanating upwards from the second row. Either player can move to $*$ by placing a stone at E2, or to 0 by going to E1. Given a game, let n be the smallest nonnegative integer n such that if a position of the game has value $*k$ then $k < 2^n$. Carlos dos Santos calls n the *nim-dimension* of the game. He shows, this volume, that **Konane** has infinite nim-dimension and ask for the nim-dimension of other games.

E7. Partial ordering of games. David Wolfe lets $g(n)$ be the number of games born by day n, notes that an upper bound is given by $g(n+1) \le g(n) + 2^{g(n)} + 2$, and a lower bound for each $\alpha < 0$ is given by $g(n+1) \ge 2^{g(n)^\alpha}$, for n sufficiently large, and asks us to tighten these bounds.

He also asks what group is generated by the all-small games (or — much harder — of all games) born by day 3. Describe the partial order of games born by day 3, identifying all the largest "hypercubes" (Boolean sublattices) and how they are interconnected. These questions have been answered for day 2, see this volume, pp.??.

Berlekamp suggests other possible definitions for games born by day n, $\mathscr{G}_n$, depending on how one defines $\mathscr{G}_0$. Our definition is 0-based, as $\mathscr{G}_0 = \{0\}$. Other natural definitions are integer-based (where $\mathscr{G}_0$ are integers) or number-based. These two alternatives do not form a lattice, for if G_1 and G_2 are born by day k, then the games

$$H_n := \{G_1, G_2 \parallel G_1, \{G_2|-n\}\}$$

form a decreasing sequence of games born by day $k+2$ exceeding any game $G \geq G_1, G_2$, and the day $k+2$ join of G_1 and G_2 cannot exist. What is the structure of the partial order given by one of these alternative definitions of birthday?

The set of all short games does not form a lattice, but Calistrate, Paulhus and Wolfe [2002] have shown that the games born by day n form a distributive lattice $\mathscr{L}_n$ under the usual partial order. They ask for a description of the exact structure of $\mathscr{L}_3$. Siegel describes $\mathscr{L}_4$ as "truly gigantic and exceedingly difficult to penetrate" but suggests that it may be possible to find its dimension and the maximum **longitude**, $\operatorname{long}_4(G)$, of a game in $\mathscr{L}_4$, which he defines as

$$\operatorname{long}_n(G) = \operatorname{rank}_n(G \vee G^\bullet) - \operatorname{rank}_n(G),$$

where $\operatorname{rank}_n(G)$ is the rank of G in $\mathscr{L}_n$ and $G^\bullet$ is the **companion** of G,

$$G^\bullet = \begin{cases} * & \text{if } G = 0, \\ \{0, (G^L)^\bullet \mid (G^R)^\bullet\} & \text{if } G > 0, \\ \{(G^L)^\bullet \mid 0, (G^R)^\bullet\} & \text{if } G < 0, \\ \{(G^L)^\bullet \mid (G^R)^\bullet\} & \text{if } G \parallel 0. \end{cases}$$

Albert and Nowakowski [2011] show that starting with any set of games, instead of just $0 = \{\cdot|\cdot\}$, then the games born on the next day form a complete but not necessarily distributive lattice. They ask which sets of games will give a distributive lattice? Is there a set which will give a nondistributive but modular lattice? Carvalho, dos Santos, Dias, Coelho, Neto, Nowakowski and Vinagre [4] answered this by showing that for any lattice L, there is a set of games which generates L on the next day.

The set of all-small games does not form a lattice, but Siegel forms a lattice $\mathscr{L}_n^0$ by adjoining least and greatest elements $\triangle$ and $\triangledown$ and asks: do the elements of $\mathcal{L}_n^0$ have an intrinsic "handedness" that distinguishes, say, $(n-1)\cdot\uparrow$ from $(n-1)\cdot\uparrow + *$?

A game is *option-closed* (Nowakowski and Ottaway [2011]) if, recursively, each G^{LL} is also $\mathcal{G}^L$ and the same for Right. For example, **Hackenbush** strings are option-closed. Nowakowski and (Angela) Siegel [16] show that the option-closed games born on day n form a planar, nondistributive lattice, but the question of *how many?* remains unanswered. Are there other natural families of games that form planar lattices?

E8. Aaron Siegel asks, given a group or monoid, $\mathcal{K}$, of games, to specify a technique for calculating the simplest game in each $\mathcal{K}$-equivalence class. He notes that some restriction on $\mathcal{K}$ might be needed; for example, $\mathcal{K}$ might be the monoid of games absorbed by a given idempotent.

E9. Siegel also would like to investigate how search methods might be integrated with a canonical-form engine.

E10. (9) Develop a **misère theory** for unions of partizan games (WW, p. 312).

E11. Four-outcome-games. Guy [2007] has given a brute force analysis of a parity subtraction game which didn't allow the use of Sprague–Grundy theory because it wasn't impartial, nor the Conway theory, because it was not last-player-winning. Is there a class of games in which there are four outcomes, $\mathcal{N}$ext, $\mathcal{P}$revious, $\mathcal{L}$eft and $\mathcal{R}$ight, and for which a general theory can be given?

E12. Impartial Misère Analysis. See **A16.** From the works of Plambeck and Siegel: Let $\mathcal{A}$ be some set of games (the *universe*) played under misère rules. Typically, $\mathcal{A}$ is the set of positions that arise in a particular game, such as **Dawson's Chess**. Games $H, K \in \mathcal{A}$ are said to be equivalent, denoted by $H \equiv K(\text{mod}\mathcal{A}$, if $H+X$ and $K+X$ have the same outcome for all games $X \in \mathcal{A}$. The relation $\equiv$ is an equivalence relation, and a set of representatives, one from each equivalence class, forms the **misère quotient**, $\mathcal{Q} = \mathcal{A}/{\equiv}$. A **quotient map** $\Phi : \mathcal{A} \to \mathcal{Q}$ is defined, for $G \in \mathcal{A}$, by $\Phi : \mathcal{G} = [G]_{\equiv}$.

(1) A quotient map $\Phi : \mathcal{A} \to \mathcal{Q}$ is said to be **faithful** if, whenever $\Phi(G) = \Phi(H)$, then G and H have the same normal-play Grundy value. Is every quotient map faithful?

(2) Let $(\mathcal{Q}, \mathcal{P})$ be a quotient and $\mathcal{S}$ a maximal subgroup of $\mathcal{Q}$. Must $\mathcal{S} \cap \mathcal{P}$ be nonempty? (Note: it's easy to get a "yes" answer in the special case when $\mathcal{S}$ is the kernel.)

(3) Extend the classification of impartial misère quotients. We have preliminary results on the number of quotients of order $n \leq 18$ but believe that this can be pushed far higher.

(4) Exhibit an impartial misère quotient with a period-5 element. Same question for period 8, etc. We've detected quotients with elements of periods 1, 2, 3, 4, 6, and infinity, and we conjecture that there is no restriction on the periods of quotient elements.

(5) In the flavor of both (3) and (4): What is the smallest quotient containing a period 4 (or 3 or 6) element?

E13. Inverses in Partizan misère games. Siegel [Misère Canonical Forms of Partizan Games, this volume] and Allen [this volume] extends the misère monoid concepts to partizan games. In normal play, games in both players have a move or neither does are called *all-small* because their values are infinitesimal. This is not true in misère play and the term *dicot* has been coined to refer to these games in all play conventions. (See Section E14 as well.)

Recall that $-G$ is obtained from the game G by reversing the roles of Left and Right — "turning the board around". In normal play, $G - G = 0$ which is decidedly not true in misère play. To avoid confusion, when dealing with misère play $-G$ is frequently written as $\overline{G}$ and is called the *conjugate* of G.

(1) Allen asks, in the dicot universe, $\mathcal{D}$, when is it true that $G + \overline{G} \equiv 0 (\mathrm{mod}\ \mathcal{D})$? For example, $* + * \equiv 0 (\mathrm{mod}\ \mathcal{D})$. McKay, Milley and Nowakowski [11] show that this is true if $G = * : x$ where x is a number in canonical form (regarded as a game tree) and more generally: if $o^{-}(H + \overline{H}) = \mathcal{N}$ for every H of G, including G, then $G + \overline{G} \equiv 0 (\mathrm{mod}\ \mathcal{D})$.

(2) Rebecca Milley asks, in a universe, $\mathcal{U}$, closed under sums, conjugation and subpositions if $G + H \equiv 0 (\mathrm{mod}\ \mathcal{U})$ is it necessarily true that $H = \overline{G}$? Jason Brown had asked the question about arbitrary universes. Milley [13] has the counterexample: on a finite strip of squares, Left places a 1×1 piece and Right places a 1×2 (domino). In the universe of sums of strips a strip of length 1 plus a strip of length 2 is equivalent to 0.

E14. Scoring games. Instead of "the last play determines the winner", another natural way is to have "whichever player has the higher score wins". Both **Dots-and-Boxes** and **Go**, for example, are scoring games.

Milnor looked at dicot scoring games with nonnegative incentives. Milnor [1953] defines a "positional game" to be what we would call a dicot (Section E13) scoring game, but only looks at positional games with nonnegative incentives. Hanner [1959] looked at the same class of games as Milnor, and invented thermography. Ettinger [2000] looked at dicot scoring games (possibly with negative incentives), which he calls "positional games" following Milnor. Johnson [9] looked at the dicot scoring games that are "well-tempered" in some sense (a subset of Ettinger's games). Stewart [23; 22] looks at general scoring games.

Since the games are no longer dicot, a choice has to be made about how to handle the situation where the current player can't move but their opponent can.

There is much work yet to be done in this field. Johnson mentions one problem worthy of a place to start: is there a general theory that covers **Dots-and-Boxes** (Section C1)?

E15. Find a formula for the number of $\mathcal{P}$-positions for **Nim** played with $2n$ tokens. there are none with an odd number of tokens. The On-line Encyclopedia of Integer Sequences (OEIS) has the value for small values of n.

In general, given a game, enumerate the number of $\mathcal{P}$-positions of a given size. We only know of **Timber** (Section D5) where the number of $\mathcal{P}$-positions, both normal and misère are related to Dyck paths with certain properties (also to Catalan numbers and Fine numbers) and Heytei [2009; 2010] which relates the number of $\mathcal{P}$-positions to the Bernoulli numbers of the second kind.

Acknowledgement

I have had help in compiling this collection from all those mentioned, and from others. We would especially like to mention Elwyn Berlekamp, Aviezri Fraenkel, Will Johnson, Adam Landsberg, Urban Larsson, Neil McKay, Thane Plambeck, Carlos Santos, Aaron Siegel and David Wolfe. All mistakes are deliberate and designed to keep the reader alert.

References

[1] M. H. Albert, J.P Grossman, S. McCurdy, R. J. Nowakowski, and D. Wolfe. Cherries. preprint, 2005. [Problem **A5**].

[2] Saúl A. Blanco and Aviezri S. Fraenkel. Triomineering, tridomineering and l-tridomineering. preprint, 2006. [Problem **B11**].

[3] N. Bullock. Domineering: solving large combinatorial search spaces. *ICGA J.*, 25:67–85, 2002. [Problem **B11**].

[4] Alda Carvalho, Carlos Pereira dos Santos, Catia Dias, Francisco Coelho, Joao Pedro Neto, Richard Nowakowski, and Sandra Vinagre. On lattices from combinatorial game theory modularity and a representation theorem: finite case. *Theoretical Computer Science*, 527:37–49, 2014. [Problem **E7**].

[5] N. Comeau, J. Cullis, R. J. Nowakowski, and J. Paek. Personal communication (class project). [Problem **A12**].

[6] Carlos dos Santos. *Alguns resultados sobre jogos imparciais e dimensao NIM*. PhD thesis, University of Lisbon, 2010. [Problem **A1**].

[7] Paul Erdős and George Szekeres. A combinatorial problem in geometry. *Compositio Math.*, 2:463–470, 1935. [Problem **C7**].

[8] P. A. Grillet. *Commutative semigroups*. Number 2 in Advances in Mathematics (Dordrecht). Kluwer Academic Publishers, Dordrecht, 2001. [Problem **A15**].

[9] William Johnson. Well-tempered scoring games. *Int. J. Game Theory*, 43(2):415–438, 2014. [Problem **E14**].

[10] Sarah McCurdy. *Two combinatorial games*. Master's thesis, Dalhousie University, 2004. [Problem **A5**].

[11] Neil A. McKay, Rebecca Milley, and Richard J. Nowakowski. Misère-play Hackenbush sprigs. preprint, 2012. to appear in *Intl. J. Game Theory*, [Problem **E13**].

[12] Neil A. McKay and Richard J. Nowakowski. Outcomes of partizan Euclid. *Integers*, 12B:Paper A9, 15, 2012/13. [Problem **A9**].

[13] Rebecca Milley. *Restricted universes of partizan misère games*. PhD thesis, Dalhousie University, 2013. [Problem **E13**].

[14] M. Müller. Solving 5×5 amazons. In *The 6th Game Programming Workshop*, number 14 in IPSJ Symposium Series, pages 64–71, 2001. [Problem **B6**].

[15] Richard J. Nowakowski, Gabriel Renault, Emily Lamoureux, Stephanie Mellon, and Timothy Miller. The game of timber! *J. of Combin. Math. and Combin. Comput.*, 85:213–225, 2013.

[16] Richard J. Nowakowski and Angela A. Siegel. The lattices of option-closed games. preprint. [Problem **E7**].

[17] László Rédei. *The theory of finitely generated commutative semigroups*. Pergamon Press, Oxford, 1965. [Problem **A15**].

[18] C. Schensted. Longest increasing and decreasing subsequences. *Canad. J. Math.*, 13:179–191, 1961. Problem **C7**].

[19] Aaron Nathan Siegel. *Loopy games and computation*. PhD thesis, University of California, Berkeley, Ann Arbor, MI, 2005. [Problem **B7, B9**].

[20] Angela Siegel. *Finite excluded subtraction sets and infinite geography*. Master's thesis, Dalhousie University, 2005. [Problem **A1**].

[21] W. L. Spight. Evaluating kos in a neutral threat environment: preliminary results. In *Computers and games: third international conference, CG 2002*, number 2883 in Lect. Notes in Comput. Sci., pages 413–428. Springer, 2003. [Problem **E4**].

[22] Fraser Stewart. Impartial scoring games. preprint, 2012. Submitted, [Problem **E14**].

[23] Fraser Stewart. Scoring play combinatorial games. preprint, 2012. Submitted, [Problem **E14**].

[24] Thea van Roode. *Partizan forms of Hackenbush*. Master's thesis, The University of Calgary, 2002. [Problem **C5**].

rjn@mathstat.dal.ca

Department of Mathematics and Statistics, Dalhousie University, Halifax, NS B3H 3J5, Canada

Games of No Chance 4
MSRI Publications
Volume **63**, 2015

Combinatorial Games: selected short bibliography with a succinct gourmet introduction

AVIEZRI S. FRAENKEL AND RICHARD J. NOWAKOWSKI

1. Short bibliography

In this instalment of the Games Bibliography, Richard Nowakowski joined as a coauthor. Unlike previous instalments, this one is restricted, mainly, to new entries of the last few years that did not appear in previous versions, such as in "Games of No Chance 3" and the Electronic J. of Combinatorics (Surveys). We apologize profusely that due to lack of time, this version does not include all the new game papers we are aware of, and also lacks some unifying editing and polishing.

2. What are combinatorial games?

Roughly speaking, the family of *combinatorial games* consists of two-player games with perfect information (no hidden information as in some card games), no chance moves (no dice) and outcome restricted to (lose, win), (tie, tie) and (draw, draw) for the two players who move alternately. Tie is an end position such as in tic-tac-toe, where no player wins, whereas draw is a dynamic tie: any position from which a player has a nonlosing move, but cannot force a win. Both the easy game of nim and the seemingly difficult chess are examples of combinatorial games. And so is go. The shorter terminology *game*, *games* is used below to designate combinatorial games.

3. Why are games intriguing and tempting?

Amusing oneself with games may sound like a frivolous occupation. But the fact is that the bulk of interesting and natural mathematical problems that are hardest in complexity classes beyond *NP*, such as *P*space, *E*xptime and *E*xpspace, are two-player games; occasionally even one-player games (puzzles) or even zero-player games (Conway's "Life"). Some of the reasons for the high complexity of two-player games are outlined in the next section. Before that we note that in addition to a natural appeal of the subject, there are applications or connections to various areas, including complexity, logic, graph and matroid theory, networks, error-correcting codes, surreal numbers, on-line algorithms, biology — and analysis and design of mathematical and commercial games!

But when the chips are down, it is this "natural appeal" that lures both amateurs and professionals to become addicted to the subject. What is the essence of this appeal? Perhaps the urge to play games is rooted in our primal beastly instincts; the desire to corner, torture, or at least dominate our peers. A common expression of these vile cravings is found in the passions roused by local, national and international tournaments. An intellectually refined version of these dark desires, well hidden beneath the façade of scientific research, is the consuming drive "to beat them all", to be more clever than the most clever, in short — to create the tools to *Math*ter them all in hot *comb*inatorial *comb*at! Reaching this goal is particularly satisfying and sweet in the context of combinatorial games, in view of their inherent high complexity.

With a slant towards artificial intelligence and classical-economic games, Judea Pearl wrote that games "offer a perfect laboratory for studying complex problem-solving methodologies. With a few parsimonious rules, one can create complex situations that require no less insight, creativity, and expertise than problems actually encountered in areas such as business, government, scientific, legal, and others. Moreover, unlike these applied areas, games offer an arena in which computerized decisions can be evaluated by absolute standards of performance and in which proven human experts are both available and willing to work towards the goal of seeing their expertise emulated by a machine. Last, but not least, games possess addictive entertaining qualities of a very general appeal. That helps maintain a steady influx of research talents into the field and renders games a convenient media for communicating powerful ideas about general methods of strategic planning."

To further explore the nature of games, we consider, informally, two subclasses.

(i) Games People Play (*playgames*): games that are challenging to the point that people will purchase them and play them.

(ii) Games Mathematicians Play (*mathgames*): games that are challenging to mathematicians or other scientists to play with and ponder about, but not necessarily to "the man in the street".

Examples of playgames are chess, go, hex, reversi; of mathgames: Nim-type games, Wythoff games, annihilation games, octal games.

Some "rule of thumb" properties, which seem to hold for the majority of playgames and mathgames are listed below.

I. Complexity. Both playgames and mathgames tend to be computationally intractable. There are a few tractable mathgames, such as Nim, but most games still live in *Wonderland*: we are wondering about their as yet unknown complexity. Roughly speaking, however, NP-hardness is a necessary but not a sufficient condition for being a playgame! (Some games on Boolean formulas are Exptime-complete, yet none of them seems to have the potential of commercial marketability.)

II. Boardfeel. None of us may know an exact strategy from a midgame position of chess, but even a novice, merely by looking at the board, gets some feel who of the two players is in a stronger position – even what a strong or weak next move is. This is what we loosely call *boardfeel*. Our informal definition of playgames and mathgames suggests that the former do have a boardfeel, whereas the latter don't. For many mathgames, such as Nim, a player without prior knowledge of the strategy has no inkling whether any given position is "strong" or "weak" for a player. Even when defeat is imminent, only one or two moves away, the player sustaining it may be in the dark about the outcome, which will startle and stump him. The player has no boardfeel. (Many mathgames, including Nim-type games, can be played, equivalently, on a board.)

Thus, in the boardfeel sense, simple games are complex and complex games are simple! This paradoxical property also doesn't seem to have an analog in the realm of decision problems. The boardfeel is the main ingredient which makes PlayGames interesting to play.

III. Math Appeal. Playgames, in addition to being interesting to play, also have considerable mathematical appeal. This has been exposed recently by the theory of partizan games established by Conway and applied to endgames of go by Berlekamp, students and associates. On the other hand, mathgames have their own special combinatorial appeal, of a somewhat different flavor. They are created by mathematicians and appeal to them, since they experience special intellectual challenges in analyzing them. As Peter Winkler called a subset of them: "games people don't play". We might also call them, in a more positive vein, "games mathematicians play". Both classes of games have applications to areas outside game theory. Examples: surreal numbers (playgames), error correcting codes (mathgames). Both provide enlightenment through bewilderment, as David Wolfe and Tom Rodgers put it.

IV. Existence. There are relatively few commercially successful playgames around. It seems to be hard to invent a playgame that catches the masses. In contrast, mathgames abound. They appeal to a large subclass of mathematicians and other scientists, who cherish producing them and pondering about them. The large proportion of mathgames-papers in the games bibliography below reflects this phenomenon.

We conclude, inter alia, that for playgames, high complexity is desirable. Whereas in all respectable walks of life we strive towards solutions or at least approximate solutions which are polynomial, there are two "less respectable" human activities in which high complexity is savored. These are cryptography (covert warfare) and games (overt warfare). The desirability of high complexity in cryptography — at least for the encryptor! — is clear. We claim that it is also desirable for playgames.

It's no accident that games and cryptography team up: in both there are adversaries, who pit their wits against each other! But combinatorial game theory is, in general, considerably harder than cryptography. For the latter, the problem whether the designer of a cryptosystem has a safe system can be expressed with two quantifiers only: $\exists$ a cryptosystem such that $\forall$ attacks on it, the cryptosystem remains unbroken? In contrast, the decision problem whether White can win if White moves first in a chess game, has the form: "$\exists\forall\exists\forall\cdots$ move: White wins?", expressing the question whether White has an opening winning move — with an unbounded number of alternating quantifiers. This makes games the more challenging and fascinating of the two, besides being fun! See also the next section.

Thus, it's no surprise that the skill of playing games, such as checkers, chess, or go has long been regarded as a distinctive mark of human intelligence.

4. Why are combinatorial games hard?

Existential decision problems, such as graph hamiltonicity and Traveling Salesperson (Is there a round tour through specified cities of cost $\leq C$?), involve a single existential quantifier ("Is there... ?"). In mathematical terms an existential problem boils down to finding a path—sometimes even just verifying its existence—in a large "decision-tree" of all possibilities, that satisfies specified properties. The above two problems, as well as thousands of other interesting and important combinatorial-type problems are NP-*complete*. This means that they are *conditionally intractable*, i.e., the best way to solve them seems to require traversal of most if not all of the decision tree, whose size is exponential in the input size of the problem. No essentially better method is known to date at any rate, and, roughly speaking, if an efficient solution will ever be found for any NP-complete problem, then all NP-complete problems will be solvable efficiently.

The decision problem whether White can win if White moves first in a chess game, on the other hand, has the form: Is there a move of White such that for *every* move of Black there is a move of White such that for *every* move of Black there is a move of White ... such that White can win? Here we have a large number of alternating existential and universal quantifiers rather than a single existential one. We are looking for an entire subtree rather than just a path in the decision tree. Because of this, most nonpolynomial games are at least *P*space-hard. The problem for generalized chess on an $n \times n$ board, and even for a number of seemingly simpler mathgames, is, in fact, Exptime-complete, which is a *provable intractability*.

Put in simple language, in analyzing an instance of Traveling Salesperson, the problem itself is passive: it does not resist your attempt to attack it, yet it is difficult. In a game, in contrast, there is your opponent, who, at every step, attempts to foil your effort to win. It's similar to the difference between an autopsy and surgery. Einstein, contemplating the nature of physics said, "Der Allmächtige ist nicht boshaft; Er ist raffiniert" (The Almighty is not mean; He is sophisticated). NP-complete existential problems are perhaps sophisticated. But your opponent in a game can be very mean!

Another manifestation of the high complexity of games is associated with a most basic tool of a game: its *game-graph*. It is a directed graph Γ whose vertices are the positions of the game, and (u, v) is an edge if and only if there is a move from position u to position v. As an example, consider a directed graph G with some tokens distributed on its vertices. A move consists of selecting a token on vertex w and shifting it to z if and only if (w, z) is an edge of G. Since every subset of vertices of G (those occupied by tokens) is a position of the game, and thus a *single* vertex in Γ, the game graph Γ of the game on G has clearly exponential size in the size of G. This kind of behavior holds, in particular, for both nim-type and chess-type games. Analyzing a game means reasoning about its game-graph. We are thus faced with a problem that is *a priori* exponential, quite unlike many present-day interesting existential (optimization) problems.

A fundamental notion is the *sum* (disjunctive compound) of games. A sum is a finite collection of disjoint games; often very basic, simple games. Each of the two players, at every turn, selects one of the games and makes a move in it. If the outcome is not a draw, the sum-game ends when there is no move left in any of the component games. If the outcome is not a tie either, then in *normal* play, the player first unable to move loses and the opponent wins. The outcome is reversed in *misère* play.

If a game decomposes into a *disjoint* sum of its components, either from the beginning (Nim) or after a while (domineering), the potential for its tractability increases despite the exponential size of the game graph. As Elwyn Berlekamp remarked, the situation is similar to that in other scientific endeavors, where we often attempt to decompose a given system into its functional components. This approach may yield improved insights into hardware, software or biological systems, human organizations, and abstract mathematical objects such as groups.

If a game doesn't decompose into a sum of disjoint components, it is more likely to be intractable (Geography or Poset Games). Intermediate cases happen when the components are not quite fixed (which explains why misère play of sums of games is much harder than normal play) or not quite disjoint (Welter). Thane Plambeck and Aaron Siegel have recently revolutionized the theory of misère play, as reflected in the bibliography below.

The hardness of games is eased somewhat by the efficient freeware package "Combinatorial Game Suite", courtesy of Aaron Siegel.

5. Breaking the rules

As the experts know, some of the most exciting games are obtained by breaking some of the rules for combinatorial games, such as permitting a player to pass a bounded or unbounded number of times, i.e., relaxing the requirement that players play alternately; or permitting a number of players other than two.

But permitting a payoff function other than (0,1) for the outcome (lose, win) and a payoff of $(\frac{1}{2}, \frac{1}{2})$ for either (tie, tie) or (draw, draw) usually, but not always, leads to *classical* games that are not combinatorial but undetermined, notably games with applications in economics. Bernhard von Stengel has recently initiated an important effort to establish a bridge between these and combinatorial games (see alse section 8 below). Though there are a few individuals who work in both disciplines, an organic connection between the two disciplines has yet to be established.

6. Why is the bibliography vast?

In the realm of existential problems, such as sorting or Traveling Salesperson, most present-day interesting decision problems can be classified into tractable, conditionally intractable, and provably intractable ones. There are exceptions, to be sure, such as graph isomorphism (are two graphs isomorphic?), whose complexity is still unknown. But the exceptions are very few. In contrast, most games are still in Wonderland, as pointed out in section 2(I) above. Only a few games have been classified into the complexity classes they belong to. Despite recent impressive progress, the tools for reducing Wonderland are still few and inadequate.

To give an example, many interesting games have a very succinct input size, so a polynomial strategy is often more difficult to come by (Richard Guy and Cedric Smith's octal games; Grundy's game). Succinctness and non-disjointness of games in a sum may be present simultaneously (Poset games). In general, the alternating quantifiers, and, to a smaller measure, "breaking the rules", add to the volume of Wonderland. We suspect that the large size of Wonderland, a fact of independent interest, is the main contributing factor to the bulk of the bibliography on games.

7. Why isn't it larger?

The bibliography below is a *partial* list of books and articles on combinatorial games and related material. It is partial not only because I constantly learn of additional relevant material I did not know about previously, but also because of certain self-imposed restrictions. The most important of these is that only papers with some original and nontrivial mathematical content are considered. This excludes most historical reviews of games and most, but not all, of the work on heuristic or artificial intelligence approaches to games, especially the large literature concerning computer games, especially computer chess. I have, however, included the compendium edited by David Levy [2009], which, with its extensive bibliography, can serve as a first guide to this world. Also some papers on chess-endgames and clever exhaustive computer searches of some games have been included.

On the other hand, papers on games that break some of the rules of combinatorial games are included liberally, as long as they are interesting and retain a combinatorial flavor. These are vague and hard to define criteria, yet combinatorialists usually recognize a combinatorial game when they see it. Besides, it is interesting to break also this rule sometimes! We have included some references to one-player games, e.g., towers of Hanoi, n-queen problems, 15-puzzle, peg-solitaire, pebbling and others, but only few zero-player games (such as Life and some games on "sand piles"). We have also included papers on various applications of games, especially when the connection to games is substantial or the application is interesting or important.

8. Meetings and publications

Conferences on combinatorial games, such as an AMS short course in Columbus, OH (1990), workshops at MSRI (1994, 2000), at BIRS (2005, 2011) resulted in books, or a special issue of a journal – the latter for the Dagstuhl seminar (2002).

During 1990–2001, *Theoretical Computer Science* ran a special Mathematical Games Section whose main purpose was to publish papers on combinatorial games. TCS still solicits papers on games. In 2002, *Integers—Electronic J. of Combinatorial Number Theory* began publishing a Combinatorial Games Section. The semi-regular 'Games at Dal' (8 meetings between 2001 and 2014) and the regular Integers Conference have generated papers to be found in the above journals. The combinatorial games community is growing in quantity and quality! In 2011 the "Game Theory Society" (classical games) posted on its web site http://www.gametheorysociety.org/ a call for papers inviting submissions of high-class papers in combinatorial game theory for its flagship publication *Intern. J. Game Theory*. Indeed, we witness a recent large influx of combinatorial game papers submissions to IJGT. Whether this will lead to fruitful scientific cooperation between the two communities remains to be seen.

9. The dynamics of the bibliography

The game bibliography below is very dynamic in nature. Previous versions have been circulated to colleagues, intermittently, since the early 1980's (surface mail!). Prior to every mailing updates were prepared, and usually also afterwards, as a result of the comments received from several correspondents. Naturally, the listing can never be "complete".

Because of its dynamic nature, it is natural that the bibliography became a "Dynamic Survey" in the Dynamic Surveys (DS) section of the *Electronic Journal of Combinatorics* (ElJC) http://www.combinatorics.org/
(click on "Surveys"). The ElJC has mirrors at various locations. Furthermore, the European Mathematical Information Service (EMIS) mirrors this Journal, as do all of its mirror sites (currently forty of them). See
http://www.emis.de/tech/mirrors/index.html

10. An appeal

I ask readers to continue sending to me corrections and comments; and inform me of significant omissions, remembering, however, that it is a *selected* bibliography. I prefer to get reprints, preprints or URLs, rather than only titles — whenever possible.

Material on games is mushrooming on the Web. The URLs can be located using a standard search engine, such as Google.

11. Idiosyncrasies

Most of the bibliographic entries refer to items written in English, though there is a sprinkling of Danish, Dutch, French, German, Japanese, Slovakian and Russian, as well as some English translations from Russian. The predominance of English may be due to certain prejudices, but it also reflects the fact that nowadays the *lingua franca* of science is English. In any case, I'm soliciting also papers in languages other than English, especially if accompanied by an abstract in English.

On the administrative side, Technical Reports, submitted papers and unpublished theses have normally been excluded; but some exceptions have been made. Abbreviations of book series and journal names usually follow the *Math Reviews* conventions. Another convention is that de Bruijn appears under D, not B; von Neumann under V, not N, McIntyre under M not I, etc.

Earlier versions of this bibliography have appeared, under the title "Selected bibliography on combinatorial games and some related material", as the master bibliography for the book *Combinatorial Games*, AMS Short Course Lecture Notes, Summer 1990, Ohio State University, Columbus, OH, *Proc. Symp. Appl. Math.* **43** (R. K. Guy, ed.), AMS 1991 with 400 items; and in the *Dynamic Surveys* section of the *Electronic J. of Combinatorics* in November 1994 with 542 items, updated there at odd times. It also appeared as the master bibliography in *Games of No Chance*, Proc. MSRI Workshop on Combinatorial Games, July, 1994, Berkeley, CA (R. J. Nowakowski, ed.), MSRI Publ. Vol. 29, Cambridge University Press, Cambridge, 1996, under the present title, containing 666 items. The version published in the palindromic year 2002 contained the palindromic number 919 of references (*More Games of No Chance*, Proc. MSRI Workshop on Combinatorial Games, July, 2000, Berkeley, CA, R. J. Nowakowski, ed., MSRI Publ. Vol. 42, Cambridge University Press, Cambridge. In GONC 3 (Games Of No Chance) it contained 1360 items.

12. Acknowledgments

Many people have suggested additions to the bibliography, or contributed to it in other ways. Ilan Vardi distilled my *Math-mas*ter (§2) into *Math*ter. Among those that contributed more than two or three items are: Akeo Adachi, Ingo Althöfer, Thomas Andreae, Eli Bachmupsky, the late Adriano Barlotti, József Beck, the late Claude Berge, Gerald E. Bergum, the late H. S. MacDonald Coxeter, Thomas S. Ferguson, James A. Flanigan, Fred Galvin, the late Martin Gardner, Alan J. Goldman, Solomon W. Golomb, Richard K. Guy, Shigeki Iwata, David S. Johnson, Victor Klee, Donald E. Knuth, the late Anton Kotzig, Jeff C. Lagarias,

Michel Las Vergnas, Hendrik W. Lenstra, Hermann Loimer, F. Lockwood Morris, Richard J. Nowakowski, Judea Pearl, J. Michael Robson, David Singmaster, Wolfgang Slany, Cedric A. B. Smith, Rastislaw Telgársky, Mark D. Ward, Yōhei Yamasaki and others. Thanks to all and keep up the game! Special thanks to Mark Ward who went through the entire file with a fine comb in late 2005, when it contained 1,151 items, correcting errors and typos. Many thanks also to various anonymous helpers who assisted with the initial TEX file, to Silvio Levy, who has edited and transformed it into LATEX2e in 1996, and to Wolfgang Slany, who has transformed it into a BIBTeX file at the end of the previous millenium, and solved a "new millenium" problem encountered when the bibliography grew beyond 999 items. Keen users of the bibliography will notice that there is a beginning of MR references, due to Richard Guy's suggestion, facilitated by his former student Alex Fink. I have learned from Kevin O'Bryant's paper "Fraenkel's partition and Brown's decomposition", *Integers* 3 (2003), A11, 17 pp., that Lord Rayleigh had anticipated the so-called 'Beatty's Theorem" many years before Beatty, but without proof. See the Rayleigh and Beatty entries below.

13. The Bibliography

1. S. Abbasi and N. Sheikh [2008], Complexity of question/answer games, *Theoret. Comput. Sci.* **409**, 364–381, 2010i:68049.
http://dx.doi.org/10.1016/j.tcs.2008.08.034
2. S. Abbasi and N. Sheikh [2008], Question/answer games on towers and pyramids, in: *Mathematical foundations of computer science 2008*, Vol. 5162 of *Lecture Notes in Comput. Sci.*, Springer, Berlin, pp. 83–95, 2011c:68157.
http://dx.doi.org/10.1007/978-3-540-85238-4_6
3. M. H. Albert and R. J. Nowakowski [2011], Lattices of games, *Order* **16**, 1–10.
4. M. R. Allen [2015], Peeking at partizan misère quotients, in: *Games of No Chance 4* (R. Nowakowski, ed.), No. 63 in Mathematical Sciences Research Institute Publications, Cambridge University Press, New York, pp. 1–12.
5. N. Alon and A. Mehrabian [2011], On a generalization of Meyniel's conjecture on the Cops and Robbers game, *Electron. J. Combin.* **18**(1), Paper 19, 7, 2012c:05205.
6. V. Andrejić [2009], On a combinatorial game, *Publ. Inst. Math. (Beograd) (N.S.)* **86(100)**, 21–25, 2011a:91085.
http://dx.doi.org/10.2298/PIM0900021A
7. S. D. Andres [2010], Directed defective asymmetric graph coloring games, *Discrete Appl. Math.* **158**, 251–260, 2011c:05223.
http://dx.doi.org/10.1016/j.dam.2009.04.025
8. S. D. Andres [2010], Erratum to: The incidence game chromatic number [Discrete Appl. Math. 157 (9) (2009) 1980–1987] [MR2522464], *Discrete Appl. Math.*

158(6), 728, 2011c:05120.
http://dx.doi.org/10.1016/j.dam.2009.11.017

9. S. D. Andres [2010], Note on the number of rooted complete N-ary trees, *Ars Combin.* **94**, 465–469, 2010m:05150.
10. S. D. Andres [2009], Asymmetric directed graph coloring games, *Discrete Math.* **309**(18), 5799–5802, 2010m:05194.
http://dx.doi.org/10.1016/j.disc.2008.03.022
11. S. D. Andres [2009], Game-perfect graphs, *Math. Methods Oper. Res.* **69**(2), 235–250, 2011a:05124.
http://dx.doi.org/10.1007/s00186-008-0256-3
12. S. D. Andres [2009], Lightness of digraphs in surfaces and directed game chromatic number, *Discrete Math.* **309**(11), 3564–3579, 2011a:05126.
http://dx.doi.org/10.1016/j.disc.2007.12.060
13. S. D. Andres [2009], The incidence game chromatic number, *Discrete Appl. Math.* **157**(9), 1980–1987, 2010i:05108.
http://dx.doi.org/10.1016/j.dam.2007.10.021
14. S. D. Andres and W. Hochstättler [2011], The game chromatic number and the game colouring number of classes of oriented cactuses, *Inform. Process. Lett.* **111**(5), 222–226, 2012b:05109.
http://dx.doi.org/10.1016/j.ipl.2010.11.020
15. S. D. Andres, W. Hochstättler and C. Schallück [2011], The game chromatic index of wheels, *Discrete Appl. Math.* **159**(16), 1660–1665, 2825608.
http://dx.doi.org/10.1016/j.dam.2010.05.003
16. J. C. S. Araújo and C. L. Sales [2009], Grundy number of P_4-classes, in: *LAGOS '09—V Latin-American Algorithms, Graphs and Optimization Symposium*, Vol. 35 of *Electron. Notes Discrete Math.*, Elsevier Sci. B. V., Amsterdam, pp. 21–27, 2579402.
http://dx.doi.org/10.1016/j.endm.2009.11.005
17. M. Asté, F. Havet and C. Linhares-Sales [2010], Grundy number and products of graphs, *Discrete Math.* **310**(9), 1482–1490, 2011g:05090.
http://dx.doi.org/10.1016/j.disc.2009.09.020
18. J. Balogh and W. Samotij [2011], On the Chvátal-Erdős triangle game, *Electron. J. Combin.* **18**(1), Paper 72, 15, 2012c:05206.
19. J. Barát and M. Stojaković [2010], On winning fast in avoider-enforcer games, *Electron. J. Combin.* **17**(1), Research Paper 56, 12, 2011d:91058.
http://www.combinatorics.org/Volume_17/Abstracts/v17i1r56.html
20. T. Bartnicki, B. Brešar, J. Grytczuk, M. Kovše, Z. Miechowicz and I. Peterin [2008], Game chromatic number of Cartesian product graphs, *Electron. J. Combin.* **15(1)**, Research Paper 72, 13, MR2411449 (2009c:05069).
http://www.combinatorics.org/Volume_15/Abstracts/v15i1r72.html
21. S. Beatty [1926], Problem 3173, *Amer. Math. Monthly* **33**, 159, Solution [to Problem 3177, numbering in **33** was erroneous] in **34** (1927) 159-160.

22. L. Beaudou, E. Duchêne and S. Gravier [2015], A survey about Solitaire Clobber, in: *Games of No Chance 4*, Cambridge University Press.
23. R. A. Beeler and D. Paul Hoilman [2011], Peg solitaire on graphs, *Discrete Math.* **311**(20), 2198–2202, 2825664.
http://dx.doi.org/10.1016/j.disc.2011.07.006
24. G. I. Bell [2009], The shortest game of Chinese checkers and related problems, *Integers* **9**, G01, 22, MR2475631 (2010a:91035).
http://dx.doi.org/10.1515/INTEG.2009.003
25. G. I. Bell [2008], Solving triangular peg solitaire, *J. Integer Seq.* **11**, Article 08.4.8, 22, 2009h:91050.
26. J. Bell and B. Stevens [2009], A survey of known results and research areas for n-queens, *Discrete Math.* **309**, 1–31, 2010a:05002.
http://dx.doi.org/10.1016/j.disc.2007.12.043
27. J. Bell and B. Stevens [2008], Results for the n-queens problem on the Möbius board, *Australas. J. Combin.* **42**, 21–34, 2009h:05049.
28. E. Berkove, J. Van Sickle, B. Hummon and J. Kogut [2008], An analysis of the (colored cubes)^3 puzzle, *Discrete Math.* **308**(7), 1033–1045, MR2382343.
29. A. M. Bersani [2010], Reformed permutations in Mousetrap and its generalizations, *Integers* **10**, G01, 575–622, 2012d:05021.
http://dx.doi.org/10.1515/INTEG.2010.101
30. S. Bhattacharya, G. Paul and S. Sanyal [2010], A cops and robber game in multidimensional grids, *Discrete Appl. Math.* **158**(16), 1745–1751, 2012b:05178.
http://dx.doi.org/10.1016/j.dam.2010.06.014
31. J. Björklund and C. Holmgren [2012], Counterexamples to a monotonicity conjecture for the threshold pebbling number, *Discrete Math.* **312**, 2401–2405.
32. S. A. Blanco and A. S. Fraenkel [2011], Tromping games: tiling with trominoes, *Integers* **11**(3), 333–362, 2988066.
http://dx.doi.org/10.1515/integ.2011.026
33. V. D. Blondel, J. M. Hendrickx and R. M. Jungers [2008], Solitaire Clobber as an optimization problem on words, *Integers* **8**, G04, 12, MR2425633 (2009d:91038).
34. G. Bloom, M. Lampis, F. A. Muntaner-Batle and M. Rius-Font [2011], Queen labelings, *AKCE Int. J. Graphs Comb.* **8**(1), 13–22, 2012f:05252.
35. H. L. Bodlaender, S. Thomassé and A. Yeo [2011], Kernel bounds for disjoint cycles and disjoint paths, *Theoret. Comput. Sci.* **412**(35), 4570–4578, 2850415.
http://dx.doi.org/10.1016/j.tcs.2011.04.039
36. T. Bohman, O. Pikhurko, A. Frieze and D. Sleator, Hat problems, The Puzzle Toad.
http://www.cs.cmu.edu/puzzle/solution15.pdf
37. A. Bonato, E. Chiniforooshan and P. Prałat [2010], Cops and Robbers from a distance, *Theoret. Comput. Sci.* **411**(43), 3834–3844, 2012d:05251.
http://dx.doi.org/10.1016/j.tcs.2010.07.003

38. A. Bonato, P. Golovach, G. Hahn and J. Kratochvíl [2009], The capture time of a graph, *Discrete Math.* **309**(18), 5588–5595, 2010j:05273. http://dx.doi.org/10.1016/j.disc.2008.04.004

39. A. Bonato, G. Hahn and C. Tardif [2010], Large classes of infinite k-cop-win graphs, *J. Graph Theory* **65**(4), 334–342, 2011k:05157. http://dx.doi.org/10.1002/jgt.20484

40. A. Bonato, G. Kemkes and P. Prałat [2012], Almost all cop-win graphs contain a universal vertex, *Discrete Math.* **312**, 1652–1657.

41. A. Bonato and R. J. Nowakowski [2011], *The game of cops and robbers on graphs*, Vol. 61 of *Student Mathematical Library*, American Mathematical Society, Providence, RI, 2830217.

42. E. Boros, V. Gurvich and V. Oudalov [2013], A polynomial algorithm for a two parameter extension of Wythoff NIM based on the Perron-Frobenius theory, *Internat. J. Game Theory* **42**(4), 891–915, 3111666. http://dx.doi.org/10.1007/s00182-012-0338-6

43. B. Brešar, S. Klavžar and D. F. Rall [2010], Domination game and an imagination strategy, *SIAM J. Discrete Math.* **24**(3), 979–991, 2011g:05202. http://dx.doi.org/10.1137/100786800

44. K. W. Burke and S.-H. Teng [2008], Atropos: a PSPACE-complete Sperner triangle game, *Internet Math.* **5**(4), 477–492 (2009), 2011d:05240. http://projecteuclid.org/getRecord?id=euclid.im/1265033176

45. S. Butler, M. T. Hajiaghayi, R. D. Kleinberg and T. Leighton [2008], Hat guessing games, *SIAM J. Discrete Math.* **22**(2), 592–605, Reprinted in SIAM Rev. 51 (2009), 399–413, MR2399367 (2009f:91016). http://dx.doi.org/10.1137/060652774

46. G. Cairns and N. B. Ho [2012], A restriction of Euclid, *Bull. Aust. Math. Soc* **86**, 506–509.

47. G. Cairns and N. B. Ho [2011], Some remarks on End-Nim, *Int. J. Comb.* pp. Art. ID 824742, 9, 2861127.

48. G. Cairns and N. B. Ho [2010], Min, a combinatorial game having a connection with prime numbers, *Integers* **10**, G03, 765–770, 2797778. http://dx.doi.org/10.1515/INTEG.2010.103

49. G. Cairns and N. B. Ho [2010], Ultimately bipartite subtraction games, *Australas. J. Combin.* **48**, 213–220, 2011m:05199.

50. G. Cairns, N. B. Ho and T. Lengyel [2011], The Sprague-Grundy function of the real game Euclid, *Discrete Math.* **311**(6), 457–462, 2799898. http://dx.doi.org/10.1016/j.disc.2010.12.011

51. N. J. Calkin, K. James, J. E. Janoski, S. Leggett, B. Richards, N. Sitaraman and S. M. Thomas [2010], Computing strategies for graphical Nim, *Proceedings of the Forty-First Southeastern International Conference on Combinatorics, Graph Theory and Computing*, Vol. 202, pp. 171–185, 2011j:05222.

52. A. Carvalho and C. Pereira dos Santos [2010], Chess and mathematics, in: *Recreational Mathematics Colloquium I*, Assoc. Ludus, Lisbon, pp. 3–26, 2807390.
53. A. Carvalho, C. Santos, C. Dias, F. Coelho, J. Neto, and S. Vinagre [2012], A recursive process related to a partizan version of Wythoff, *Integers, Electr. J. of Combinat. Number Theory* **12**, #G3, 15 pp., Comb. Games Sect. http://www.integers-ejcnt.org/vol12.html
54. T. Cazenave [2015], Monte-Carlo approximation of temperature, in: *Games of No Chance 4* (R. Nowakowski, ed.), No. 63 in Mathematical Sciences Research Institute Publications, Cambridge University Press, New York, pp. 41–46.
55. T. Cazenave and R. J. Nowakowski [2015], Retrograde analysis of Woodpush, in: *Games of No Chance 4* (R. Nowakowski, ed.), No. 63 in Mathematical Sciences Research Institute Publications, Cambridge University Press, New York, pp. 47–56.
56. S. Chaiken, C. R. H. Hanusa and T. Zaslavsky [2010], Nonattacking queens in a rectangular strip, *Ann. Comb.* **14**(4), 419–441, 2776757 (2012d:05034). http://dx.doi.org/10.1007/s00026-011-0068-7
57. J. Chalopin, V. Chepoi, N. Nisse and Y. Vaxès [2011], Cop and robber games when the robber can hide and ride, *SIAM J. Discrete Math.* **25**(1), 333–359, 2801232. http://dx.doi.org/10.1137/100784035
58. M. Chan and A. P. Godbole [2008], Improved pebbling bounds, *Discrete Math.* **308**, 2301–2306, MR2404560.
59. H. Chang and X. Zhu [2009], Colouring games on outerplanar graphs and trees, *Discrete Math.* **309**(10), 3185–3196, 2011a:05097. http://dx.doi.org/10.1016/j.disc.2008.09.015
60. V. Chatelain and J. L. Ramírez Alfonsín [2012], The switching game on unions of oriented matroids, *European J. Combin.* **33**(2), 215–219, 2854642. http://dx.doi.org/10.1016/j.ejc.2011.08.006
61. L.-J. Chen, J.-J. Lin, M.-Z. Shieh and S.-C. Tsai [2011], More on the Magnus-Derek game, *Theoret. Comput. Sci.* **412**(4-5), 339–344, 2012c:91059. http://dx.doi.org/10.1016/j.tcs.2010.09.031
62. R. W. Chen, A. Zame and B. Rosenberg [2009], On the first occurrence of strings, *Electron. J. Combin.* **16**(1), Research Paper 29, 16, MR2482097 (2010h:91058). http://www.combinatorics.org/Volume_16/Abstracts/v16i1r29.html
63. X. Chen, X. Deng and S.-H. Teng [2009], Settling the complexity of computing two-player Nash equilibria, *J. ACM* **56**(3), Art. 14, 57, 2010i:68040. http://dx.doi.org/10.1145/1516512.1516516
64. S.-H. Chiang, I.-C. Wu and P.-H. Lin [2011], Drawn k-in-a-row games, *Theoret. Comput. Sci.* **412**(35), 4558–4569, 2850414. http://dx.doi.org/10.1016/j.tcs.2011.04.033
65. P. Ciancarini and G. P. Favini [2010], Monte Carlo tree search in Kriegspiel, *Artificial Intelligence* **174**(11), 670–684, 2011j:91060. http://dx.doi.org/10.1016/j.artint.2010.04.017

66. P. Ciancarini and G. P. Favini [2010], Playing the perfect Kriegspiel endgame, *Theoret. Comput. Sci.* **411**(40-42), 3563–3577, 2011h:91055. http://dx.doi.org/10.1016/j.tcs.2010.05.019

67. J. Cibulka, J. Kynčl, V. Mészáros, R. Stolař and P. Valtr [2013], Graph sharing games: complexity and connectivity, *Theoret. Comput. Sci.* **494**, 49–62, Preliminary version in: Theory and applications of models of computation, Springer, 2010, Lecture Notes in Comput. Sci, vol. 6108, pp. 340–349, 2010, 3066015. http://dx.doi.org/10.1016/j.tcs.2012.12.029

68. J. Cibulka, V. Mészáros, R. Stolar and P. Valtr [2010], Solution of Peter Winkler's pizza problem, in: *Fete of combinatorics and computer science*, Vol. 20 of *Bolyai Soc. Math. Stud.*, János Bolyai Math. Soc., Budapest, pp. 63–93, 2797961. http://dx.doi.org/10.1007/978-3-642-13580-4_4

69. A. Cincotti [2014], Lattices of Multi-player Games, *Proceedings of the International MultiConference of Engineers and Computer Scientists*, Vol. II, pp. 3–6.

70. A. Cincotti [2014], Counting the Number of Multi-player Partizan Cold Games Born by Day d, *Lecture Notes in Electrical Engineering*, Vol. 275, pp. 289–304.

71. A. Cincotti [2011], The game of n-player Cutcake, *Theoret. Comput. Sci.* **412**(41), 5678–5683, 2866564. http://dx.doi.org/10.1016/j.tcs.2011.06.036

72. A. Cincotti [2011], The lattice structure of three-player games, in: *Computers and games*, Vol. 6515 of *Lecture Notes in Comput. Sci.*, Springer, Berlin, pp. 230–237, 2804004. http://dx.doi.org/10.1007/978-3-642-17928-0_21

73. A. Cincotti [2010], n-player partizan games, *Theoret. Comput. Sci.* **411**(34-36), 3224–3234, 2011f:91035. http://dx.doi.org/10.1016/j.tcs.2010.05.018

74. A. Cincotti [2010], The game of synchronized quadromineering, *Int. J. Comput. Math. Sci.* **4**(4), 183–187, 2011f:91034.

75. A. Cincotti [2009], Further results on three-player domineering, *Int. J. Comput. Math. Sci.* **3**(6), 286–288, 2011j:91061.

76. A. Cincotti [2009], On the complexity of N-player Hackenbush, *Integers* **9**, G03, 621–627, 2011b:91078. http://dx.doi.org/10.1515/INTEG.2009.049

77. A. Cincotti [2008], The game of Cutblock, *Integers* **8**, G06, 12, MR2425635 (2009g:91039).

78. A. Cincotti, S. Komori and H. Iida [2008], The game of synchronized triomineering and synchronized tridomineering, *Int. J. Comput. Math. Sci.* **2**(3), 143–148, 2011a:91087.

79. S. Clark, P. Johnson and M. Walsh [2011], Edge-sensitivity of the Grundy number, *Proceedings of the Forty-Second Southeastern International Conference on Combinatorics, Graph Theory and Computing*, Vol. 207, pp. 171–180, 2856318.

80. N. E. Clarke [2009], A witness version of the cops and robber game, *Discrete Math.* **309**, 3292–3298, 2011a:91080. http://dx.doi.org/10.1016/j.disc.2008.09.032

81. N. E. Clarke, S. L. Fitzpatrick, A. Hill and R. J. Nowakowski [2010], Edge critical cops and robber, *Discrete Math.* **310**, 2299–2309, 2011k:05057. http://dx.doi.org/10.1016/j.disc.2010.05.004

82. S. Collette, E. D. Demaine, M. L. Demaine and S. Langerman [2015], Narrow misère Dots-and-Boxes, in: *Games of No Chance 4* (R. Nowakowski, ed.), No. 63 in Mathematical Sciences Research Institute Publications, Cambridge University Press, New York, pp. 57–64.

83. D. Collins and T. Lengyel [2008], The game of 3-Euclid, *Discrete Math.* **308**(7), 1130–1136, 2008k:91050. http://dx.doi.org/10.1016/j.disc.2007.03.064

84. D. Craft, Z. Miller and D. Pritikin [2009], A solitaire game played on 2-colored graphs, *Discrete Math.* **309**, 188–201, 2010a:91030. http://dx.doi.org/10.1016/j.disc.2007.12.069

85. A. Csernenszky, C. I. Mándity and A. Pluhár [2009], On Chooser-Picker positional games, *Discrete Math.* **309**, 5141–5146, 2011e:05171. http://dx.doi.org/10.1016/j.disc.2009.03.051

86. A. Csernenszky, R. R. Martin and A. Pluhár [2011], On the complexity of chooser-picker positional games, *Integers, Electr. J. of Combinat. Number Theory* **11**, #G02, 16 pp., Comb. Games Sect. http://www.integers-ejcnt.org/vol11.html

87. D. Curtis, T. Hines, G. Hurlbert and T. Moyer [2009], On pebbling graphs by their blocks, *Integers* **9**, G02, 411–422, 2011b:05169. http://dx.doi.org/10.1515/INTEG.2009.033

88. C. Daskalakis, P. W. Goldberg and C. H. Papadimitriou [2009], The complexity of computing a Nash equilibrium, *SIAM J. Comput.* **39**(1), 195–259, 2010e:91006. http://dx.doi.org/10.1137/070699652

89. C. Daskalakis, A. Mehta and C. Papadimitriou [2009], A note on approximate Nash equilibria, *Theoret. Comput. Sci.* **410**(17), 1581–1588, 2010g:68303. http://dx.doi.org/10.1016/j.tcs.2008.12.031

90. E. D. Demaine, M. L. Demaine, R. Fleischer, R. A. Hearn and T. von Oertzen [2009], The complexity of the Dyson Telescopes puzzle, in: *Games of No Chance 3,* Proc. BIRS Workshop on Combinatorial Games, July, 2005, Banff, Alberta, Canada, MSRI Publ. (M. H. Albert and R. J. Nowakowski, eds.), Vol. 56, Cambridge University Press, Cambridge, pp. 271–285.

91. E. D. Demaine and R. A. Hearn [2009], Playing games with algorithms: algorithmic combinatorial game theory, in: *Games of No Chance 3,* Proc. BIRS Workshop on Combinatorial Games, July, 2005, Banff, Alberta, Canada, MSRI Publ. (M. H. Albert and R. J. Nowakowski, eds.), Vol. 56, Cambridge University

Press, Cambridge, pp. 3–56, an updated version of E.D. Demaine's paper of the same title, 2001.

92. E. D. Demaine and R. A. Hearn [2008], Constraint logic: a uniform framework for modeling computation as games, in: *Twenty-Third Annual IEEE Conference on Computational Complexity*, IEEE Computer Soc., Los Alamitos, CA, pp. 149–162, 2010d:68039.
http://dx.doi.org/10.1109/CCC.2008.35

93. J. DeMaio and B. Mathew [2011], Which chessboards have a closed knight's tour within the rectangular prism?, *Electron. J. Combin.* **18**(1), Paper 8, 14, 2012d:05210.

94. D. Dereniowski [2011], The complexity of node blocking for dags, *J. Combin. Theory Ser. A* **118**, 248–256, 2737198.
http://dx.doi.org/10.1016/j.jcta.2010.03.011

95. D. Dereniowski [2010], Phutball is PSPACE-hard, *Theoret. Comput. Sci.* **411**, 3971–3978, 2011h:91056.
http://dx.doi.org/10.1016/j.tcs.2010.08.019

96. M. Develin and S. Payne [2010], Discrete bidding games, *Electron. J. Combin.* **17**(1), Research Paper 85, 40, 2011j:91062.
http://www.combinatorics.org/Volume_17/Abstracts/v17i1r85.html

97. R. G. Donnelly [2008], Eriksson's numbers game and finite Coxeter groups, *European J. Combin.* **29**(7), 1764–1781, 2009d:05259.
http://dx.doi.org/10.1016/j.ejc.2007.06.029

98. P. Dorbec, E. Duchêne and S. Gravier [2008], Solitaire Clobber played on Hamming graphs, *Integers, Electr. J of Combinat. Number Theory* **8**, G3, 21, MR2393376 (2009b:91037).

99. I. Douglas and N. Sieben [2012], Edge Animal Weak (1,2)-Achievement Games, *Integers* **12**, G5.

100. G. C. Drummond-Cole [2104], An Update on Domineering on Rectangular Boards, *Integers* **14**, G3.

101. Y. Du [2008], On the complexity of deciding bimatrix games similarity, *Theoret. Comput. Sci.* **407**(1-3), 583–586, 2010g:91008.
http://dx.doi.org/10.1016/j.tcs.2008.07.021

102. E. Duchêne, A. S. Fraenkel, S. Gravier and R. J. Nowakowski [2009], Another bridge between NIM and WYTHOFF, *Australas. J. Combin.* **44**, 43–56, 2010h:91056.

103. E. Duchêne, A. S. Fraenkel, R. J. Nowakowski and M. Rigo [2010], Extensions and restrictions of Wythoff's game preserving its *P* positions, *J. Combin. Theory Ser. A* **117**(5), 545–567, 2011d:91073.
http://dx.doi.org/10.1016/j.jcta.2009.07.010

104. E. Duchêne and S. Gravier [2009], Geometrical extensions of Wythoff's game, *Discrete Math.* **309**, 3595–3608, 2010k:91051.
http://dx.doi.org/10.1016/j.disc.2007.12.089

105. E. Duchêne, S. Gravier and M. Mhalla [2009], Combinatorial graph games, *Ars Combin.* **90**, 33–44, 2010b:05123.

106. E. Duchêne, S. Gravier and J. Moncel [2009], New results about impartial solitaire clobber, *RAIRO Oper. Res.* **43**, 463–482, 2011b:91079. http://dx.doi.org/10.1051/ro/2009029

107. E. Duchêne and M. Rigo [2010], Invariant games, *Theoret. Comput. Sci.* **411**, 3169–3180, 2011i:91038. http://dx.doi.org/10.1016/j.tcs.2010.05.007

108. E. Duchêne and M. Rigo [2008], A morphic approach to combinatorial games: the Tribonacci case, *Theor. Inform. Appl.* **42**(2), 375–393, 2009b:91038. http://dx.doi.org/10.1051/ita:2007039

109. E. Duchêne and M. Rigo [2008], Cubic Pisot unit combinatorial games, *Monatsh. Math.* **155**(3-4), 217–249, 2009m:68199. http://dx.doi.org/10.1007/s00605-008-0006-x

110. A. Duffy, G. Kolpin and D. Wolfe [2009], Ordinal partizan End Nim, in: *Games of No Chance 3,* Proc. BIRS Workshop on Combinatorial Games, July, 2005, Banff, Alberta, Canada, MSRI Publ. (M. H. Albert and R. J. Nowakowski, eds.), Vol. 56, Cambridge University Press, Cambridge, pp. 419–425.

111. S. Dyrkolbotn and M. Walicki [2012], Kernels in digraphs that are not kernel perfect, *Discrete Math.* **312**, 2498–2505.

112. R. B. Ellis and K. L. Nyman [2009], Two-batch liar games on a general bounded channel, *J. Combin. Theory Ser. A* **116**(8), 1253–1270, 2011j:91054. http://dx.doi.org/10.1016/j.jcta.2009.03.005

113. P. L. Erdős and L. Soukup [2009], Quasi-kernels and quasi-sinks in infinite graphs, *Discrete Math.* **309**(10), 3040–3048, 2010i:05243. http://dx.doi.org/10.1016/j.disc.2008.08.006

114. L. Esperet and X. Zhu [2009], Game colouring of the square of graphs, *Discrete Math.* **309**(13), 4514–4521, 2010j:05134. http://dx.doi.org/10.1016/j.disc.2009.02.014

115. K. Etessami and M. Yannakakis [2010], On the complexity of Nash equilibria and other fixed points, *SIAM J. Comput.* **39**(6), 2531–2597, 2011i:91013. http://dx.doi.org/10.1137/080720826

116. C. D. A. Evans and J. D. Hamkins [2014], Transfinite Game Values in Infinite Chess, *Integers* **14**, G2.

117. G. Fülep and N. Sieben [2010], Polyiamonds and polyhexes with minimum site-perimeter and achievement games, *Electron. J. Combin.* **17**(1), Research Paper 65, 14, 2011d:05078. http://www.combinatorics.org/Volume_17/Abstracts/v17i1r65.html

118. G. Farr and J. Schmidt [2008], On the number of Go positions on lattice graphs, *Inform. Process. Lett.* **105**(4), 124–130, MR2381402. http://dx.doi.org/10.1016/j.ipl.2007.08.018

119. A. Ferber and D. Hefetz [2011], Winning strong games through fast strategies for weak games, *Electron. J. Combin.* **18**(1), Paper 144, 13, 2817794.

120. A. Fink [2012], Lattice games without rational strategies, *J. Combin. Theory Ser. A* **119**(2), 450–459, 2860605.
http://dx.doi.org/10.1016/j.jcta.2011.10.005

121. A. Fink, A. S. Fraenkel and C. Santos [2014], LIM is not slim, *Internat. J. Game Theory* **43**(2), 269–281, 3197293.
http://dx.doi.org/10.1007/s00182-013-0380-z

122. A. Fink and R. Guy [2009], Rick's tricky six puzzle: S_5 sits specially in S_6, *Math. Mag.* **82**, 83–102, 2512593.
http://dx.doi.org/10.4169/193009809X468896

123. A. Fink, R. J. Nowakowski, A. N. Siegel and D. Wolfe [2015], Toppling conjectures, in: *Games of No Chance 4* (R. Nowakowski, ed.), No. 63 in Mathematical Sciences Research Institute Publications, Cambridge University Press, New York, pp. 65–76.

124. E. Fisher and N. Sieben [2008], Rectangular polyomino set weak (1, 2)-achievement games, *Theoret. Comput. Sci.* **409**(3), 333–340, MR2473908 (2010a:91036).
http://dx.doi.org/10.1016/j.tcs.2008.05.024

125. F. V. Fomin, P. A. Golovach, J. Kratochvíl, N. Nisse and K. Suchan [2010], Pursuing a fast robber on a graph, *Theoret. Comput. Sci.* **411**(7-9), 1167–1181, 2011f:68055.
http://dx.doi.org/10.1016/j.tcs.2009.12.010

126. F. V. Fomin, P. A. Golovach and J. Kratochvil [2008], On tractability of cops and robbers game, in: *Fifth IFIP International Conference on Theoretical Computer Science—TCS 2008*, Vol. 273 of *IFIP Int. Fed. Inf. Process.*, Springer, New York, pp. 171–185, 2012b:91044.
http://dx.doi.org/10.1007/978-0-387-09680-3_12

127. F. V. Fomin, P. A. Golovach and D. Lokshtanov [2011], Guard games on graphs: keep the intruder out!, *Theoret. Comput. Sci.* **412**(46), 6484–6497, 2883020.
http://dx.doi.org/10.1007/978-3-642-12450-1_14

128. D. Forge and A. Vieilleribière [2009], The directed switching game on Lawrence oriented matroids, *European J. Combin.* **30**(8), 1833–1834, 2011a:05066.
http://dx.doi.org/10.1016/j.ejc.2008.12.002

129. A. Fraenkel, R. Hearn and A. Siegel [In Press], Theory of combinatorial games, in: *Handbook of Game Theory* (H. P. Young and S. Zamir, eds.), Vol. 4.

130. A. Fraenkel and Y. Tanny [2012/13], A class of Wythoff-like games, *Integers* **12B** (Proceedings of the Integers Conference 2011), Paper No. A7, 18, 3055681.

131. A. S. Fraenkel [2015], The Rat Game and the Mouse Game, in: *Games of No Chance 4* (R. Nowakowski, ed.), No. 63 in Mathematical Sciences Research Institute Publications, Cambridge University Press, New York, pp. 95–113.

132. A. S. Fraenkel and R. J. Nowakowski [2015], Combinatorial Games: selected short bibliography with a succinct gourmet introduction, No. 63 in Mathematical

Sciences Research Institute Publications, Cambridge University Press, New York, pp. 309–339.

133. A. S. Fraenkel [2012], The vile, dopey, evil and odious game players, *Discrete Math.* **312**, 42–46, special volume in honor of the 80th birthday of Gert Sabidussi, 2852506.
http://dx.doi.org/10.1016/j.disc.2011.03.032

134. A. S. Fraenkel [2012], RATWYT, *College Math. J.* **43**(2), 160–164, In memory of Martin Gardner, 2897480.
http://dx.doi.org/10.4169/college.math.j.43.2.160

135. A. S. Fraenkel [2011], Aperiodic subtraction games, *Electron. J. Combin.* **18**(2), Paper 19, 12, 2012g:91039.

136. A. S. Fraenkel [2010], Complementary iterated floor words and the Flora game, *SIAM J. Discrete Math.* **24**(2), 570–588, 2011g:91033.
http://dx.doi.org/10.1137/090758994

137. A. S. Fraenkel [2010], From enmity to amity, *Amer. Math. Monthly* **117**(7), 646–648, 2681526.
http://dx.doi.org/10.4169/000298910X496787

138. A. S. Fraenkel [2009], *The cyclic Butler University game,* in: Mathematical Wizardry for a Gardner, Volume honoring Martin Gardner, A K Peters, Natick, MA, pp. 97–105, E. Pegg Jr, A. H. Schoen, and T. Rodgers, eds.

139. A. Fraenkel [2013], Beating Your Fractional Beatty Game Opponent and: What's the Question to Your Answer?, *Advances in Combinatorics, in Memory of Herbert S. Wilf* (I. Kotsireas and E. Zima, eds.).

140. A. S. Fraenkel and U. Peled [2015], Harnessing the unwieldy MEX function, in: *Games of No Chance 4* (R. Nowakowski, ed.), No. 63 in Mathematical Sciences Research Institute Publications, Cambridge University Press, New York, pp. 77–94.

141. A. S. Fraenkel and E. Reisner [2009], The game of End-Wythoff, in: *Games of No Chance 3,* Proc. BIRS Workshop on Combinatorial Games, July, 2005, Banff, Alberta, Canada, MSRI Publ. (M. H. Albert and R. J. Nowakowski, eds.), Vol. 56, Cambridge University Press, Cambridge, pp. 329–347.

142. T. Gagie [2012], Bounds from a card trick, *J. Discrete Algorithms* **10**, 2–4.

143. H. Galeana-Sánchez [2008], Kernels in edge-coloured orientations of nearly complete graphs, *Discrete Math.* **308**(20), 4599–4607, 2009e:05115.
http://dx.doi.org/10.1016/j.disc.2007.08.079

144. H. Galeana-Sánchez and M.-k. Guevara [2009], Some sufficient conditions for the existence of kernels in infinite digraphs, *Discrete Math.* **309**, 3680–3693, 2010h:05211.
http://dx.doi.org/10.1016/j.disc.2008.01.025

145. H. Galeana-Sánchez and M. Olsen [2011], Kernels by monochromatic paths in digraphs with covering number 2, *Discrete Math.* **311**(13), 1111–1118, 2012d:

05166.
http://dx.doi.org/10.1016/j.disc.2010.09.022

146. H. Galeana-Sánchez and R. Rojas-Monroy [2008], Kernels and some operations in edge-coloured digraphs, *Discrete Math.* **308**(24), 6036–6046, 2009j:05077. http://dx.doi.org/10.1016/j.disc.2007.11.022

147. S. Ganzell and A. Meadows [2009], Untangle: knots in combinatorial game theory, *Geombinatorics* **18**(3), 101–108, 2479639.

148. S. Ganzell, A. Meadows and J. Ross [2014], Twist Untangle and Related Knot Games, *Integers* **14**, G4.

149. S. M. Garrabranta, E. J. Friedman and A. S. Landsberg [2013], Cofinite Induced Subgraphs of Impartial Combinatorial Games: An Analysis of CIS-Nim, *INTEGERS* **13**, G2.

150. Q. R. Gashi, T. Schedler and D. E. Speyer [2012], Looping of the numbers game and the alcoved hypercube, *J. Combin. Theory Ser. A* **119**(3), 713–730, 2871758. http://dx.doi.org/10.1016/j.jcta.2011.11.012

151. T. Gavenčiak [2010], Cop-win graphs with maximum capture-time, *Discrete Math.* **310**(10-11), 1557–1563, 2011d:05241.
http://dx.doi.org/10.1016/j.disc.2010.01.015

152. H. Gebauer [2012], On the clique-game, *European J. Combin.* **33**(1), 8–19, 2854626.
http://dx.doi.org/10.1016/j.ejc.2011.07.005

153. A. Georgakopoulos and P. Winkler [2012], Two-color Babylon, *Integers, Electr. J of Combinat. Number Theory* **12**, #G1, 7 pp., Comb. Games Sect.
http://www.integers-ejcnt.org/vol12.html

154. M. Goadrich and M. Schlatter [2011], Analyzing two-color Babylon, *Integers* **11**, G01, 12, 2012c:91060.
http://dx.doi.org/10.1515/INTEG.2011.001

155. W. Goddard, A. Sinko, P. Slater and H. Xu [2011], The graph distance game, *AKCE Int. J. Graphs Comb.* **8**(1), 85–96, 2012e:05252.

156. J. Goldwasser, X. Wang and Y. Wu [2011], Minimum light numbers in the σ-game and lit-only σ-game on unicyclic and grid graphs, *Electron. J. Combin.* **18**(1), Paper 214, 27, 2853071.

157. J. Goldwasser, X. Wang and Y. Wu [2009], Does the lit-only restriction make any difference for the σ-game and σ^+-game?, *European J. Combin.* **30**, 774–787, 2010c:05087.
http://dx.doi.org/10.1016/j.ejc.2008.09.020

158. J.P. Grossman and R. J. Nowakowski [2015], A ruler regularity in hexadecimal games, in: *Games of No Chance 4* (R. Nowakowski, ed.), No. 63 in Mathematical Sciences Research Institute Publications, Cambridge University Press, New York, pp. 115–128.

159. A. Guignard and É. Sopena [2009], Compound Node-Kayles on paths, *Theoret. Comput. Sci.* **410**, 2033–2044, 2010g:91030.
http://dx.doi.org/10.1016/j.tcs.2008.12.053

160. A. Guo [2012], Winning strategies for aperiodic subtraction games, *Theoret. Comput. Sci.* **421**, 70–73.

161. A. Guo and E. Miller [2011], Lattice point methods for combinatorial games, *Adv. in Appl. Math.* **46**(1-4), 363–378, 2794028.
http://dx.doi.org/10.1016/j.aam.2010.10.004

162. A. Guo and E. Miller [2012], Corrigendum to "Lattice point methods for combinatorial games" [Adv. in Appl. Math. 46 (1–4) (2011) 363–378] [MR2794028], *Adv. in Appl. Math.* **48**(1), 269–271, 2845518.
http://dx.doi.org/10.1016/j.aam.2011.09.001

163. A. Guo and E. Miller [2013], Algorithms for lattice games, *Internat. J. Game Theory* **42**(4), 777–788, 3111660.
http://dx.doi.org/10.1007/s00182-012-0319-9

164. V. Gurvich [2012], Further generalizations of the Wythoff game and the minimum excludant, *Discrete Appl. Math.* **160**, 941–947.

165. G. Gutowski [2011], Mr. Paint and Mrs. Correct go fractional, *Electron. J. Combin.* **18**(1), Paper 140, 8, 2817790.

166. I. Halupczok and J.-C. Schlage-Puchta [2008], Some strategies for higher dimensional animal achievement games, *Discrete Math.* **308**(16), 3470–3478, 2009g: 91040.
http://dx.doi.org/10.1016/j.disc.2007.07.005

167. P. Harding and P. Ottaway [2014], Edge Deletion Games with Parity Rules, *Integers* **14**, G1.

168. F. Havet and L. Sampaio [2010], On the Grundy number of a graph, in: *Parameterized and exact computation*, Vol. 6478 of *Lecture Notes in Comput. Sci.*, Springer, Berlin, pp. 170–179, 2012e:05134.
http://dx.doi.org/10.1007/978-3-642-17493-3_17

169. R. B. Hayward, B. Toft and P. Henderson [2012], How to play Reverse Hex, *Discrete Math.* **312**(1), 148–156, 2852517.
http://dx.doi.org/10.1016/j.disc.2011.06.033

170. R. A. Hearn and E. D. Demaine [2009], *Games, puzzles, and computation*, A K Peters Ltd., Wellesley, MA, 2537584.

171. G. Helleloid, M. Khalid, D. P. Moulton and P. M. Wood [2009], Graph pegging numbers, *Discrete Math.* **309**, 1971–1985, 2010h:05247.
http://dx.doi.org/10.1016/j.disc.2008.03.029

172. P. Henderson and R. B. Hayward [2015], A handicap strategy for Hex, in: *Games of No Chance 4* (R. Nowakowski, ed.), No. 63 in Mathematical Sciences Research Institute Publications, Cambridge University Press, New York, pp. 129–136.

173. P. Henderson and R. B. Hayword [2013], Captured-Reversible Moves and Star Decomposition Domination in Hex, *Integers* **13**, G1.

174. C. Hernández-Cruz and H. Galeana-Sánchez [2012], k-kernels in k-transitive and k-quasi-transitive digraphs, *Discrete Math.* **312**, 2522–2530.

175. D. S. Herscovici [2010], On graph pebbling numbers and Graham's conjecture, *Graph Theory Notes N. Y.* **59**, 15–21, 2849399.

176. D. S. Herscovici [2008], Graham's pebbling conjecture on products of many cycles, *Discrete Math.* **308**(24), 6501–6512, 2010b:05177.
http://dx.doi.org/10.1016/j.disc.2007.12.045

177. D. S. Herscovici, B. D. Hester and G. H. Hulbert [2012], Generalization of Graham's pebbling conjecture, *Discrete Math.* **312**, 2286–2293.

178. D. S. Herscovici, B. D. Hester and G. H. Hurlbert [2011], Optimal pebbling in products of graphs, *Australas. J. Combin.* **50**, 3–24, 2829275.

179. G. Hetyei [2010], Enumeration by kernel positions for strongly Bernoulli type truncation games on words, *J. Combin. Theory Ser. A* **117**(8), 1107–1126, 2011j: 05007.
http://dx.doi.org/10.1016/j.jcta.2010.01.003

180. G. Hetyei [2009], Enumeration by kernel positions, *Adv. in Appl. Math.* **42**(4), 445–470, 2010f:05013.
http://dx.doi.org/10.1016/j.aam.2008.11.001

181. N. Ho [2012], Two variants of Wythoff's game preserving its P-positions, *J. Combin. Theory Ser. A* **119**, 1302–1314.

182. N. Ho [2012], Variations of the game 3-Euclid, *Int. J. Comb.* **Article ID 406250**.

183. N. B. Ho [2014], The Max-Welter game, *Discrete Math.* **318**, 41–47.

184. M. Hoffmann, J. Matoušek, Y. Okamoto and P. Zumstein [2011], The t-pebbling number is eventually linear in t, *Electron. J. Combin.* **18**(1), Paper 153, 4, 2823374.

185. M. Hoffmann, J. Matoušek, Y. Okamoto and P. Zumstein [2010], Minimum and maximum against k lies, in: *Algorithm theory—SWAT 2010*, Vol. 6139 of *Lecture Notes in Comput. Sci.*, Springer, Berlin, pp. 139–149, 2678492.

186. S. Hofmann, G. Schuster and J. Steuding [2008], Euclid, Calkin & Wilf—playing with rationals, *Elem. Math.* **63**(3), 109–117, 2009e:91046.
http://dx.doi.org/10.4171/EM/95

187. C. W. Hong [2010], The biased, distance-restricted n-in-a-row game for small p, *Theoret. Comput. Sci.* **411**(16-18), 1895–1897, 2011b:91080.
http://dx.doi.org/10.1016/j.tcs.2010.02.002

188. D. Horrocks [2010], Winning positions in simplicial Nim, *Electron. J. Combin.* **17**(1), Research Paper 84, 13, 2012a:91047.
http://www.combinatorics.org/Volume_17/Abstracts/v17i1r84.html

189. D. G. Horrocks and M. A. Trenton [2008], Subtraction with a Muller twist, *Integers, Electr. J of Combinat. Number Theory* **8**, G05, 12, MR2425634 (2009h: 91051).

190. S. C. Huneke [2014], An inductive proof of hex uniqueness, *Amer. Math. Monthly* **121**(1), 78–80, 3139586.
http://dx.doi.org/10.4169/amer.math.monthly.121.01.078

191. S. C. Huneke, R. Hayward and B. Toft [2014], A winning strategy for $3 \times n$ Cylindrical Hex, *Discrete Math.* **331**, 93–97, 3225310.
http://dx.doi.org/10.1016/j.disc.2014.05.002

192. Y.-C. Hung, R. W. Chen, A. Zame and M.-R. Chen [2010], A note on the first occurrence of strings, *Electron. J. Combin.* **17**(1), Note 5, 8, 2011d:05013.
http://www.combinatorics.org/Volume_17/Abstracts/v17i1n5.html

193. S. Huntemann and R. J. Nowakowski [2014], Doppelganger Placement Games, *Recreational Mathematics Magazine* **1**, 55–61.

194. P. Hunter and S. Kreutzer [2008], Digraph measures: Kelly decompositions, games, and orderings, *Theoret. Comput. Sci.* **399**(3), 206–219, 2009g:05169.
http://dx.doi.org/10.1016/j.tcs.2008.02.038

195. E. J. Ionascu, D. Pritikin and S. E. Wright [2008], k-dependence and domination in kings graphs, *Amer. Math. Monthly* **115**, 820–836, 2009i:05180.

196. V. Isler and N. Karnad [2008], The role of information in the cop-robber game, *Theoret. Comput. Sci.* **399**(3), 179–190, 2009f:91017.
http://dx.doi.org/10.1016/j.tcs.2008.02.041

197. Y. Iwasaki [2009], Solution for the Tower of Hanoi problem with four pegs, *Adv. Appl. Discrete Math.* **4**(2), 95–104, 2590297.

198. J. James and M. Schlatter [2008], Some observations and solutions to short and long global nim, *Integers, Electr. J of Combinat. Number Theory* **8**, G01, 14, MR2373090 (2008k:91051).

199. W. Johnson [2014], The combinatorial game theory of well-tempered scoring games, *Internat. J. Game Theory* **43**(2), 415–438, 3197299.
http://dx.doi.org/10.1007/s00182-013-0386-6

200. B. Jones, L. Taalman and A. Tongen [2013], Solitaire Mancala games and the Chinese remainder theorem, *Amer. Math. Monthly* **120**(8), 706–724, 3096479.
http://dx.doi.org/10.4169/amer.math.monthly.120.08.706

201. D. M. Kane [2010], On solving games constructed using both short and long conjunctive sums, *Integers* **10**, G04, 849–878, 2797777.
http://dx.doi.org/10.1515/INTEG.2010.104

202. T. Khovanova and J. B. Lewis [2011], Baron Münchhausen redeems himself: bounds for a coin-weighing puzzle, *Electron. J. Combin.* **18**(1), Paper 37, 14, 2776813.

203. H. A. Kierstead and A. V. Kostochka [2009], Efficient graph packing via game colouring, *Combin. Probab. Comput.* **18**(5), 765–774, 2010j:05322.
http://dx.doi.org/10.1017/S0963548309009973

204. H. A. Kierstead, C.-Y. Yang, D. Yang and X. Zhu [2012], Adapted game colouring of graphs, *European J. Combin.* **33**(4), 435–445, 2864427.
http://dx.doi.org/10.1016/j.ejc.2011.11.001

205. H. Kierstead, B. Mohar, S. Špacapan, D. Yang and X. Zhu [2009], The two-coloring number and degenerate colorings of planar graphs, *SIAM J. Discrete Math.* **23**(3), 1548–1560, 2011b:05073.
http://dx.doi.org/10.1137/070703697

206. J. Y. Kim [2011], The incidence game chromatic number of paths and subgraphs of wheels, *Discrete Appl. Math.* **159**(8), 683–694, 2012b:05121.
http://dx.doi.org/10.1016/j.dam.2010.01.001

207. K. Knauer, P. Micek and T. Ueckerdt [2011], How to eat 4/9 of a pizza, *Discrete Math.* **311**(16), 1635–1645, 2012e:91185.
http://dx.doi.org/10.1016/j.disc.2011.03.015

208. S. C. Kontogiannis, P. N. Panagopoulou and P. G. Spirakis [2009], Polynomial algorithms for approximating Nash equilibria of bimatrix games, *Theoret. Comput. Sci.* **410**(17), 1599–1606, 2010h:68217.
http://dx.doi.org/10.1016/j.tcs.2008.12.033

209. W. O. Krawec [2012], Analyzing n-player impartial games, *Int. J. Game Theory* **41**, 345–367.

210. M. Krivelevich and T. Szabó [2008], Biased positional games and small hypergraphs with large covers, *Electron. J. Combin.* **15**(1), Research Paper 70, 13, 2009h:91047.
http://www.combinatorics.org/Volume_15/Abstracts/v15i1r70.html

211. K. Kruczek and E. Sundberg [2010], Potential-based strategies for tic-tac-toe on the integer lattice with numerous directions, *Electron. J. Combin.* **17**(1), Research Paper 5, 15, 2011g:05204.
http://www.combinatorics.org/Volume_17/Abstracts/v17i1r5.html

212. K. Kruczek and E. Sundberg [2008], A pairing strategy for tic-tac-toe on the integer lattice with numerous directions, *Electron. J. Combin.* **15**(1), Note 42, 6, 2010c:91029.
http://www.combinatorics.org/Volume_15/Abstracts/v15i1n42.html

213. D. Kunkle and G. Cooperman [2009], Harnessing parallel disks to solve Rubik's cube, *J. Symbolic Comput.* **44**(7), 872–890, 2010g:05015.
http://dx.doi.org/10.1016/j.jsc.2008.04.013

214. U. Larsson [2015], Restrictions of m-Wythoff Nim and p-complementary Beatty sequences, in: *Games of No Chance 4* (R. Nowakowski, ed.), No. 63 in Mathematical Sciences Research Institute Publications, Cambridge University Press, New York, pp. 137–160.

215. U. Larsson [2014], Wythoff Nim Extensions and Splitting Sequences, *Journal of Integer Sequences* **17**, Article 14.5.7.

216. U. Larsson [2012], A Generalized Diagonal Wythoff Nim, *Integers, Electr. J. of Combinat. Number Theory* **12**, #G02, 24 pp., Comb. Games Sect.
http://www.integers-ejcnt.org/vol12.html

217. U. Larsson [2012], The *-operator and invariant subtraction games, *Theoret. Comput. Sci.* **422**(1), 52–58.

218. U. Larsson [2011], Blocking Wythoff nim, *Electron. J. Combin.* **18**(1), Paper 120, 18, 2811089.

219. U. Larsson [2009], 2-pile Nim with a restricted number of move-size imitations, *Integers* **9**, G04, 671–690, With an appendix by Peter Hegarty, 2011e:91054. http://dx.doi.org/10.1515/INTEG.2009.053

220. U. Larsson, P. Hegarty and A. S. Fraenkel [2011], Invariant and dual subtraction games resolving the Duchêne-Rigo conjecture, *Theoret. Comput. Sci.* **412**(8-10), 729–735, 2012d:91053.
http://dx.doi.org/10.1016/j.tcs.2010.11.015

221. U. Larsson and J. Wästlund [2014], Maharaja Nim: Wythoff's Queen meets the Knight, *Integers* **14**, G5.

222. E. L. Leiss [2013], The worst Hanoi graphs, *Theoret. Comput. Sci.* **498**, 100–106, 3083516.
http://dx.doi.org/10.1016/j.tcs.2013.05.020

223. J. Lemoine and S. Viennot [2015], Computer Analysis of Sprouts with Nimbers, in: *Games of No Chance 4* (R. Nowakowski, ed.), No. 63 in Mathematical Sciences Research Institute Publications, Cambridge University Press, New York, pp. 161–182.

224. W. Li, D. Wu and F. Pan [2012], A construction for doubly pandiagonal magic squares, *Discrete Math.* **312**(2), 479–485, 2852609.
http://dx.doi.org/10.1016/j.disc.2011.09.031

225. W. Lin and X. Zhu [2009], Circular game chromatic number of graphs, *Discrete Math.* **309**(13), 4495–4501, 2010i:05127.
http://dx.doi.org/10.1016/j.disc.2009.02.011

226. W. A. Liu and H. Li [2014], A restricted version of Wythoff's Game, *Electronic Journal of Combinatorics* .

227. W. A. Liu, H. Li and B. Li [2011], A restricted version of Wythoff's game, *Electron. J. Combin.* **18**(1), Paper 207, 17, 2853064.

228. W. A. Liu, H. Li, Y. Ma and B. Li [2012], On Supplements of 2 x m Board in Toppling Towers, *Integers* **12**, G4.

229. L. Longpré and P. McKenzie [2009], The complexity of solitaire, *Theoret. Comput. Sci.* **410**(50), 5252–5260, Appeared first in Mathematical foundations of computer science 2007, 182193, Lecture Notes in Comput. Sci., 4708, Springer, Berlin, 2007, 2011b:91082.
http://dx.doi.org/10.1016/j.tcs.2009.08.027

230. A. M. Loosen [2008], Directional phutball, *Atl. Electron. J. Math.* **3**, 30–45, 2827227.

231. T. Mütze and R. Spöhel [2011], On the path-avoidance vertex-coloring game, *Electron. J. Combin.* **18**(1), Paper 163, 33, 2831099.

232. P. Manuel, I. Rajasingh, B. Rajan and J. Punitha [2009], Kernel in oriented circulant graphs, in: *Combinatorial algorithms*, Vol. 5874 of *Lecture Notes in Comput. Sci.*, Springer, Berlin, pp. 396–407, 2577955.

233. N. A. McKay [2011], Canonical forms of uptimals, *Theoret. Comput. Sci.* **412**(52), 7122–7132, 2895141.
http://dx.doi.org/10.1016/j.tcs.2011.09.021

234. N. A. McKay, R. J. Nowakowski and A. A. Siegel [2015], Navigating the MAZE, in: *Games of No Chance 4* (R. Nowakowski, ed.), No. 63 in Mathematical Sciences Research Institute Publications, Cambridge University Press, New York, pp. 183–193.

235. N. A. McKay, R. J. Nowakowski and A. A. Siegel [2010], The gods play maze, in: *Recreational Mathematics Colloquium I*, Assoc. Ludus, Lisbon, pp. 153–159, 2807403.

236. A. Mehrabian [2011], Lower bounds for the cop number when the robber is fast, *Combin. Probab. Comput.* **20**(4), 617–621, 2012e:05253.
http://dx.doi.org/10.1017/S0963548311000101

237. A. Mehrabian [2011], The capture time of grids, *Discrete Math.* **311**(1), 102–105, 2011i:05154.
http://dx.doi.org/10.1016/j.disc.2010.10.002

238. N. Mehta and Á. Seress [2011], Connected, bounded degree, triangle avoidance games, *Electron. J. Combin.* **18**(1), Paper 193, 37, 2836828.

239. K. Meng, C. Lin, W. A. Liu and Y. Yang [2011], Q-ary Rényi-Ulam pathological liar game with one lie, *Discrete Appl. Math.* **159**(6), 478–483, 2012c:91063.
http://dx.doi.org/10.1016/j.dam.2010.12.021

240. P. Micek and B. Walczak [2012], Parity in graph sharing games, *Discrete Math.* **312**, 1788–1795.

241. P. Micek and B. Walczak [2011], A graph-grabbing game, *Combin. Probab. Comput.* **20**(4), 623–629, 2012d:05252.
http://dx.doi.org/10.1017/S0963548311000071

242. E. Miller [2013], Affine stratifications from finite misère quotients, *J. Algebraic Combinatorics* **37**, 1–9.

243. R. Milley, R. J. Nowakowski and P. Ottaway [2012], Misère moniod of one-handed alternating games, *Integers 2011*, to appear.

244. R. Milley and G. Renault [2013], Dead ends in misère play: the misère monoid of canonical numbers, *Discrete Math.* **313**(20), 2223–2231, 3084267.
http://dx.doi.org/10.1016/j.disc.2013.05.023

245. M. Montassier, A. Pêcher, A. Raspaud, D. B. West and X. Zhu [2010], Decomposition of sparse graphs, with application to game coloring number, *Discrete Math.* **310**(10-11), 1520–1523, 2011g:05244.
http://dx.doi.org/10.1016/j.disc.2010.01.008

246. R. E. Morrison, E. J. Friedman and A. S. Landsberg [2011], Combinatorial games with a pass: A dynamical systems approach, *Chaos 21, 043108 (2011); doi:10.1063/1.3650234* **21**.

247. H. Nagamochi [2011], Cop-robber guarding game with cycle robber-region, *Theoret. Comput. Sci.* **412**(4-5), 383–390, 2012b:05181. http://dx.doi.org/10.1016/j.tcs.2010.04.014

248. T. Nakamura [2015], Evaluating territories of Go positions with capturing races, in: *Games of No Chance 4* (R. Nowakowski, ed.), No. 63 in Mathematical Sciences Research Institute Publications, Cambridge University Press, New York, pp. 195–205.

249. Z. Nedev [2010], An $O(n)$-round strategy for the Magnus-Derek game, *Algorithms (Basel)* **3**(3), 244–254, 2725450. http://dx.doi.org/10.3390/a3030244

250. Z. Nedev [2008], Universal sets and the vector game, *Integers* **8**, A45, 5, 2010b: 91053.

251. Z. Nedev and S. Muthukrishnan [2008], The Magnus-Derek game, *Theoret. Comput. Sci.* **393**(1-3), 124–132, 2009h:91052. http://dx.doi.org/10.1016/j.tcs.2007.11.016

252. Z. Nedev and A. Quas [2008], Balanced sets and the vector game, *Int. J. Number Theory* **4**, 339–347, MR2424326.

253. R. J. Nowakowski [2015], Unsolved problems in Combinatorial Games, in: *Games of No Chance 4* (R. Nowakowski, ed.), No. 63 in Mathematical Sciences Research Institute Publications, Cambridge University Press, New York, pp. 279–308.

254. R. J. Nowakowski and P. Ottaway [2011], Option-closed games, *Contrib. Discrete Math.* **6**(1), 142–153, 2012e:91080.

255. I. Pak and J. Yang [2013], The complexity of generalized domino tilings, *Electron. J. Combin.* **20**(4), Paper 12, 23, 3139397.

256. M. B. Paterson and D. R. Stinson [2010], Yet another hat game, *Electron. J. Combin.* **17**(1), Research Paper 86, 12, 2011e:91055. http://www.combinatorics.org/Volume_17/Abstracts/v17i1r86.html

257. S. Payne and E. Robeva [2015], Artificial intelligence for Bidding Hex, in: *Games of No Chance 4* (R. Nowakowski, ed.), No. 63 in Mathematical Sciences Research Institute Publications, Cambridge University Press, New York, pp. 207–214.

258. T. Plambeck and G. Whitehead [2014], The Secrets of Notatko: Winning at X-Only Tic-Tac-Toe, *Recreational Mathematics Magazine* **1**, 49–54.

259. T. E. Plambeck and A. N. Siegel [2008], Misère quotients for impartial games, *J. Combin. Theory Ser. A* **115**, 593–622, 2009m:91036.

260. J. W. S. Rayleigh [1937], *The Theory of Sound*, MacMillan, London, 1st ed. 1877, 2nd ed. 1894, reprinted 1928,1929,1937. There is also a Dover 1945 publication, which is a reprint of the 2nd. The "sound" equivalent of Beatty's theorem, is stated, without proof, in sect. 92a, pp. 122-123, in the 1937 reprint of the 2nd ed., the only one I saw. Kevin O'Bryant told me that he checked the 1st edition and the statement is not there, MR0016009 (7,500e).

261. C. P. dos Santos [2011], Embedding processes in combinatorial game theory, *Discrete Appl. Math.* **159**(8), 675–682, 2012d:91054. http://dx.doi.org/10.1016/j.dam.2010.11.019

262. C. P. dos Santos and J. N. Silva [2015], Nimbers in Partizan Games, in: *Games of No Chance 4* (R. Nowakowski, ed.), No. 63 in Mathematical Sciences Research Institute Publications, Cambridge University Press, New York, pp. 215–224.

263. C. P. d. Santos and J. N. Silva [2008], Konane has infinite Nim-dimension, *Integers* **8**, G2, 6, 2008k:91052.

264. U. Schauz [2011], Colorings and nowhere-zero flows of graphs in terms of Berlekamp's switching game, *Electron. J. Combin.* **18**(1), Paper 65, 33, 2012e:05254.

265. U. Schauz [2009], Mr. Paint and Mrs. Correct, *Electron. J. Combin.* **16**(1), Research Paper 77, 18, 2010i:91064. http://www.combinatorics.org/Volume_16/Abstracts/v16i1r77.html

266. A. Scott and U. Stege [2010], Parameterized pursuit-evasion games, *Theoret. Comput. Sci.* **411**(43), 3845–3858, 2012c:68102. http://dx.doi.org/10.1016/j.tcs.2010.07.004

267. A. Scott and U. Stege [2008], Parameterized chess, in: *Parameterized and exact computation*, Vol. 5018 of *Lecture Notes in Comput. Sci.*, Springer, Berlin, pp. 172–189, 2010f:68062. http://dx.doi.org/10.1007/978-3-540-79723-4_17

268. A. Shapovalov [2011], Occupation games on graphs in which the second player takes almost all vertices, *Discrete Appl. Math.* **159**(15), 1526–1527, 2012f:05202. http://dx.doi.org/10.1016/j.dam.2011.03.002

269. N. Sieben [2008], Polyominoes with minimum site-perimeter and full set achievement games, *European J. Combin.* **29**(1), 108–117, MR2368619 (2008j:05092). http://dx.doi.org/10.1016/j.ejc.2006.12.008

270. N. Sieben [2008], Proof trees for weak achievement games, *Integers, Electr. J of Combinat. Number Theory* **8**, G07, 18, MR2472080 (2010b:91054).

271. A. N. Siegel [2015], The structure and classification of misère quotients, in: *Games of No Chance 4* (R. Nowakowski, ed.), No. 63 in Mathematical Sciences Research Institute Publications, Cambridge University Press, New York, pp. 241–266.

272. A. N. Siegel [2015], Misère canonical forms of partizan games, in: *Games of No Chance 4* (R. Nowakowski, ed.), No. 63 in Mathematical Sciences Research Institute Publications, Cambridge University Press, New York, pp. 225–239.

273. A. N. Siegel [2013], *Combinatorial Game Theory*, American Math. Society.

274. A. Sinko and P. J. Slater [2008], Queen's domination using border squares and (A, B)-restricted domination, *Discrete Math.* **308**(20), 4822–4828, 2009f:05208. http://dx.doi.org/10.1016/j.disc.2007.08.065

275. M. Soltys and C. Wilson [2011], On the complexity of computing winning strategies for finite poset games, *Theory Comput. Syst.* **48**(3), 680–692, 2012e:91016. http://dx.doi.org/10.1007/s00224-010-9254-y

276. M. Z. Spivey [2009], Staircase rook polynomials and Cayley's game of Mousetrap, *European J. Combin.* **30**(2), 532–539, 2010f:05016. http://dx.doi.org/10.1016/j.ejc.2008.04.005

277. F. Stewart [2013], Pirates and Treasure, *Integers* **13**, G3.

278. Z. Szaniszlo, M. Tomova and C. Wyels [2009], The N-queens problem on a symmetric Toeplitz matrix, *Discrete Math.* **309**, 969–974, 2010d:05023. http://dx.doi.org/10.1016/j.disc.2008.01.042

279. W. Szumny, I. Włoch and A. Włoch [2008], On the existence and on the number of (k, l)-kernels in the lexicographic product of graphs, *Discrete Math.* **308**(20), 4616–4624, 2009e:05235. http://dx.doi.org/10.1016/j.disc.2007.08.078

280. X. L. Tan and M. D. Ward [2014], On Kotzig's Nim, *Integers* **14**, G6.

281. T. Thanatipanonda [2011], Further hopping with Toads and Frogs, *Electron. J. Combin.* **18**(1), Paper 67, 12, 2788684.

282. T. Thanatipanonda [2008], How to beat Capablanca, *Adv. in Appl. Math.* **40**, 266–270, 2388615. http://dx.doi.org/10.1016/j.aam.2006.11.005

283. T. Thanatipanonda and D. Zeilberger [2009], A symbolic finite-state approach for automated proving of theorems in combinatorial game theory, *J. Difference Equ. Appl.* **15**, 111–118, 2010b:91055. http://dx.doi.org/10.1080/10236190802006423

284. B. Torrence [2011], The easiest lights out games, *College Math. J.* **42**, 361–372.

285. X. Wang and Y. Wu [2010], Lit-only σ-game on pseudo-trees, *Discrete Appl. Math.* **158**(17), 1945–1952, 2011i:05180. http://dx.doi.org/10.1016/j.dam.2010.08.009

286. M. Weimerskirch [2015], An algorithm for computing indistinguishability quotients in misère impartial combinatorial games, in: *Games of No Chance 4* (R. Nowakowski, ed.), No. 63 in Mathematical Sciences Research Institute Publications, Cambridge University Press, New York, pp. 267–277.

287. M. Wildon [2010], Knights, spies, games and ballot sequences, *Discrete Math.* **310**(21), 2974–2983, 2011k:05014. http://dx.doi.org/10.1016/j.disc.2010.07.005

288. J. Wu and X. Zhu [2008], Lower bounds for the game colouring number of partial k-trees and planar graphs, *Discrete Math.* **308**(12), 2637–2642, MR2410475 (2009c:05093). http://dx.doi.org/10.1016/j.disc.2007.05.023

289. J. Wu and X. Zhu [2008], The 6-relaxed game chromatic number of outerplanar graphs, *Discrete Math.* **308**(24), 5974–5980, MR2464888 (2009j:91044). http://dx.doi.org/10.1016/j.disc.2007.11.015

290. Y. Wu [2009], Lit-only sigma game on a line graph, *European J. Combin.* **30**(1), 84–95, 2009k:91039. http://dx.doi.org/10.1016/j.ejc.2008.02.003

291. B. Yang, D. Dyer and B. Alspach [2009], Sweeping graphs with large clique number, *Discrete Math.* **309**(18), 5770–5780, 2010k:05324. http://dx.doi.org/10.1016/j.disc.2008.05.033

292. C. Yang [2011], Sliding puzzles and rotating puzzles on graphs, *Discrete Math.* **311**(14), 1290–1294, 2012f:05139. http://dx.doi.org/10.1016/j.disc.2011.03.011

293. D. Yang [2012], Coloring games on squares of graphs, *Discrete Math.* **312**, 1400–1406.

294. D. Yang [2009], Activation strategy for relaxed asymmetric coloring games, *Discrete Math.* **309**, 3323–3335, 2010g:05247. http://dx.doi.org/10.1016/j.disc.2008.09.047

295. D. Yang [2009], Relaxed very asymmetric coloring games, *Discrete Math.* **309**, 1043–1050, 2010d:05064. http://dx.doi.org/10.1016/j.disc.2007.11.058

296. D. Yang and H. A. Kierstead [2008], Asymmetric marking games in line graphs, *Discrete Math.* **308**(9), 1751–1755, MR2392616 (2009a:05081). http://dx.doi.org/10.1016/j.disc.2007.03.082

297. D. Yang and X. Zhu [2010], Game colouring directed graphs, *Electron. J. Combin.* **17**(1), Research Paper 11, 19, 2011d:05150. http://www.combinatorics.org/Volume_17/Abstracts/v17i1r11.html

298. D. Yang and X. Zhu [2008], Activation strategy for asymmetric marking games, *European J. Combin.* **29**, 1123–1132, MR2419216 (2009b:91039). http://dx.doi.org/10.1016/j.ejc.2007.07.004

299. X. Zhu [2008], Game coloring the Cartesian product of graphs, *J. Graph Theory* **59**, 261–278, MR2463181 (2009h:05095). http://dx.doi.org/10.1002/jgt.20338

300. X. Zhu [2008], Refined activation strategy for the marking game, *J. Combin. Theory Ser. B* **98**(1), 1–18, 2008k:91049. http://dx.doi.org/10.1016/j.jctb.2007.04.004

aviezri.fraenkel@weizmann.ac.il *Department of Computer Science and Applied Mathematics, Weizmann Institute of Science, 76100 Rehovot, Israel*

rjn@mathstat.dal.ca *Mathematics and Statistics, Dalhousie University, Halifax B3H 3J5, Canada*